TROPICAL RAIN FORESTS

Tropical Rain Forests

An Ecological and Biogeographical Comparison

Richard T. Corlett and
Richard B. Primack

Second edition

A John Wiley & Sons, Ltd., Publication

Blackwell Publishing was acquired by John Wiley & Sons in February 2007. Blackwell's publishing program has been merged with Wiley's global Scientific, Technical and Medical business to form Wiley-Blackwell.

Registered office: John Wiley & Sons Ltd, The Atrium, Southern Gate, Chichester, West Sussex, PO19 8SQ, UK

Editorial offices: 9600 Garsington Road, Oxford, OX4 2DQ, UK
The Atrium, Southern Gate, Chichester, West Sussex, PO19 8SQ, UK
111 River Street, Hoboken, NJ 07030-5774, USA

For details of our global editorial offices, for customer services and for information about how to apply for permission to reuse the copyright material in this book please see our website at www.wiley.com/wiley-blackwell

Library of Congress Cataloguing-in-Publication Data
Corlett, Richard.
  Tropical rain forests: an ecological and biogeographical comparison /
Richard Corlett and Richard Primack. – 2nd ed.
    p. cm.
  Rev. ed. of: Tropical rain forests : an ecological and biogeographical comparison /
Richard Primack & Richard Corlett. c2005.
  Includes bibliographical references and index.
  ISBN 978-1-4443-3254-4 (cloth) – ISBN 978-1-4443-3255-1 (pbk.)
  1. Rain forests. 2. Rain forest ecology. I. Primack, Richard B., 1950– II. Title.
  QH86.P75 2011
  577.34–dc22
                                                                    2010037701

A catalogue record for this book is available from the British Library.

This book is published in the following electronic formats: eBook 9781444392272; Wiley Online Library 9781444392296; ePub 9781444392289

Set in 9/11.5pt Meridien by Graphicraft Limited, Hong Kong

Printed and bound in Malaysia by Vivar Printing Sdn Bhd

1   2011

Contents

Preface to the first edition

In the popular imagination, the topical rain forest consists of giant trees towering above a tangle of vines and beautiful orchids below, with colorful birds, tree frogs, and monkeys everywhere abundant. Scientists and visitors quickly realize that this image is not accurate: animal life, while highly diverse, is not necessarily strikingly abundant, and flowers are often very small and hard to find. But beyond the difference between perception and reality, there are tremendous differences among regions. Biologists working in one area rapidly recognize the special features of the biological community in their area, yet they would find themselves in highly unfamiliar terrain should they move, for example, from their accustomed study site in Borneo to a seemingly similar location in New Guinea. Indeed, even in locations within the same overall zone, such as the Amazon and Central American forests, there can be differences both dramatic and subtle from place to place. There are unique plants and animals in every community, and even those organisms common to other regions are part of a distinctive mixture of species that interact in ways readily distinguishable from other forests. Thus, the tendency of both popular media and science to make sweeping statements about "the rain forest" is highly misleading.

On a larger scale, rain forests on different continents have fundamentally different characteristics that make each of them unique. In many earlier books on rain forests, authors such as Paul Richards in *The Tropical Rainforest* and Tim Whitmore in *An Introduction to Tropical Rain Forests* tried to describe the unifying properties of rain forests on each continent – features such as the high diversity of tree species and the low nutrient status of the soils. They took comparable examples from each region and emphasized certain principles of tropical ecology that are true on all continents. However, this emphasis on commonalities meant that readers could – and often did – easily overlook the fact that each of these rain forests has its own unique features of plants, animals, climate, topography, and past history. Our goal in writing this book is therefore to redress this oversight by emphasizing the ways in which the major rain forest areas are special.

We believe this approach can suggest new research questions that can be investigated in comparative studies of rain forests in different regions. At the end of each chapter, we suggest specific new approaches, sometimes involving experimental methods, that could be used to develop new research questions. Finally, in the last chapter we consider the unique threats faced by rain forests in each area of the world and suggest strategies for conservation. Such topics may have relevance to policy initiatives aimed at protecting rain forest habitats – initiatives that currently are based upon a misunderstanding that the various communities would respond in the same manner to the same methods of management.

We hope that readers will come away with an appreciation that our planet is host to not one monolithic tract of rain forest, but many unique tropical rain forest habitats, all worth of study and protection.

Richard B. Primack and Richard T. Corlett

Preface to the second edition

This new book is more than just an update of the first addition, published six years ago. The chapter on island rain forests is completely new and fills a significant gap in the first edition. The total area of these island rain forests worldwide is small, but their biogeographical isolation means that together they support a significant proportion of all rain forest species. Also, many of these rain forests, such as those on the Hawaiian Islands and in the Caribbean, are both intensively studied and frequently visited by tropical biologists. The final chapter, on the future of tropical rain forests, has also been completely re-written, reflecting the rapid pace of change in the major rain forest countries, such as Brazil, Indonesia, the Democratic Republic of the Congo, and Papua New Guinea. The bad news is even worse, with climate change looming large over everything. However, there are also reasons for cautious optimism and there has never been a more exciting time to be involved in tropical conservation. The other chapters have been brought up to date. Major changes in our understanding of the evolutionary history of plants and bats have been incorporated and new ideas and new examples have been added throughout. Reptiles now get more space than they did in the first edition and a special effort has been made to include more examples from Africa rain forests, which are less studied than rain forests elsewhere. Finally, the photographs are largely new and all in color.

Richard T. Corlett and Richard B. Primack

Acknowledgments

Anyone attempting to write a book that makes simple clear statements about the whole range of tropical rain forest ecology, biogeography, and conservation, quickly reaches the limits of their own knowledge. Books can be studied, articles examined and the Internet searched. But in the end, it is the tropical ecologists themselves who must be consulted for the accuracy of impressions and ideas, and to explain facts which seem confusing. Therefore we have many people to thank for their expertise in making this book as accurate as possible, for providing colorful examples, and also for saying when we really do not have the data to make definitive statements. Individual chapters for the first edition were read by J.R. Flenley, Geoff Hope, Chris Schneider (Introduction), Ian Turner, John Kress, Ghillean Prance, Chris Dick (Chapter 2), John Fleagle, Cheryl Knott, Colin Chapman (Chapter 3), Tom Kunz, John Hart (Chapter 4), Mercedes Foster, David Pearson, Fred Wasserman (Chapter 5), Tom Kunz (Chapter 6), James Traniello, Phil DeVries, David Roubik (Chapter 7), and Kamal Bawa and Peter Feinsinger (Chapter 8). For the second edition we received help from many of the same people and also William Laurance, Navjot Sodhi, Hugh Tan, Robert Morley, Francis Putz, Ahimsa Campos-Arceiz, Rhett Harrison, Soumya Prasad, and Vojtech Novotny. The book is greatly improved by the numerous images supplied to us from photographers throughout the world. These people are identified by name next to their photographs. A special effort to expand the representation of photographs from African rain forests resulted in numerous excellent pictures from people, many of which are included here. Finally we are grateful to our respective Universities and Departments for their support during the production of this new edition and to the staff at Wiley-Blackwell for their help at every stage in the process.

Chapter 1

Many Tropical Rain Forests

It is easy to make generalizations about tropical rain forests. Travel posters, magazine articles, and television programs give the casual observer the impression that tropical rain forests from any spot in the world are one interchangeable mass of tall, wet trees – the canopy filled with brightly colored birds, chirping tree frogs, and acrobatic monkeys, the ground level home to silent predators, tangled vines, and exotic flowers. This popular perception has been useful because it creates a compellingly attractive image of an untamed, beautiful place that is, on the one hand, a source of infinite mystery and adventure, and on the other, a fragile natural treasury that must be protected. Both conservation and ecotourism rely on this generalized image to promote the idea of tropical rain forests as a "good thing" to be preserved and enjoyed. Yet, beneficial as this image may be in encouraging travel and conservation, it obscures the fact that the world's tropical rain forests have major differences from one another in addition to their obvious similarities.

A major drawback of this generalized image is that it encourages a belief that saving "the rain forest" is a single problem with a single, universally applicable, set of answers. Nothing could be further from the truth. There are many different rain forests, all of which need action for protection, but this action must be targeted at the specific threats present in each region and adapted to the specific ecological characteristics of each rain forest. Policies, tactics, and techniques that work in one region may prove ineffective in another. The major differences between the tropical rain forests in different regions also mean that successes in one region will not compensate for losses in another. The task we are faced with is not "saving *the* rain forest," but "saving the *many* rain forests."

Scientists also have usually emphasized the common appearance of rain forests on different continents and highlighted examples of similar-looking species in separate regions of the world. This emphasis on the common features of rain forests worldwide has had the unintended effect of discouraging research that makes comparisons between regions. The assumption that all rain forests are alike has also led to a tendency to fill gaps in the scientific understanding of one rain forest region by reference to studies in other regions. This tendency

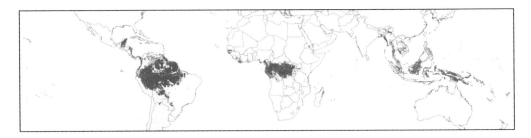

Fig. 1.1 The current global distribution of lowland tropical rain forests. (Courtesy of UNEP World Conservation Monitoring Centre, 2004.)

in turn gives the false impression that our understanding of rain forests is greater than it really is, so that scientific research that could fill the gaps is given a low priority. It also implies that differences between regions are minor, at least in comparison with the similarities.

Our principal reason for writing this book is that we do not believe that the differences between rain forests are minor. It is our contention that the various rain forest regions are sufficiently distinct from one another that they merit individual consideration. In this book, therefore, we will compare the major rain forest regions of the world. The three largest of these rain forest regions are in the Amazon basin of South America, in the Congo River basin of Central Africa, and on the everwet peninsula and islands of Southeast Asia (Fig. 1.1). There are also two smaller and very distinctive rain forest regions on the giant islands of Madagascar and New Guinea. We will show that the rain forests of these five regions are unique biogeographical and ecological entities, each with many distinctive plants, animals, and ecological interactions that are not found in the other regions.

Rain forests occur outside of these core areas as well, but they are less extensive in area and usually less diverse in species. There are rain forests in Central America and coastal Brazil that are basically similar in species composition to those found in the Amazon, but have fewer species and occupy a much smaller area. Similarly, the rain forests of Sri Lanka and the Western Ghats of India resemble those of Southeast Asia, and Australian rain forests have many similarities to the more extensive and diverse forests of New Guinea. Each of these many areas has numerous noteworthy features and unique species, which we will mention in this book, but the focus will be on the differences among the five main regions. In addition, there are also small but distinctive areas of rain forest on many tropical oceanic islands, which are the subject of a separate chapter.

Each rain forest region has different geographical, geological, and climatic features; each region supports plants and animals with separate evolutionary histories; and each region has experienced different past and present human impacts (Table 1.1). These differences have important implications for understanding how rain forests work and deciding how they should be exploited or conserved. Results from scientific research in one region may not apply in the others. At the end of each chapter, we suggest comparative investigations and experiments that could provide deeper insights into rain forest biology. Each rain forest area, and even local areas within each of these regions, must be viewed separately

Table 1.1
Some key characteristics of the main rain forest regions.

	Neotropics	Africa	Madagascar	Southeast Asia	New Guinea
Main geographical feature(s)	Amazon River basin and Andes Mountains	Congo River basin	Forests along eastern edge of island	Peninsula and islands on Sunda Shelf	Large, mountainous island
Distinctive biological features*	Bromeliad epiphytes, high bird diversity, small primates	Low plant richness, forest elephants, many forest browsers	Lemurs, low fruit abundance	Dipterocarp tree family, mast fruiting of trees, large primates	Marsupial mammals, birds of paradise
Annual rainfall (mm)†	2000–3000	1500–2500	2000–3000	2000–3000, often > 3000	2000–3000, often > 3000
Largest country	Brazil	Democratic Republic of Congo	Malagasy Republic	Indonesia	Papua New Guinea

* Unfamiliar terms are explained in the text.
† Rainfall is highly variable within each region. These are the ranges over most of the core rain forest area (1000 mm equals 40 inches).

when scientific investigation, conservation efforts, and responsible development are undertaken. The issues of human impacts, conservation, and development will be considered in detail in the final chapter.

What are tropical rain forests?

Tropical rain forests are the tall, dense, evergreen forests that form the natural vegetation cover of the wet tropics, where the climate is always hot and the dry season is short or absent. This broad definition allows for a considerable range of variation, as is necessary for any global comparison. One important variable is the proportion of deciduous trees in the forest canopy. We have excluded the predominantly deciduous tropical forests that occur in areas with a long dry season, but many forests that we and others call rain forest have some deciduous trees in the canopy. Where this proportion is large, the forests can be called semievergreen (or semideciduous) rain forest. Another important distinction is between lowland and montane rain forests. On high mountains in the wet tropics, forests extend from sea level to around 4000 m (13,000 feet), but the typical tall, lowland rain forest is confined below an altitude of 900–1200 m. Rain forests above this altitude are termed "montane" and have a distinctive ecology of their own.

Precise definitions are difficult in ecology, and there are large areas of forest in the tropics that some ecologists call rain forest while others do not. Definitions also differ between regions, with less strict, more inclusive, definitions in areas where rain forests are less extensive. Thus, on the very dry continent of Australia, almost any area of closed forest is called rain forest. Conversely, foresters familiar with the everwet rain forests of equatorial Southeast Asia might exclude much of what their African counterparts call rain forest. In this book, we focus on hot, wet, tall, and largely evergreen tropical rain forests and we have made it clear when we are referring to unusual or marginal types.

Where are the tropical rain forests?

On a simpler planet, tropical rain forests would form a broad belt around the equator, extending 5–10° to the north and south. On our untidy Earth, inter-actions between wind direction and mountain ranges, variations in sea surface temperature, and various other factors exclude rain forest from parts of this belt – notably most of East Africa – and, in other places, extend it for some distance outside (Fig. 1.1). At least, that was the situation until very recently. During the last few hundred years, and particularly in the last few decades, between one-third to one-half of this rain forest has been converted into other land uses, ranging from productive farmland or tree plantations to urban areas or unproductive grasslands. This book is mostly about the rain forests that still survive, but the devastating effects of human impacts are considered in the last chapter.

The Neotropics

Approximately half of the world's tropical rain forests are in tropical America, the region that biologists call the Neotropics (literally "new tropics" or New World tropics). The Neotropical rain forests form three main blocks. The single largest block of tropical rain forest in the world covers the adjoining basins of the Amazon and Orinoco Rivers (Fig. 1.2). The Amazon River basin is centered on northern and central Brazil. It stretches more than 3000 km (2000 miles) from the foothills of the Andes Mountains in western South America, across the entire South American continent, until the Amazon empties into the Atlantic Ocean. The Orinoco River basin drains eastern Colombia and Venezuela and adjoins the Amazon basin along the Brazilian border. This rain forest block also continues to the northeast of Brazil into the countries of Guyana, Surinam, and French Guiana.

In addition to this giant block of rain forest centered on the Amazon basin, there were two other major blocks of tropical rain forest in the Americas. The Brazilian Atlantic Forest ran along the southeast coast of Brazil from Recife south to São Paulo. This narrow band of forest was over 2000 km long and covered an area of 1.5 million square kilometers, although by no means all of that was rain forest. It was separated from the Amazon forest block by hundreds of kilometers of dry scrub and savanna. This forest has now been reduced to 11–16% of its original area, with most of what remains existing as small, widely separated, fragments (Ribeiro et al. 2009). A third block extended from the Pacific coast of northwest South America through Central America to southernmost Mexico. The rain forests of northwest South America became separated from the Amazon basin rain forest by the uplift of the Andes beginning around 25 million years ago, and then became continuous with those of Central America when the final marine barrier between them disappeared 3 million years ago (see below). Only fragments of this rain forest block survive today. There

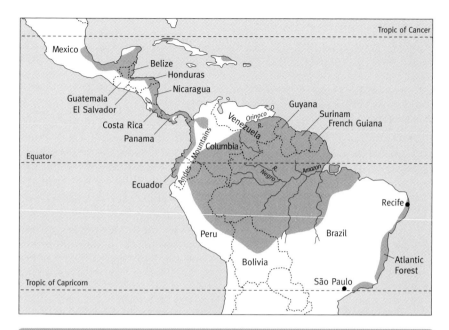

Fig. 1.2 The estimated historical extent of tropical rain forest in South and Central America. (From multiple sources.)

were also smaller areas of rain forest on many of the Caribbean islands, where very little now remains.

Africa

The second largest block of tropical rain forest is in Africa, centered on the Congo River basin (Fig. 1.3). About half of this rain forest is in the Democratic Republic of the Congo (formerly Zaire), with most of the rest divided between the Republic of the Congo, Gabon, and Cameroon. This Central African rain forest block formerly extended northwest into southern Nigeria, but little of this now remains. Rain forest also extended until recently as a belt, up to 350 km (200 miles) wide, along the coast of West Africa, from Ghana through Côte D'Ivoire (Ivory Coast) and Liberia to the eastern margin of Sierra Leone. Most of this has now gone and the remaining fragments are under threat. The larger Central and smaller West African rain forest blocks were separated by 300 km of dry woodland and savanna at the Dahomey Gap, in Togo, Benin, and eastern Ghana. There are also outlying "islands" of rain forest in East Africa, mostly centered on mountains, and surrounded by a "sea" of dry woodland. Although the total area of these East African rain forest patches is small – approximately 3,500 km² – some of them, on older mountains, are apparently of great age and have been isolated from the forests of West and Central Africa for millions of years (Burgess et al. 2007). On a longer timescale, all the rain forests of modern Africa can be seen as remnants of the much more extensive rain forest that spanned the entire continent until around 30 million years ago (Plana 2004).

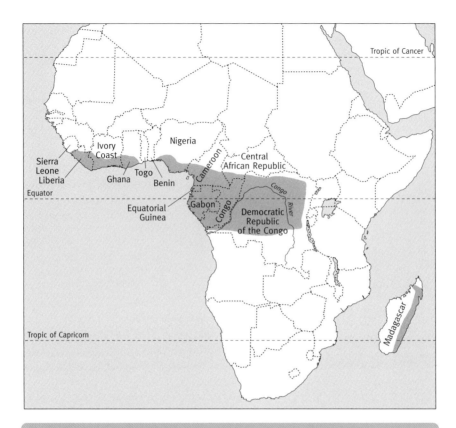

Fig. 1.3 The estimated historical extent of tropical rain forest in Africa and Madagascar. (From multiple sources.)

Asia

The third largest rain forest area until recently occupied most of the Malay Peninsula and the large islands of Borneo, Sumatra, and Java (Fig. 1.4). Ecologists call this region "Sundaland," after the surrounding Sunda continental shelf, and we will follow the convention in this book. Despite the large expanses of sea between the major islands, the Sundaland rain forests are surprisingly uniform and can be considered as a single block. Similar forests also covered much of the Philippines, as well as Sulawesi and many of the smaller Indonesian islands east of Borneo. Rain forest also extended north from Sundaland into the more seasonal climates of mainland Southeast Asia, including most of Cambodia, Laos, and Vietnam, and much of Thailand and Myanmar (formerly Burma). However, large areas in the interior of Myanmar and Thailand did not support rain forest as a result of rainshadows caused by several long north–south mountain chains. Rain forest also once covered much of tropical southern China, in a mosaic with drier forest types, east to the southern tip of Taiwan. To the north, this tropical rain forest merged gradually into the subtropical and warm temperate forests (Corlett 2009a). Today a billion people inhabit the East Asian tropics, so it is not

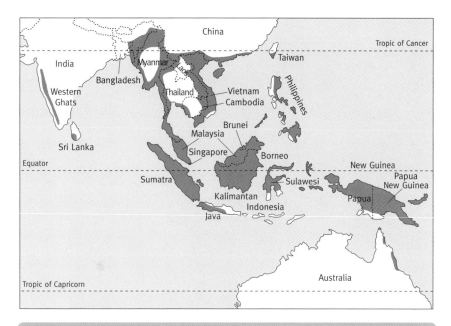

Fig. 1.4 The estimated historical extent of tropical rain forest in Asia, New Guinea, and Australia. (From multiple sources.)

surprising that rain forest has already been eliminated from much of the region and is under threat almost everywhere where it still survives.

The rain forests of Southeast Asia extend westward through Myanmar into northeastern India. India also had a completely separate rain forest area as a long narrow strip, 50–100 km wide, running parallel to the west coast for 1500 km along the crest of the Western Ghats. This now fragmented rain forest band occurs on the tops and sides of these hills, where sea mists can drench the plants even during the dry season. Apart from this isolated ridge, the remainder of India south of the Himalayas is too dry to support rain forest. Just across the Palk Strait lies the island of Sri Lanka. Formerly much of the southwest of the island supported rain forest, but now only small fragments remain.

New Guinea and Australia

The fourth largest block of rain forest covers most of the large island of New Guinea, except for the dry southern and eastern margins, and the highest mountain peaks (Fig. 1.4). Although biologically uniform, the island is divided politically into two halves: the western half forms the Indonesian province of Papua and the eastern half is the independent country of Papua New Guinea. The neighboring and largely dry continent of Australia also supports a small area of rain forest in the northeast. The largest block of rain forest occurs along the coast between Cooktown and Townsville, but there are also numerous smaller patches.

Fig. 1.5 Reconstruction of an early Miocene (~20 million years old) rain forest community from the Riversleigh World Heritage site in northwestern Queensland, Australia. (By artist Dorothy Dunphy from the book *Riversleigh* by Michael Archer, Suzanne J. Hand, and Henk Godthelp; Reed, Chatswood NSW, 1994.) The site supported a far wider range of marsupial forms than inhabit Australia's rain forests today, including marsupial lions, carnivorous kangaroos, and browsing cow-sized diprotodontids.

Australia's tiny rain forest area has many similarities to the much more extensive rain forests of New Guinea – similarities that were even greater in the recent past (Hocknull et al. 2007) – but also many differences, including several distinctive endemic plant genera. These differences reflect the very different histories of the two regions, despite their proximity and intermittent contacts. Rain forest covered much of northern Australia during the early to middle Miocene (23 to 15 million years ago) (Fig. 1.5), but has become restricted to the northeast because of the subsequent drying of the continent (Long et al. 2002). Most of the rain forest in New Guinea, by contrast, occupies land that was uplifted above sea level only in the last 10–15 million years, making this the youngest of the major rain forest blocks.

Madagascar

The final major area of rain forest is on the island of Madagascar (now the Malagasy Republic) (see Fig. 1.3). Although Madagascar has about two-thirds the land area of New Guinea, most of the island is very dry, and rain forest was confined to a 120 km (75 miles) wide band along the eastern edge. Humans first arrived in Madagascar around 2000–2500 years ago, resulting in mass extinctions among the larger and more vulnerable vertebrates (Burney et al. 2003). Rain forests seem to have been the last habitat to be settled by humans, but most of the rain forest band has now been cleared, and what remains is fragmented and in many places highly degraded.

Rain forest environments

Rainfall

The very name "tropical rain forest" suggests a steamy jungle that is unfailingly hot and wet, every day of the year. In reality, there is no place on Earth where it rains every day, and rain forests are found in a surprisingly wide range of climates. In the tropical lowlands, these forests grow on almost all soil types where the annual rainfall is well distributed and greater than about 1800 mm (70 inches). Mean annual rainfall can be as low as 1500 mm on sites with soils that can hold water well or where dry-season water stress is moderated by cloud or low temperatures. At the other extreme, the mean annual rainfall is greater than 10,000 mm (33 feet!) at Cherrapunji in northeast India, Ureka in Equatorial Guinea, and in parts of the Chocó region of western Colombia.

To a first approximation, the amount and timing of rainfall in the tropics is controlled by the seasonal movements of the Intertropical Convergence Zone (ITCZ) – a band of low pressure, cloudiness, and rainfall that migrates north and south a month or two behind the overhead sun. The ITCZ results from rising warm air masses over regions where the sun is most directly overhead at midday. In this simple model the equatorial region is continuously influenced by the proximity of the ITCZ, so it is wet all year, with two rainfall peaks, a month or so after the equinoxes. Away from the equator, rainfall is concentrated in the periods when the ITCZ is present and there are dry periods when it moves away. With increasing distance from the equator, the two rainfall peaks move together and the "winter" period of low rainfall becomes longer and drier. Near the margins of the tropics, there is only one, relatively short, wet season, and a long, rainless dry season.

This simple model works fairly well over the oceans, but many factors introduce complications over land. While all rainfall results from upward movements of moist air, surface heating by the overhead sun is by no means the only mechanism that can cause this. The most important disruptions to the general pattern described above occur when moist air is forced to rise over a mountain range, producing rainfall on the windward slopes. Where mountains face a sustained flow of moist air throughout the year, this effect can produce an everwet climate well away from the equator. This explains the presence of rain forest in eastern Madagascar, at latitudes where we might expect a seasonally dry climate. Other examples of high rainfall resulting from such "orographic" uplift include parts of the Caribbean coast of Central America, eastern Brazil, West Africa, the west coast of India, coastal Queensland, and many tropical islands. There are also anomalies in the opposite direction: dry climates at latitudes we expect to be wet. The most striking interruption to the equatorial belt of rain forest climates is in East Africa, where a combination of relatively dry monsoon air flows and large latitudinal movements of the ITCZ makes most of the region too dry for rain forest. Cold ocean currents produce anomalously dry climates in western Ecuador, while other dry areas are in the lee of mountain ranges. As a result of these and other factors, the overall patterns of rainfall and rainfall seasonality in the tropics can be very complex.

All the major rain forest regions have relatively dry and relatively wet areas, but the amount of rainfall in the most extensive forest type varies between regions (Fig. 1.6). In general, the wettest rain forests are those of equatorial Southeast Asia, centered on the core Sundaland region of western Indonesia and Malaysia, and those on the island of New Guinea. In both these regions, the mean annual rainfall exceeds 3000 mm over large areas. In contrast, most Madagascan and American rain forest receives 2000–3000 mm, although there are wetter areas receiving 3000 mm or more in the upper Amazon basin, western Colombia, and the eastern slopes of Central America. The Atlantic Coastal Forest of Brazil is mostly somewhat drier, with less than 2000 mm of rain. Most African rain forests are distinctly drier than rain forests elsewhere, with an annual rainfall of only 1500–2000 mm, except in narrow fringes along the coast where the rainfall can exceed 4000 mm. Within particular regions, there is often considerable variation in rainfall, determined by distance from the coast, elevation, land use patterns, and other climatic factors. For example, in the Amazon River basin, rainfall varies from less than 1200 mm per year to over 6400 mm, with higher rainfall along the coast and in the northwestern edge of the basin.

Even in regions with very high total rainfall, most rain forests experience some dry months, when the water lost by evaporation and transpiration – around 100 mm per month – is greater than the amount of rain that falls. In tropical Asia outside the everwet Sundaland core, in almost all African rain forests, and in most of those in tropical America, there is an annual dry season lasting 1–4 months. This dry season is harsher in tropical America and continental Asia, where it is usually accompanied by cloudless skies, than in much of the African rain forest, where the rainless months are often misty and overcast (McGregor & Nieuwolt 1998). The predictable rainfall seasonality in these forests is reflected in all aspects of their biology, with more or less regular annual peaks in leafing, flowering, fruiting, and animal reproduction.

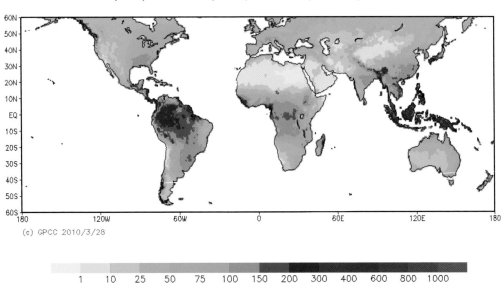

GPCC Normals 1951/2000 0.5 degree
precipitation for year (Jan – Dec) in mm/month

(c) GPCC 2010/3/28

1   10   25   50   75   100   150   200   300   400   600   800   1000

Fig. 1.6 Worldwide distribution of rainfall over land (1951–2000). The scale is in mm/month. (Global Precipitation Climatology Centre.)

In striking contrast to the typical African or American rain forest, there are large areas of Sundaland and on the island of New Guinea where there is no regular dry season, and the mean rainfall for every month is over 100 mm. In some places, dry months are very rare, and longer dry periods are unknown. In most Sundaland rain forests, however, dry periods occur every few years, with important consequences for their biology. These forests show what can only be described as "multiyear seasonality." Just as the fall in temperate zones brings on specific biological changes, such as leaf-dropping by deciduous trees and hibernation in some mammals, these less frequent events in Sundaland rain forests trigger dramatic biological changes, including mass flowering of many tree species, increased reproduction in many animal species, and large-scale migrations in others (see Chapters 2–7). The relative unpredictability of food availability in these forests has been used to explain many of their ecological characteristics.

In Borneo and the eastern side of the Malay Peninsula, these dry periods are associated with the El Niño–Southern Oscillation cycle (ENSO). The term El Niño means "Christ child" and was originally used by Peruvian fishermen to describe the warm current appearing off the western coast of Peru around Christmas time. Today, El Niño refers to the warm phase of a naturally occurring sea surface temperature oscillation in the tropical Pacific Ocean. The El Niño cycle is associated with the Southern Oscillation – a seesaw shift in surface air pressure between Darwin, Australia, and the Pacific island of Tahiti. Hence, the name El Niño–Southern Oscillation or ENSO cycle for this complex coupled cycle in the ocean–atmospheric system. There are three phases in the cycle: (i) a normal

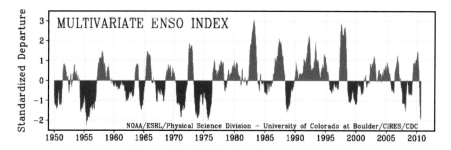

Fig. 1.7 Graph of the Multivariate ENSO Index (MEI) developed at NOAA: positive values represent El Niño events and negative values La Niña events. (From the NOAA MEI website.)

or neutral phase; (ii) the El Niño phase with unusually warm sea surface temperatures in the tropical Pacific; and (iii) the opposite La Niña ("little girl") phase with unusually cold sea surface temperatures. La Niña events occur after some, but not all, El Niños.

El Niño events generally occur at intervals of 2–8 years, although their intensity varies greatly (Fig. 1.7). The strongest recent events – and probably the strongest for the last century – were in 1982–3 and 1997–98, while there were weaker events in 1986–7, 1991–2, 1993, 1994, 2002–3, 2004–5, 2006–7, and 2009–10. A strong El Niño brings low rainfall to large areas of Indonesia and Malaysia, and coincides with mass flowering, then synchronized fruiting, by most individuals of the dominant tree family, the dipterocarps (Dipterocarpaceae), across huge areas (see Chapter 2 for more details). Many nondipterocarps flower and fruit at the same time so, for a few weeks only, the usually green and monotonous rain forest experiences a burst of colorful reproductive activity. The superabundance of flowers and fruits, in turn, leads to an inflow of nomadic animals, including giant honeybees, parakeets, and bearded pigs, and a burst of reproduction in the resident animal species.

Strong El Niño episodes lead to dry periods of increased severity in most other rain forests as well, including those of New Guinea, Africa, Central America, and Amazonia. However, the impact of these extreme events on the forest seems to be much greater in the Sundaland rain forests, which lack a regular dry season, than in forests that are adapted to an annual shortage of water. At Lambir in Sarawak, the long-term average rainfall of the driest month is 168 mm, which is not low enough to cause any water stress. During the exceptional El Niño-associated drought of 1998, in contrast, the total for the 3 months from January to March was only 139 mm, resulting in extensive leaf loss and a large increase in tree mortality (Potts 2003). The same El Niño episode in central Amazonia merely intensified the annual dry season, causing wilting and leaf fall but only a modest increase in tree mortality (Williamson et al. 2000).

El Niño episodes in the Neotropics also increase fruit production, but there is no equivalent of the mass, synchronized fruiting at multiyear intervals seen in the Sundaland rain forests. High fruit production during El Niño episodes on Barro Colorado Island, Panama, is followed by low fruit production in the wet, cloudy years that often follow, leading to increased mortality among primates and other fruit-eating animals (Wright et al. 1999). The same pattern occurs widely across Central and South America, leading to synchronous declines in the populations of frugivorous muriqui, spider and woolly monkeys in the year following El Niño events (Wiederholt & Post 2010). This Neotropical pattern of a regular fruit supply, with famines at multiyear intervals, is in striking contrast to the situation in the rain forests of Sundaland, where fruit famines are the normal situation and feasts occur at multiyear intervals. We will see in later chapters how these contrasting patterns of fruit (and flower) supply have apparently resulted in contrasting adaptations in the animals of these forests.

Before all differences between rain forests are attributed to the varying effects of the ENSO cycle, a note of caution is necessary. Reliable historical records of El Niño episodes go back only a century or so, but a variety of indirect sources of information suggest that both the frequency and intensity of El Niño events has varied considerably on timescales ranging from thousands to millions of years (Abram et al. 2007; Bush 2007). This raises the interesting possibility that the ENSO-associated patterns of plant and animal reproduction that can be observed today, particularly in the Sundaland rain forests, are not necessarily typical of even the last ten thousand years. Variability on longer timescales is likely to be even greater, so it is unlikely that plant and animal responses to the ENSO cycle are as finely tuned as they sometimes appear to be.

Temperature

Wetness is only half of the tropical rain forest equation: the other half is warmth. Typical equatorial lowland rain forests have a mean annual temperature of 25–26°C (77°F) and very little seasonal variation. At Danum in Sabah, Malaysia, for instance, the difference between the highest daytime temperature and the lowest night-time temperature is 8–9°C (14–16°F), but the difference between the average temperature of the hottest and coolest months is less than 2°C (36°F) (Walsh & Newbery 1999). Even near the equator, however, brief incursions of cold air cause surprisingly low minimum temperatures in some rain forest areas. In the upper Amazon basin, cold air from temperate South America moves northwards along the Andes and can bring temperatures as low as 11°C to Iquitos, just north of the equator (Walsh 1996). Even lower minimum temperatures (8°C) have been recorded in the lowland rain forest at Cocha Cashu, in Manu National Park, Peru, 12° south of the equator. Annual temperature ranges increase with distance from the equator because of reduced solar radiation in winter, as well as an increased impact of these "cold waves," known as "friagems" in Brazil. Near the southern margins of the tropical rain forest in South America and Australia, and the northern margins in Asia, frosts (i.e. subzero temperatures) can occur down to sea level, resulting in selective defoliation and shoot dieback in sensitive tropical plant species.

Nevertheless, the latitudinal limits of tropical rain forest are, in most places, set by drought rather than cold. Only in East Asia is there a continuous belt of lowland forest climates from the equator to the Arctic, without an intervening belt of climates too dry to support forest. Forests that closely resemble the typical tropical rain forests of Southeast Asia in terms of structure, floristics, and diversity extend north of the Tropic of Cancer in southwestern China, northern Myanmar, and northeast India (Corlett 2009a). The climate in southwest China is extreme for tropical rain forest, with a mean annual rainfall as low as 1500 mm and a long and very dry winter, with minimum temperatures regularly falling below 10°C (Zhu 1997). The low winter temperatures and the frequent thick fog reduce water stress, so the forest is still largely evergreen (Liu et al. 2008). An unforgettable experience for a tropical rain forest ecologist in China is to watch the fog clear on a wintry morning in Xishuangbanna, to reveal a rain forest with emergent dipterocarp trees rising to 60 m (200 feet). The dipterocarps all belong to one species – *Shorea wantianshuea* (*Parashorea chinensis*) – in contrast to the dozens of coexisting dipterocarp species in Bornean rain forests, but the whole appearance of the forest is distinctly tropical.

Temperatures also decline with increasing altitude above sea level, but in this case there is no associated increase in seasonality. On equatorial mountains, such as Mount Wilhelm in Papua New Guinea, the mean annual temperature at the altitudinal tree limit, at 4000 m (13,000 feet) above sea level, is only around 5°C (41°F), but there is very little seasonal variation (Hnatiuk et al. 1976). The temperature falls to near zero every night and rises above 10°C during the day: a climate that has been aptly termed "summer every day, winter every night." With increasing altitude above the lowlands, the rain forest becomes shorter, tree heights more even, the crowns and leaves smaller, rooting more shallow, and cold-intolerant plant families, such as dipterocarps and figs, progressively drop out. The direct effects of declining temperature may, however, be less important than changes in other factors, such as soil conditions, and a marked increase in soil organic matter is the most consistent environmental change at the upper limits of the lowland rain forest (Ashton 2003). The most dramatic vegetation changes often coincide with the zone of persistent cloud cover, where trunks and branches become gnarled and bryophytes – mosses and liverworts – cover all surfaces. This vegetation is often referred to as "cloud forest" or "mossy forest," although the bryophytes are mostly liverworts rather than mosses. In this book we focus our attention on lowland rain forests.

Wind

Another climatic factor with a major influence on the structure, if not the distribution, of tropical rain forest is wind. The combination of very tall trees and shallow root systems makes rain forests particularly vulnerable to strong winds. All rain forests are subject to occasional squalls of strong wind that may blow down single trees or, more rarely, fell large swathes of forest. Indeed, the ecological importance of these rare but widespread blowdown events may have been underestimated (Whitmore and Burslem 1998; Proctor et al. 2001). The most dramatic effects of wind, however, are in the rain forest areas subject to tropical cyclones (Fig. 1.8). Such cyclones are absent from the region approximately 10° either side of the equator that contains most tropical rain forest, but these storms affect with varying frequency the rain forests of the Caribbean, much of Central America, Madagascar, northern Southeast Asia (particularly the northern Philippines), northeastern Australia, and many oceanic islands. Sustained wind speeds during a major cyclone can exceed 70 m per second (150 mph), with brief gusts of much higher speeds.

The short-term impact of a single, severe, hurricane-strength cyclone is dramatic, with a large proportion of canopy trees uprooted or snapped off in the

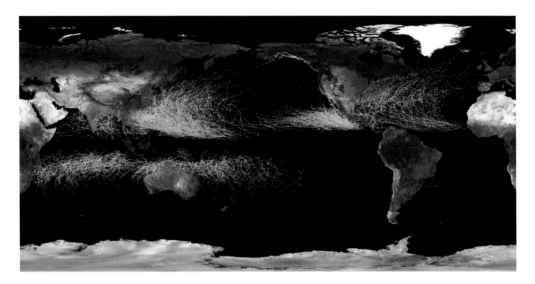

Fig 1.8 The tracks of all tropical cyclones from 1985 to 2005. Image created by Nilfanion, Wikimedia Commons (http://commons.wikimedia.org/wiki/File:Global_tropical_cyclone_tracks-edit2.jpg).

worst affected areas and almost complete defoliation in less damaged areas (Lugo 2008). Most of these damaged and defoliated trees will soon put out a new crop of leaves. However, "super cyclones" of extreme intensity occur at longer intervals in some regions, and may kill trees over a large area. In the longer term, repeated cyclone damage may allow an increased proportion of light-demanding tree species to persist in the forest. In areas with a very high frequency of cyclones, such as the islands of Mauritius and Fiji, and parts of Queensland, a distinct "cyclone forest" may develop. Such a forest is dominated by short-lived, rapidly reproducing, light-demanding species that are either less easily damaged by strong winds or able to complete their life cycle between successive hurricanes. Thus, the impact of a single cyclone will depend not only on its severity but also on the time since the previous one. Even rare cyclones eliminate the advantages for a tree of being taller than its neighbors, so rain forests in the cyclone belt tend to be relatively short (de Gouvenain & Silander 2003; Keppel et al. 2010).

Sunlight

In the equatorial region the sun is high in the sky throughout the year, but cloudiness and the high water vapor content of the air greatly reduce the amount of solar radiation reaching the forest canopy. A perhaps surprising consequence of this is that the availability of light – rather than water, temperature, or soil nutrients – can limit plant growth at certain times of the year. This was neatly demonstrated by installing high-intensity lamps above a canopy tree species, *Luehea seemannii*, in semievergreen rain forest in Panama (Graham et al. 2003). Trees given extra lighting during the cloudiest periods of the wet season grew more than those receiving only natural light. Light availability in the rain forest canopy is known to vary between sites, between seasons, and between phases of the ENSO cycle, but the potential consequences of this variation have not yet been investigated. It has been suggested, for instance, that above-average light intensities during El Niño events, as a result of reduced cloud cover, may be at least partly responsible for the enhanced fruit production observed in many rain forests (Wright and Calderón 2006).

Soils

There is an increasing amount of evidence that soil factors control plant distributions in tropical rain forests on both local and regional scales. Soil characteristics can also strongly influence plant biomass and, indirectly, animal biomass (Meiri et al. 2008). Exactly which soil factors are most important is still not certain, however, since soil texture, drainage, nutrients, and surface topography are all usually correlated and few studies in lowland rain forests have looked at the full range of possible factors.

Soil properties depend, in part, on the nature of the geological substrate from which they are formed. Over time, however, soil depth increases and the weatherable minerals in the soil are lost, leaving only quartz and clays. Those soil nutrients that were derived from weathering of the parent rock are either leached out of the soil (calcium, magnesium, potassium) or, in the case of phosphorus, form insoluble compounds that are unavailable to plants. High temperatures and rainfall in the humid tropics speed up these processes, but even then it can take several million years before the final stages are reached (Hedin et al. 2003). Soils this old are found only in geologically stable areas, such as the Amazon basin, Central Africa, and those parts of tropical Asia that are furthest from the margins of tectonic plates. Elsewhere, as in much of Central America, Southeast Asia, and New Guinea, tectonic movements or volcanic eruptions reset the clock at intervals, keeping the soils relatively young (Vitousek et al. 2010). The annual influx of river-borne sediments has the same effect in the floodplains of major rivers.

Deep, old, highly leached, and weathered soils are acid and infertile, with very low levels of plant-available phosphorus, calcium, potassium, and magnesium, and high levels of potentially toxic aluminum (Nortcliff 2010). Such soils are unsuitable for most forms of permanent agriculture, yet can support tall, dense, hyperdiverse rain forests. This apparent paradox reflects the ability of undisturbed rain forests on poor soils to recycle nutrients with very little loss. Most nutrients are withdrawn before leaves are dropped and the nutrients released in the litter layer are rapidly taken up by a dense mat of roots and their associated mycorrhizal fungi. If there is no unweathered parent material left within the root zone, the inevitable small losses of nutrients from the forest ecosystem must be replenished from the atmosphere, in dust and rain, and by biological nitrogen fixation.

Tropical rain forests occur on a wide range of soil types, by no means all of which are unsuitable for permanent agriculture. Relatively fertile soils occur in a variety of situations, such as in the volcanic areas of Java and on the flood-plains of whitewater rivers in the Amazon region. Unsurprisingly, rain forests on these more fertile soils are particularly prone to clearance, while long-term protection is most likely for forests on the least fertile sites. Deforestation is thus concentrated in the areas that support the highest plant and animal biomass, so the impact on biodiversity and carbon storage is even greater than crude estimates of percentage area loss imply.

Variations in soil texture, drainage, and chemistry affect the botanical composition of the rain forest, but only the most extreme soil types support distinctly different vegetation types. Most distinctive are the heath forests, which are also known by a variety of different local names, such as *caatinga* in Amazonia and *kerangas* in Southeast Asia. Heath forests develop on infertile, drought-prone, sandy soils derived from coastal deposits or the weathering of sandstone. Compared with typical tropical lowland rain forests, they are lower in stature and the trees have smaller, harder leaves. The streams that drain these forests are blackish or dark brown as a result of the presence of particulate and colloidal organic matter. Heath forests are found in all the major rain forest regions but they are most extensive in the upper reaches of the appropriately named Rio Negro (Black River) in South America. Other distinctive, but more variable, forest types occur on soils derived from limestone, as well as those on ultramafic (iron- and nickel-rich) rocks. Forests on these soils are also typically low in stature, with many distinctive plant species.

Different forest types also develop on sites where peat, consisting largely of partly decomposed woody plant material, has built up to such a depth that the forest is isolated from the ground water. These peat swamp forests are totally dependent on nutrient input from the rainfall, which also saturates and preserves the peat, and both the height and species diversity of the vegetation decrease with increasing peat depth. Raised, deep peat beds are found only in areas with high rainfall and without a long dry season, and are particularly extensive on the islands of Borneo, Sumatra, and New Guinea. Deep peat also occurs in parts of the Amazon region, but its extent is currently unknown (Lähteenoja et al. 2009).

Flooding

Flooding by river water produces an array of different forest types depending on whether the floods are permanent or periodic, and whether the periodicity is daily, monthly, or annual. Freshwater swamp forests, known locally as várzea, are most extensive along the Amazon River, which has annual floods and is also influenced by tides up to 900 km (600 miles) from its mouth (Goulding 1989). Extensive, but little studied, swamp forests also occur in the Congo River basin, and there are smaller areas in New Guinea and Southeast Asia. Freshwater swamp forests generally support a lower diversity of plant species than dryland forests, presumably because of the problems of dispersal, germination, establishment, and growth in an environment that experiences such seasonal extremes (Lopez & Kursar 2007). Várzea swamp forests in Amazonia support a lower diversity of mammals than adjacent terra firme (unflooded) forests, but a higher density of arboreal primates (Haugaasen & Peres 2005) and the same is likely to be true in other rain forest regions. Near the mouths of major rivers, freshwater swamp forests are replaced by brackish-water swamp forests and then mangrove forest, which has a much simpler structure and lower plant diversity than other tropical forest types.

Rain forest histories

Plate tectonics and continental drift

To understand the similarities and differences between modern rain forests, it is necessary to learn about their pasts. Most of the land masses that currently support tropical rain forest have a common origin in the ancient southern supercontinent of Gondwana (Fig. 1.9) (Morley 2007). Gondwana means "land of the Gonds," and is named after a tribe from southern India whose land provided the first evidence that India had been part of the supercontinent. The core of modern Southeast Asia is made up of continental blocks that broke away

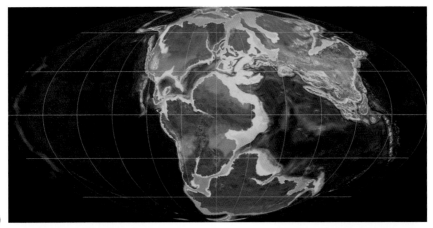

(a)

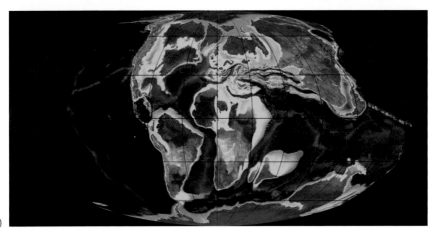

(b)

Fig. 1.9 Movements of the continents have had a big influence on the biogeography of the tropics. (a) In the later Jurassic (150 million years ago), the southern continents were connected into the supercontinent of Gondwana, and species could migrate among these areas. (b) By the late Cretaceous (90 million years ago), South America, Africa, India, and Australia had separated from one another, and connections with northern continents were severed.

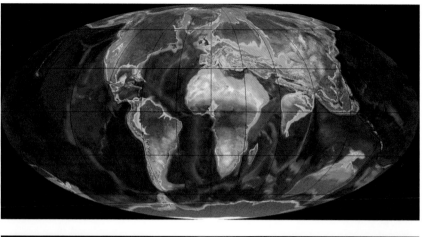

(c)

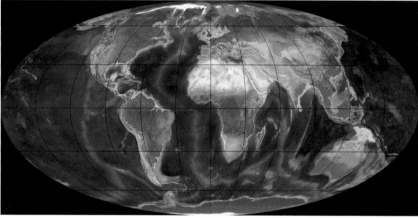

(d)

Fig. 1.9 (*cont'd*) (c) In the middle Eocene (50 million years ago), Madagascar had separated from Africa, India was moving toward Asia, and New Guinea had started to emerge. (d) By the early Miocene (20 million years ago), India had merged with Asia; Borneo, Sumatra, and neighboring islands had emerged; and Africa was connected to Asia.

from Gondwana between 400 and 160 million years ago, which is too early for them to have carried modern groups of plants and animals. Australia, India, Madagascar, Africa, and South America, in contrast, separated from each other and drifted north during the late Jurassic, Cretaceous, and early Tertiary (160 to 30 million years ago). This is the period during which many modern groups of plants and animals originated, so the sequence and timing of the break-up has had a significant influence on modern biogeographical patterns. New Caledonia, an island east of Australia, and the Seychelles, an archipelago east of Africa in the Indian Ocean, are also fragments of Gondwana.

India broke away early (130 million years ago) and, although the timing of India's collision with Eurasia is still uncertain (Rust et al. 2010), its modern flora and fauna are Asian. Africa's longer period of isolation in the late Cretaceous

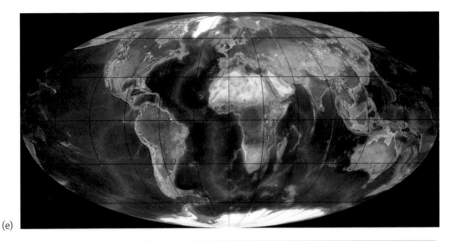

(e)

Fig. 1.9 *(cont'd)* (e) The modern world, showing the establishment of Central America and the broad contact of Africa and Asia. (Maps copyright Ron Blakey, Northern Arizona University, Geology.)

and early Tertiary produced a spectacular radiation of forms in an endemic clade of mammals, the Afrotheria, represented today by fewer than a hundred species in six orders: Proboscidea (elephants), Hyracoidea (hyraxes), Macroscelidea (elephant shrews or sengis), Tubulidentata (aardvark), Afrosoricida (golden moles and Madagascan tenrecs), and Sirenia (dugong and manatees). Although Africa has now been physically connected to Asia for at least 20 million years, climatic barriers such as large arid zones (represented today by the Sahara Desert) have allowed only limited exchange of rain forest taxa between the two continents for most of this period. Madagascar has remained isolated for 90 million years, and its relatively small rain forest area has developed on its own unique evolutionary path.

South America remained isolated from other continents for over 70 million years. This presented the opportunity for peculiar animal forms to evolve that are found nowhere else in the world, including the sloths, armadillos, and anteaters in the endemic order Xenarthra, the New World monkeys (see Chapter 3), and endemic radiations of marsupials, rodents, and ungulates (see Chapter 4). South America's long isolation finally ended when the Isthmus of Panama rose 3 million years ago, connecting the two American continents and allowing the intermingling of North American and South American faunas (Fig. 1.10). This dramatic event has become known as the Great American Interchange (Webb 1997). Some movement of animals between continents via rafting, long-distance dispersal, and island hopping appears to have begun as early as 8–10 million years ago and the formation of the land bridge seems to have been of less significance for plants (Cody et al. 2010). The North American invaders, which included such familiar mammalian groups as the cats, tapirs, deer, and squirrels, underwent explosive diversification in South America and, today, their descendents make up more than half the mammal fauna. Despite their ability to fly, bird exchanges increased greatly at the same time (Weir et al. 2009). This merging of previously separated faunas resulted in many extinctions, but may also have contributed to the exceptional diversity of the modern Neotropical biota.

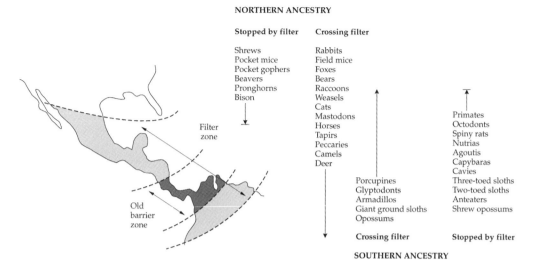

NORTHERN ANCESTRY

Stopped by filter	Crossing filter
Shrews	Rabbits
Pocket mice	Field mice
Pocket gophers	Foxes
Beavers	Bears
Pronghorns	Raccoons
Bison	Weasels
	Cats
	Mastodons
	Horses
	Tapirs
	Peccaries
	Camels
	Deer

Primates
Octodonts
Spiny rats
Nutrias
Agoutis
Capybaras
Cavies
Three-toed sloths
Two-toed sloths
Anteaters
Shrew opossums

Porcupines
Glyptodonts
Armadillos
Giant ground sloths
Opossums

Crossing filter **Stopped by filter**

SOUTHERN ANCESTRY

Fig. 1.10 The Central American land bridge, which was formed 3 million years ago, acted as a filter, preventing some North American animal families from crossing the barrier and allowing many others to pass on to South America. In contrast, many South American families did not disperse to North America. (From Lomolino et al. 2006.)

Central America is a composite of geological units of different ages and origins whose biota, before the Interchange, was dominated by North American organisms. The rain forests of Central America today, however, are overwhelmingly dominated by plants and animal groups that are shared with South America, and the effects of the deep-sea barrier that separated Central and South America until 3 million years ago have been almost totally erased.

Australia and New Guinea are still isolated from Southeast Asia by marine barriers, but the northward movement of Australia in the Miocene – continuing to this day – gave rise to the Indonesian archipelago, which has permitted an increasing interchange with Southeast Asia for organisms that can disperse from island to island, such as birds, bats, insects, and many plants. The modern rain forest biotas of Australia and New Guinea are thus a mixture of lineages from two very different origins: ancient Gondwana and post-Miocene Asia (Oliver & Sanders 2009). The distinctive differences in animal communities between Southeast Asia and New Guinea were first noted in the 19th century by the great naturalist Alfred Russel Wallace (Wallace 1859). In recognition of his discovery, the line of separation between these two biotic regions is now called "Wallace's line" and the region between Borneo and New Guinea, with its numerous islands, is known as Wallacea (Fig. 1.11). Geologists estimate that Australia and New Guinea will become connected to Asia in about 40 million years time, allowing their biotas to mix more completely in a "Great Australasian Interchange."

Changes in climate and sea level

Over the last few million years, rain forests expanded and contracted in area depending on the climate of the times (Morley 2007). Fossil evidence suggests

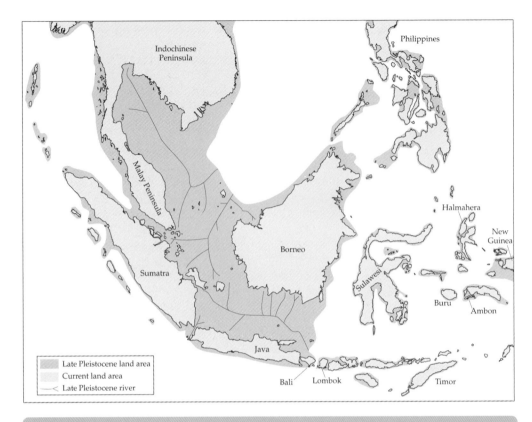

Fig. 1.11 The Sunda Shelf (shaded) was exposed during the last glacial maximum, allowing the movement of animals and plants between Borneo, Sumatra, Java, the Malay Peninsula, and many smaller islands. Note that the Philippines and Sulawesi were still separated from the Sunda Shelf. The area between the Sunda Shelf and the Sahul Shelf, which surrounds New Guinea and Australia, is known as Wallacea; the western boundary of Wallacea was described by Wallace as the eastern limit of distribution for many species of Asian animals, and is known today as Wallace's line. (From Lomolino et al. 2006; after Heaney 2004.)

that the tropical lowlands were both substantially cooler and, in large areas of the tropics, a lot drier during the glacial periods (ice ages) that occupied most of the last 2 million years. Lower atmospheric carbon dioxide levels during the glacial periods may also have been an important factor and may complicate the interpretation of the climatic records. These changes altered the composition of tropical rain forests and reduced their ranges. Pollen records from many sites show that montane or savanna plants were more widespread in glacial times and that rain forests disappeared from marginal areas. Rainfall remained high in many upland areas, but lower temperatures probably made these areas unsuitable as "refuges" for lowland rain forest organisms.

Australia and Africa show the strongest evidence for glacial drying and rain forest contraction. In Africa, rain forest was reduced to perhaps 10% of its area during the glacial maximum, persisting only in a few areas with high rainfall and as gallery forests along river margins (Morley 2000). This desiccation of African rain forests is a major reason for the far lower current diversity of palms, orchids, epiphytic species, and amphibians in African rain forests in comparison with rain forests in the Amazon and Asia (see Chapter 2). The impact of glacial changes was less severe in the Asian and American tropics, and on the islands of Madagascar and New Guinea, due to higher levels of rainfall associated with mountain ranges. Such upland areas are notably lacking in the Congo basin, and are found only as isolated mountains in East Africa.

Lower sea levels (up to 130 m (430 feet) below the present level) during glacial maxima linked the major land masses of insular Southeast Asia – Sumatra, Java, and Borneo – to the Malay Peninsula and Asian mainland, and greatly increased the exposed land area (Fig. 1.11). Although much of Southeast Asia seems to have been too dry to support rain forest during glacial times, both modeling studies and a limited amount of fossil evidence show that large areas of lowland rain forest persisted (Cannon et al. 2009). Much of this rain forest was on the exposed Sunda Shelf so, in striking contrast to the other rain forest regions, the total extent of rain forest in Southeast Asia was larger in glacial times than in the interglacials. Past land connections between the major modern land areas of Southeast Asia are reflected in the similarities of their modern forest biotas, although recent studies suggest that migration across these glacial land bridges was more limited than previously assumed, presumably because the lowest sea levels coincided with the driest climates and greatest fragmentation of the forest (e.g. den Tex et al. 2010). Connections among the islands could have occurred via migration along river valleys that extended from the coastal plains of northwestern Borneo, Sumatra, western Java, and the east coast of the Malay Peninsula onto the exposed Sunda Shelf. In contrast, most of the Philippines, Sulawesi, and the smaller islands between Sulawesi and New Guinea remained as islands, albeit often larger and connected among themselves, even at the lowest sea levels.

New Guinea had land connections to Australia during the glacial maxima, with the most recent connection being interrupted only 8000 years ago. However, the cooler, drier climate of those periods probably restricted interchange to those species that could migrate along the forested margins of waterways, and recent molecular studies suggest that most dispersal events pre-date the Pleistocene (e.g. Malekian et al. 2010). New Guinea and Australia share rain forest species of tree kangaroo, possums, birds, snakes, frogs, and even fishes – and shared more as recently as 300,000 years ago (Hocknull et al. 2007) – but New Guinea, with its larger rain forest area, has also developed its own unique biota, which is far richer than that now found in Australia.

A key point here is that the drier, cooler glacial episodes contracted African forests and divided them into smaller, isolated blocks, and probably had a similar effect on the Amazon basin as well. In contrast, glacial periods created land connections among the islands of Southeast Asia and the Asian mainland, and between New Guinea and Australia. As a result, the rain forests of Southeast Asia are more uniform, and those of Africa and South America less uniform, than might be expected from their present-day geography. Moreover, the cyclical changes in the total area of rain forest in response to the glacial cycles appear to have been in opposite phase in Africa (glacial minimum) and Asia (glacial maximum), with the other regions between these extremes.

Human occupation

One of the most important, but least understood, differences between rain forest regions is in their histories of human occupation. The popular idea that tropical rain forests were untouched, virgin ecosystems until the 20th century is a myth that has proved very hard to dispel, even among scientists. All rain forests have been modified by people and they cannot be understood if this fact is ignored. The broad picture of the evolution and spread of humans across the Earth's surface is now quite well documented (Fig. 1.12). Modern humans originated in Africa and spread overland to the warmer regions of Asia. New Guinea, Australia, and the Americas were colonized during the last glacial period, when low sea levels eliminated or reduced the water gaps. Larger water gaps remained a barrier until the development of improved boating technology within the last few thousand years. Madagascar was the last major tropical land mass to be reached by humans, a mere 2500–2000 years ago. Some smaller, more isolated tropical islands remained uninhabited until the last few hundred years.

The presence of people on a continent does not necessarily mean that they occupied the rain forest. Dense tropical forests are one of the least attractive environments for human occupation, since only a small proportion of the edible plant and animal material is accessible from ground level. Heavy rainfall washes away soil and mineral nutrients, requiring specialized techniques for agriculture, such as shifting cultivation and tree farming. Tropical rain forests may, however, have appeared more attractive before the first human hunters eliminated the most vulnerable ground animals, and recent archeological work

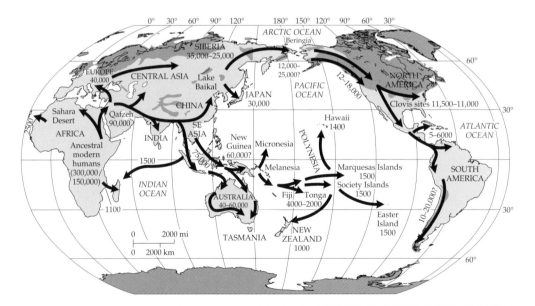

Fig. 1.12 The spread of modern humans over the world's surface. Values indicate the number of years that humans have been in a place; for example, humans arrived in Australia 40,000–60,000 years ago. (From Lomolino et al. 2006.)

has tended to push back the dates of first occupation (Mercader 2003). Some areas have been inhabited long enough for a distinct human body type to have evolved (Perry & Dominy 2009). "Pygmies", with average adult male heights < 155 cm (5 feet), occur in Africa, Southeast Asia, and South America, largely in populations that hunt and gather food in rain forest. And although hunters were the first humans to enter rain forests, evidence for cultivation, in the form of charcoal layers, broken pottery, crop remains, and modified soils, shows that most rain forest regions have also supported agricultural populations for millennia. In Amazonia, pre-Columbian agriculture left scattered patches of rich, black soil, known as *terra preta de Indio* (Indian dark earth), which are still prized by farmers for their fertility (Junqueira et al. 2010).

Origins of the similarities and differences among rain forests

The similarities and differences among the tropical rain forests in different regions can be explained in two major ways: "ecological" explanations relate the similarities and differences between rain forests to similarities and differences between their present-day environments, while "historical" explanations relate them to events that happened in the past. Each of these major types of explanation can, in turn, involve a huge range of possible factors; for example, soil nutrients and rainfall seasonality are ecological factors, while the movement of tectonic plates, changes in climate and sea level, and past human impacts are all historical factors. Further complications arise when ecological and historical factors interact as, for instance, when human impacts have been concentrated on the most fertile soils.

To distinguish between the many possible explanations for the distribution of a particular group of organisms, we need to know several things. First, we need to know the pattern of branching of lineages during evolution of the group – its phylogeny. The phylogeny for the primates, for instance, tells us that New World primates arose as a branch of the Old World primates (see Chapter 3), while that for army ants shows that the New and Old World species evolved from a common ancestor (see Chapter 7). Second, we need to know the timing of these branching events in relation to the availability of dispersal routes between rain forest regions. If, as current evidence suggests, the New World primates branched from their Old World ancestors around 30 million years ago, then they must have crossed the sea to reach the Americas, since no land route was available during this period. In contrast, the split between the two main army ant lineages may just be old enough for it to have occurred while there were still dryland connections between Africa and South America. Finally, we need to know the pattern of extinctions. Today, only Africa and Asia support really large (> 800 kg) species of mammals, but the fossil record shows that tropical America had a diverse "megafauna" when humans first arrived.

In the past, most discussions of these issues were largely speculation. In the last few decades, however, molecular techniques have greatly improved our understanding of the phylogenies of major plant and animal groups. These techniques can also provide approximate dates for branching events, by counting the number of mutations that have accumulated since, although this "molecular clock" must be calibrated from the very incomplete fossil record. The DNA evidence so far has provided a surprising result: most lowland rain forest species examined appear to have diverged from their nearest living relative prior to the Pleistocene glaciations of the last 2–3 million years, indicating that these biological communities are very old and speciation processes are slower than expected. This new understanding of phylogeny has coincided with the equally striking advances in our understanding of the Earth's history that have come from developments in plate tectonics and paleoclimatology (the study of past climates). Together, these developments have given us new insights into rain forest communities.

Similarities

Some similarities between rain forests in different parts of the world, such as the many shared families and genera of plants, may be inherited from ancient Gondwana. For a group of organisms to have reached all the major parts of Gondwana before it broke up, it would need to have originated at least 130–140 million years ago, but land connections persisted longer between some fragments than others, so such regions would be expected to have more similarities than areas that broke away completely at an earlier time. Moreover, dispersal between the fragments of Gondwana would have been relatively easy while they were still close together in the late Cretaceous and early Tertiary. A Gondwanan origin has therefore been suggested at one time or another for many groups of rain forest organisms, with a ride north on India providing a plausible route into tropical Asia. However, very few of these suggestions have been backed up by evidence – from fossils or molecular clocks – that the organisms involved existed early enough to take advantage of these opportunities. One exception is the worm-like scolecophidian snakes, where the timing of the initial diversification appears to match that of the break-up of Gondwana, but even in this ancient group several oceanic dispersal events are need to explain the current distribution (Vidal et al. 2010).

In the early Tertiary, the extension of frost-free climates to much higher latitudes than at present provided an alternative, northern, route between some of the rain forest regions. Around 50 million years ago, during the early Eocene, tropical forests grew at latitudes that today would be considered temperate, and a land bridge across the North Atlantic via southern Greenland provided a frost-free link between western Eurasia and North America (Milne 2006). Many rain forest plant genera are old enough to have used this "boreotropical connection." Cooling at the end of the Eocene broke this link forever by making the high latitudes too cold for tropical organisms.

Other similarities undoubtedly reflect later dispersal events as the Gondwanan fragments approached, and were eventually joined with, the northern continents: first India (34–55 million years ago), then Africa (20 million years ago), and finally South America (3 million years ago). Although there is still no dry-land connection between Australia/New Guinea and Southeast Asia, the largest water gap that an organism would have needed to cross during the last period of low sea level was less than 70 km (45 miles), making dispersal relatively easy for some types of organisms. Dispersal across much larger water gaps is rare, but not impossible, particularly for plants, and its significance in explaining current distribution patterns may have been underestimated (Givnish & Renner 2004).

Many other similarities between rain forest regions are not a result of shared ancestry at all, but of convergent evolution – the development of similar adaptations by unrelated organisms because they inhabit similar environments. Similarities resulting from convergent evolution have received a lot of attention in the past, but they are often superficial, as we shall see later in the book. Common ancestry and convergence are not mutually exclusive explanations for similarities, since the more closely related two organisms are, the more likely they are to evolve similar adaptations to similar environments.

Differences

The main theme of this book is not the similarities, but the many and important differences between rain forest regions. Such differences could have arisen for a variety of reasons, which are often impossible to disentangle. The simplest explanation for many of the biological differences is the ecological one: that they are a response to the differences in the physical environments of the various regions that have been outlined already. Even closely related organisms may evolve divergently in each region because the rain forest environments differ. For the same reason, some types of organisms may be able to invade and diversify in the rain forest in one region but not in another. The major differences among regions in the amount and seasonal distribution of rainfall have been called upon to explain many of the differences among rain forest regions. Rain forest organisms themselves also form an important part of the environment, and late arrivals may be unable to establish or diversify if their potential niches are already occupied. Thus the absence of specialist leaf-eaters among the New World primates may reflect their late arrival in a continent that already had leaf-eating sloths (see Chapter 3).

The different plate tectonic histories of the rain forest regions can also explain many of the differences in which groups of organisms are present or absent. Particular groups of plants and animals are shared between regions only if these regions were once connected or dispersal between them was once possible. Without the land bridge provided by the Isthmus of Panama, for instance, the rain forests of South America would be even more distinctive in comparison to rain forests elsewhere than they are now. Chance may play a major role here, particularly when dispersal on or over the open sea is involved. In many cases, the presence of a particular group of organisms in a region appears to be the result of a single, very unlikely event. The presence of primates in South American and Madagascan rain forests, and their absence in New Guinea, can perhaps be explained in this way.

Extinction is another cause of differences, and one that is often difficult to detect. The recent chance discovery of a single fossil honeybee (*Apis* sp.) in Middle Miocene rocks in Nevada, for example, has changed our understanding of honeybee biogeography, proving that their absence from the native bee fauna of the Americas reflects post-Miocene extinction, rather than a failure to get there (Engel et al. 2009). Changes in climate, as a result of plate movements, mountain uplift, or global climate change, may eliminate sensitive organisms from one region while they survive in another. Past human impacts are another, often overlooked, source of differences between regions. In the Neotropics, Madagascar, and New Guinea, the "megafauna" – the very large animals – were largely eliminated by

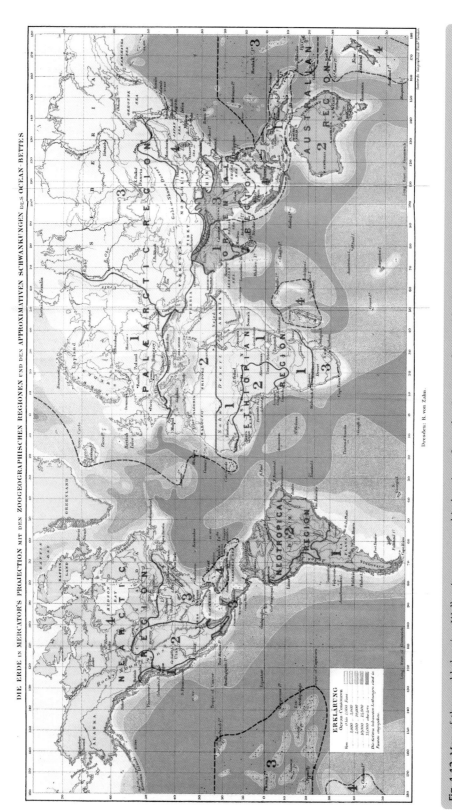

Fig. 1.13 Map of the world showing Wallace's zoogeographical regions, from the German edition of his book, *The Geographical Distribution of Animals*, published in 1876. This differs from the scheme used in this book only in the inclusion of Madagascar as a subregion in the Ethiopian (African) Region and the inclusion of Sulawesi in the Australian Region. Wallace later changed his mind on Sulawesi and transferred it to the Oriental (Asian) Region. (From Wikipedia.)

the first human arrivals, long before their ecological roles had been documented (Martin & Steadman 1999). In most cases, it is not even known if they inhabited rain forests.

Given time, evolution may fill the niches left vacant by the initial absence or subsequent extinction of a particular group of organisms. But evolution is a gradual process and can only act on the organisms that are present in the region. Particular niches may remain unfilled for a long time, or be filled in very different ways. When this happens, rain forests may differ not only in their biotas but also in the ways in which they function.

Functional consequences

The major question about tropical rain forests that we attempt to answer in this book is: why is what where? This is a very traditional approach to ecology, but one that has benefited greatly from the recent scientific advances outlined above. An alternative or complementary approach would be to look at the functioning of the whole rain forest ecosystem: such attributes as the production of biomass, the cycling of nutrients, and the rate at which these processes recover after natural and human disturbance. Combining the two approaches, we can ask what, if any, are the functional consequences of the observed differences between the biotas of the different rain forest regions? Does the presence of fungus-growing termites in Old World rain forests affect nutrient cycling? Does the absence of primates from New Guinea affect seed dispersal?

The answer to the great majority of such questions has to be that we do not know. We can and do speculate, but identifying functional differences requires comparisons between sites that have been carefully matched for the major environmental factors. These comparisons have not yet been made. Identifying which of the numerous biological differences between regions are responsible for particular differences in function will require the experimental removal or addition of the organisms in question. In some cases, such experiments could be done quite easily (e.g. the exclusion of browsing herbivores from an area of rain forest), in others they have already been done by accident (e.g. the introduction of honeybees to tropical America), while many more would be too dangerous to carry out in practice and should remain forever as "thought experiments" (e.g. the introduction of leaf-cutter ants to the Old World).

Many rain forests

For the reasons outlined above, the tropical rain forests of each region have distinctive characteristics and elements that give each a quality all its own. These differences were first formally recognized by Wallace 140 years ago (Fig. 1.13). The Neotropical rain forest is the most extensive, most diverse, and in many ways the most distinctive. The richest Neotropical rain forest sites have more tree species (see Chapter 2), more bird species (see Chapter 5), more bat species (see Chapter 6), and more butterfly species (see Chapter 7) living together than rain forests elsewhere, and the same pattern is found in many, but not all, other groups of organisms. The effects of South America's long isolation have not been erased by the influx from the north after the formation of the Panama land bridge, and many characteristic groups of plants and mammals are found in no other rain forest region. The epiphytic plant family Bromeliaceae gives an unmistakable appearance to the forest and their water tanks provide a unique canopy resource that is exploited by numerous species (see Chapter 2). Hummingbirds (see Chapter 5) and the flowers that they pollinate (see Chapter 2) show a degree of evolutionary diversification that is unparalleled in other rain forests, while other New World endemic groups of birds dominate the insectivore, frugivore, and scavenging niches. Both primates (see Chapter 3) and rodents (see Chapter 4) diversified along very different lines in the Neotropics from their ancestors in the Old World, with giant, long-legged rodents partly filling niches occupied by ungulates in Africa and Asia. The fruit bats (see Chapter 6) in Neotropical rain forests are an entirely separate evolutionary radiation from the fruit bats in all other rain forests, with different flight, sensory, and fruit-processing capabilities. Long columns of leaf-cutter ants (see Chapter 7) bringing cut wedges of leaf back to their underground nests are another distinctive feature of Neotropical rain forests that has no equivalent elsewhere.

African rain forests could hardly be more different. They are mostly drier, lower, and more open than rain forests elsewhere and have a relatively less diverse flora (see Chapter 2), apparently as a result of both present and past climates. Diversity is high in some other groups, however, including the primates (see Chapter 3) and termites (see Chapter 7). Perhaps the most distinctive feature of the African rain forests is the abundance and diversity of large, ground-living mammals, including many species of primates (see Chapter 3) and terrestrial herbivores (see Chapter 4). The African elephant is the largest of all rain forest mammals and the gorilla by far the largest primate. The bird fauna shares most major groups with Asian rain forests, but there is an endemic group of frugivores, the turacos (see Chapter 5).

Most Asian rain forests can be characterized as "dipterocarp forests," because they are dominated by large trees in the family Dipterocarpaceae (see Chapter 2). Many dipterocarps are among the tallest trees in any rain forest. Probably because of this dominance by a single family, Southeast Asian dipterocarp forests show a unique pattern of mass flowering and fruiting at 2–7-year intervals, described in more detail in Chapter 2. The short periods of "feast" separated by long periods of "famine" for any animal that eats flowers, fruits, or seeds, appear to shape the whole ecology of the forest. Another peculiar feature of these forests – the abundance and diversity of gliding animals (see Chapter 6) – may be connected to this phenomenon.

The rain forests of New Guinea are a paradox, with a basically Asian flora (see Chapter 2), but a very un-Asian fauna (see Chapters 3 and 4). This is the only rain forest region without primates or placental carnivores. Indeed, bats and rodents are the only placental mammals, and marsupials occupy most other mammalian niches. New Guinea and Australia are also notable for several unique radiations of birds (see Chapter 5), including families such as the cassowaries, birds of paradise, and bowerbirds that could probably not have survived predation by placental carnivores.

The assembly of Madagascar's unique rain forests appears to owe a great deal to chance. The entire nonflying mammal fauna can be explained by four colonization events, involving, respectively, a single ancestral species of lemur (see Chapter 3), a mongoose-like carnivore (see Chapter 4), a rodent, and an insectivore. Most of the birds also represent endemic radiations from a very small number of initial colonizers (see Chapter 5) and the same pattern is shown whenever modern molecular techniques are brought to bear on other groups of organisms.

Conclusions

In this first chapter, the major rain forest regions of the world have been introduced. The rain forests differ in their biogeographical histories and in both their past and present environments. Most notably, the presence or absence of particular groups of animals and plants gives each region a distinctive character. However, the purpose of this book is not merely to list differences in bio-geography, environment, and species, but to show how the differences impact on evolutionary and ecological relationships. And, in the end, the goal is to bring together the accumulated insights to examine how conservation strategies might use this information.

   In the next chapter, plant communities are examined first, as plants represent the building blocks of the biological community. Subsequent chapters will consider animal communities.

Chapter 2

Plants: Building Blocks of the Rain Forest

In a book about rain forests, it makes sense to cover plants before animals because plants are the foundation of any rain forest community. They not only provide the physical structure to the forest, but the diversity of plants also provides a huge variety of different food sources for animals. With a wide variety of plants, it is possible for many animal species to coexist by specializing on the flowers, fruits, seeds, leaves, twigs, bark, or roots of individual plant species or groups of related species. One beetle species, for example, could feed exclusively on the roots of ginger plants, while another species could live on the young leaves of a passionflower vine. The relationship between rain forest plants and animals is by no means one way, however. As we will see in the following chapters, most rain forest plants depend on animals to pollinate their flowers and disperse their seeds, and many have defensive or nutritional relationships with ants. It is not surprising, therefore, that the rapid rise to dominance of the angiosperms (flowering plants) around 100 million years ago appears to have coincided with the diversification of many of the groups of organisms that inhabit these forests, including amphibians, placental mammals, ants, beetles, and ferns (Wang et al. 2009).

It would seem logical to suggest that if plant communities differ among rain forest regions, for historical or environmental reasons, we should expect to see corresponding differences in animal communities. Such differences between the plant communities do exist, both in the structure of the forest and in the plant species present. Among the most striking differences are the abundance of dipterocarp tree species in Asian forests and the proliferation of epiphytic bromeliads in the Americas (see below). Both dipterocarps and bromeliads have profound effects on the animal communities in their respective rain forests, some of which are discussed below and some in later chapters. On the other hand, the lowland rain forests of Asia and New Guinea have relatively similar floras, but very different faunas. We might also predict that more diverse plant communities would support more diverse animal communities. In agreement with this, the rain forests of central and western Amazonia are exceptionally rich in both plants and animals (Fig. 2.1) and the rain forests of Southeast Asia

Tropical Rain Forests: An Ecological and Biogeographical Comparison, Second edition.
© Richard T. Corlett and Richard B. Primack. Published 2011 by Blackwell Publishing Ltd.

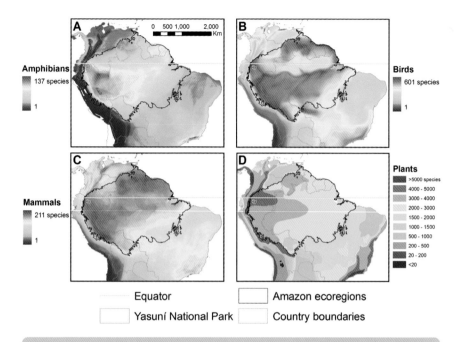

Fig. 2.1 Patterns of plant and animal diversity in northern South America, showing their partial coincidence (Bass et al. 2010). Note the position of Ecuador's huge Yasuní National Park in western Amazonia, where all four taxonomic groups reach their maximum diversity.

generally next richest in both. But there are also exceptions to this general pattern, making it unlikely that that there is a simple cause-and-effect relationship between plant and animal diversity. African rain forests combine a relatively low diversity of plant species with a high diversity in some groups of plant-dependent animals, including primates (see Chapter 3), squirrels, and terrestrial herbivores (see Chapter 4). Madagascar's rain forests, in contrast, are rich in plant species but relatively poor in animal species. In this chapter, we will describe the plant communities of each region, and in subsequent chapters, the role of these plant communities in structuring animal communities will be explored.

Plant distributions

All rain forests share the same basic growth forms of trees, shrubs, herbs, climbers, and epiphytes. In addition, they share many of the same major plant families, such as the Annonaceae (soursop family) (Fig. 2.2a), Arecaceae (palm family), Burseraceae (frankincense and myrrh family), Fabaceae (legume family), Lauraceae (avocado family), Meliaceae (mahogany family), Moraceae (fig family), Myristicaceae (nutmeg family) (Fig. 2.2b), Rubiaceae (coffee family), and Sapotaceae (the chicle family). The presence of the same major families in tropical forests throughout the world, despite continental separation for tens

(a)

(b)

Fig. 2.2 Fruits of two distinctive plant families found in tropical rain forests throughout the tropics. (a) Annonaceae fruits (*Polyalthia rumphii*) from Malaysia. This cluster of stalked fruits is the product of a single flower with many free pistils. (Courtesy of K.M. Wong.) (b) Nutmeg fruit (*Knema* sp., Myristicaceae), showing the single large seed completely covered in an orange aril, from Gunung Palung National Park, Indonesian Borneo. (Courtesy of Tim Laman.)

of millions of years, suggests that there is a conservative quality to these basic building blocks of the forest community, but the reasons for this are not at all obvious.

In general, the most ancient land plants have the most cosmopolitan distributions. Ancient plant families have not only had a longer time in which to spread around the world but, as discussed in the previous chapter, there also were overland dispersal routes between rain forest regions available in the Mesozoic and early Tertiary that did not exist in later times. Among the flowering plant families found in all the major rain forest regions, some, such as the Annonaceae, Arecaceae, Lauraceae, and Myristicaceae, may be old enough to have spread to all the southern land masses while they still formed part of Gondwana, and to have ridden north to Asia on India. Even in these families, however, some shared or related genera probably result from much later dispersal events (Pennington & Dick 2004). Other pantropical families, such as the Burseraceae, Moraceae, Meliaceae, and Sapotaceae, appear to have diversified well after the break-up of Gondwana and owe their present distributions to later migrations by various routes (Muellner et al. 2006; Smedmark and Anderberg 2007).

The presence of plants on oceanic islands that have never had overland connections shows that many species can disperse long distances, even over water. The Hawaiian Islands are 3900 km (2500 miles) from the nearest continent (North America), yet at least 272 flowering plant species have reached these specks of land in the vast Pacific Ocean (see Chapter 8). This includes species with light, fluffy, wind-dispersed seeds (such as members of the Asteraceae, the sunflower family), dust-like seeds (such as orchids, Orchidaceae), small seeds in fleshy fruits that could be retained inside the guts of long-distance seafaring birds (such as members of the Aquifoliaceae, the holly family), and floating seeds (such as some palms and legumes). Sometimes whole groups of plants and animals may be dispersed when a tangled mass of trees and vines falls into a rain-swollen river and is swept out to sea. Large expanses of water – as well as deserts and mountain ranges – are, however, barriers to the dispersal of many rain forest plant species. As discussed in Chapter 8, the rain forests of the Hawaiian Islands have no representatives at all from many important rain forest plant families. Hawaii is an extreme case, but the distributions of several major rain forest families, many genera, and almost all species show that the dispersal of plants between rain forest regions has been limited. Most such dispersal seems to have occurred when there were overland connections with an appropriate climate, but the role of long-distance dispersal across oceans has probably been underestimated (Givnish & Renner 2004).

From a floristic point of view, the rain forests of the Neotropics are the most distinctive: a pattern that we will see repeated in several animal groups in the following chapters. Not only do Neotropical rain forests have several important plant families (e.g. Bromeliaceae, Cactaceae, Cyclanthaceae, and Vochysiaceae) and many important genera that have few or no representatives in rain forests elsewhere, but they also lack some families (e.g. Pandanaceae) and genera that are important in most other rain forests. In comparison with the distinctiveness of the New World, the floras of Old World rain forests have more in common with each other. African rain forests are distinguished more by the relative poverty of their flora than by unique floristic elements. Madagascar has a very rich rain forest flora that has both African and Asian affinities, as well as unique features resulting from its prolonged isolation. The diverse flora of New Guinea's rain forest, in contrast, is relatively young and surprisingly similar at the generic and family levels to that of Asia. This similarity has encouraged botanists to treat the whole region from Southeast Asia to New Guinea as a single floristic unit, often called Malesia, with the main floristic boundary between New Guinea and Australia. This is in striking contrast to the vertebrate faunas considered in later chapters, where there are major differences between Asia and New Guinea and major similarities between New Guinea and Australia. This contrast reflects the much greater ability of most plants to disperse across multi-kilometer water gaps in comparison with most vertebrates.

Rain forest structure

An alternative to comparing the families, genera, and species of plants present in different rain forests is to compare their structure, including such characteristics as canopy height and the number of trees per hectare. An additional commonly used characteristic is the basal area of trees: that is the total area of the tree stems in a hectare, measured at 1.3 m (4.3 feet) height on the trees. Such comparisons are not as simple as might be expected, since different researchers have used different methods and made different measurements in various parts of the tropics. The most easily comparable data come from studies that describe the characteristics of a forest by censusing every tree in research plots, which are typically 0.1 or 1 ha (0.24 or 2.4 acres) in area. Large numbers of such studies have been published, and 46 have been gathered together and synthesized (de Gouvenain & Silander 2003). The main conclusions of this survey are that the typical dipterocarp forests of Southeast Asia are taller than rain forests elsewhere, followed in decreasing height by typical rain forests in South America, Africa, and Madagascar. There were not enough data available to include New Guinea and Australia in this comparison. The differences are substantial. Lowland rain forests in Southeast Asia typically have a tree canopy at 30–50 m (100–165 feet) above the ground, with emergent trees reaching 70 m or more; while, at the other extreme, most lowland rain forests in Madagascar are less than 30 m tall. African rain forests contrast with rain forests elsewhere in tending to have a lower density of trees. Across the range of densities, which vary from 300 to 1000 trees per hectare, African rain forests tend to be on the lower end of the scale, in the range of 300–600 trees per hectare. Despite the low density of trees, African forests do not have an especially low basal area of trees, suggesting that these forests are lowest in the density of small trees, which have less of an effect on total basal area.

The differences in canopy height between rain forests are largely attributed by de Gouvenain and Silander (2003) to the occurrence of tropical cyclones (see Chapter 1), which reduce the advantages that might otherwise accrue to a tree that is taller than its neighbors. All their Madagascan sites were subject to cyclones, as were particularly low forests in the Caribbean, Australia, and the Philippines. In contrast, the exceptionally tall forests of Malaysia and Indonesia are all outside the cyclone belt. Additional environmental factors are also likely to be important in some areas. Exceptionally poor or shallow soils reduce canopy height all over the tropics. It is also striking that the tallest forests are almost all dominated by trees in the family Dipterocarpaceae (see below), and it is possible that dipterocarps simply have an exceptional capacity for height growth, perhaps associated with the need to emerge above the main canopy in order to disperse their large winged seeds (Slik et al. 2010).

Structural differences between forests would be expected to have an impact on the communities of animals that climb or fly through the forest, although there has been no systematic study of this in the tropics. Tall forests may provide more opportunities for animals to specialize on a particular layer in the forest, with distinct animal communities in the understorey, subcanopy, canopy, and emergent layers. The density of trees and the degree to which adjacent trees are connected by overlapping crowns and woody climbers may also be important. High density and connectivity would be expected to make it easier for climbing animals to move through the forest, while providing more obstacles for flying animals. The effects of differences in structure will be most easily observed when adjacent forests in the same area are compared, while comparisons between rain forest regions are confounded by many other differences. One possible consequence of the distinctive structure of Southeast Asian dipterocarp forests for animal movement is considered in Chapter 6.

How many plant species?

Tropical rain forests are by far the most species-rich of all terrestrial ecosystems. Despite occupying only about 6% of the Earth's land surface area, Turner (2001a) estimated that tropical rain forests worldwide support 175,000 species of vascular plants (flowering plants, conifers, and ferns), which is around two-thirds of the estimated global total. Rain forest plant species are divided in an approximately 3 : 2 : 1 ratio between the Neotropics, the Asia–Pacific region (Asia, New Guinea, Australia, and the Pacific islands), and Africa plus Madagascar (Table 2.1). Within the Asia–Pacific region, the greatest numbers of species are found in the wettest and least seasonal areas, centered on the Sundaland region of Southeast Asia and the island of New Guinea. New Guinea could well have the richest flora in the region, but it is also the least well-collected and new species continue to be added by every expedition. Outside these everwet areas, in drier and more isolated parts of the region, the richness of species declines dramatically.

On a smaller scale, we do not have a complete species list of all the plants and animals for any area of rain forest, and complete lists of just the plants are rare. A survey of a 1 ha plot (100 m × 100 m, somewhat larger than a typical soccer pitch) in the exceptionally diverse rain forest at Cuyabeno, Ecuador, in western Amazonia, found a total of 942 vascular plant species, of which about half were trees and the remainder was divided among shrubs, climbers, ground herbs, and epiphytes (Balslev et al. 1998). Most other studies have only looked at the trees. The largest set of comparable data from around the world concerns the number of trees species more than 10 cm (about 4 inches) in diameter found in a 1 ha plot. There are still huge areas of tropical rain forest for which there are no plot data, so although some general patterns are clear, others will undoubtedly change as more data become available.

The most diverse tropical rain forest plots so far, with more than 250 trees species in a single hectare, are in central and western Amazonia, in the Pacific coast rain forests of Chocó Province, Colombia, in the Atlantic Coastal Forest

Table 2.1
The estimated numbers of plant species in the tropical rain forests of the world.

Area	Number of species	Area of rain forest (millions of ha)	Comments
Neotropics	93,500	400	Mainly Amazon basin; also includes some other forest types
African tropics	20,000	180	
African mainland	16,000		Includes West Africa, Congo region, and montane areas
Madagascar	4,000		A preliminary estimate; the flora is not well known
Asia–Pacific region	61,700	250	
Malesia	45,000		Malaysia and Indonesia, including New Guinea
Indo-China and adjacent areas	10,000		An estimate as no flora has been completed for the region
Southern India	4,000		Mainly the Western Ghats
Sri Lanka	1,000		Sinharaja and neighboring localities
Australia	700		Queensland rain forest
Pacific islands	1,000		Isolated rain forests with low species richness
Total	175,200		

Modified from Turner (2001a) and Whitmore (1998).

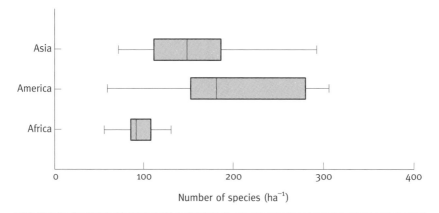

Fig. 2.3 The number of large tree species per hectare in rain forest sites in Asia, the American tropics, and Africa. The line in the box is the mean value, the horizontal line covers all values, and the box includes 50% of the values above and below the mean. (Modified from Turner 2001b.)

of Brazil, and on the Southeast Asian island of Borneo. Plots with more than 200 different tree species have also been found on the islands of New Guinea and Madagascar, but not – so far – in Africa (Fig. 2.3). The most species-rich African sites, as well many sites in Amazonia, Sundaland, and New Guinea, have 100–200 tree species per hectare.

Outside the core areas of each rain forest region, tree diversity is generally lower. Hence, in the Neotropics, forests in eastern Amazonia, the Guiana Shield (between the Amazon River and the mouth of the Orinoco), most of the Brazilian Atlantic region, Bolivia, and Central America, usually have fewer than 100 species per hectare, although there are some exceptions. Similar declines in diversity with distance from the core area occur in other rain forest regions, but in tropical Asia even the northernmost rain forest sites studied – in the Himalayan foothills of northeast India (Proctor et al. 1998), in southwest China (Lan et al. 2008), and on Hainan Island (An et al. 1999) – have more than 100 species per hectare. The isolated Asian rain forest areas of the Western Ghats and Sri Lanka, in contrast, have fewer than 100 species in a hectare. Large tree diversity is also generally lower on oceanic islands, with 84 species in one hectare on the Andaman Islands (Rasingam and Parathasarathy 2009), 42 in Puerto Rico (Thompson et al. 2002), and only 9 in Hawaii (Hawaii Permanent Plot Network, www.hippnet.hawaii.edu). Age can apparently compensate for isolation, however, with 124 species in a single hectare on Viti Levu, in the relatively ancient Fijian archipelago, with half these species endemic (Keppel et al. 2010).

An annual rainfall of less than 2000 mm (80 inches) or a severe dry season also reduces tree diversity, as do periodic flooding and extreme soil types such as sand, limestone, or peat. In the Amazon region, dry season length (i.e. the number of months with less than 100 mm of rainfall) is an excellent predictor of the maximum plot diversity within an area – with the highest diversity plots in areas with only 0–1 dry months – but a poor predictor of average plot diversity, since there are low diversity plots in all areas (ter Steege et al. 2003). The history of a rain forest area is probably as important as the current environment, particularly at the regional and global scales, with the highest diversities in areas such as the northwestern Amazon (Fig. 2.1), where rain forest plant species have had a place to persist with other organisms through the cooler, drier glacial episodes.

A network of much larger, 10–52 ha, plots has now been established at 30 forest sites across the tropics by the Center for Tropical Forest Science (CTFS) of the Smithsonian Tropical Research Institute (www.ctfs.si.edu) (Fig. 2.4). Within these plots, all plants more than 1 cm in diameter have been mapped, measured, and identified: with more than 300,000 individual plants in the largest rain forest plots! Periodic remeasuring of the plots gives information on growth, mortality, and regeneration. The results published so far show that species richness of large trees in single hectare plots is, in general, a good predictor of total tree species richness in these larger areas. African rain forests are an exception, with misleadingly low diversities at scales of 1 ha due to the tendency of one or several species to dominate a local area, but with more respectable diversities at scales of 10 ha and over. However, no African site approaches the very high diversities found in Amazonia and Southeast Asia, where the richest plots have more than 800 tree species greater than 10 cm diameter and more than 1000 species greater than 1 cm diameter in 25 ha.

To put the plant diversity of tropical rain forests in perspective, even the most species-rich temperate forests have fewer than 25 tree species in a 1 ha plot, and most have fewer than 10. In the richest tropical rain forests, on average every second tree is a new species, while many temperate forests are dominated over large areas by just one or two species. A mere 52 hectares (130 acres) of lowland rain forest at Lambir, Sarawak, supports more tree species (1175) than the

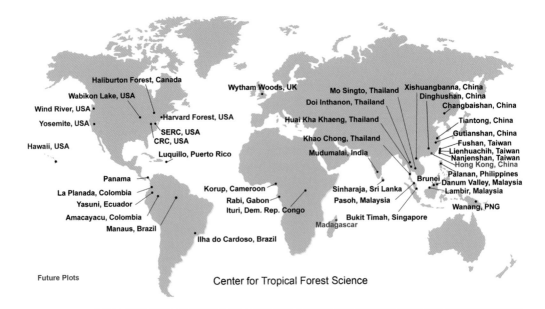

Fig. 2.4 Map of the global network of research plots coordinated by the Center for Tropical Forest Science (CTFS), showing the strong focus on tropical regions. CTFS and its partners monitor the growth and survival of approximately 3.5 million trees of 7,500 species. (Courtesy of CTFS.)

entire forests of the northern temperate zone – North America, Europe, and Asia (1166 species) (Wright 2002). Even the "common" species in tropical rain forests are "rare" by the standards of temperate forests, with important consequences for the animals that feed on them.

Low-diversity forests do occur in the wet tropics, but in most cases they reflect either extreme soil conditions, such as the "alan" peat swamp forests of Borneo dominated by *Shorea albida*, or recent disturbance. The young forests that develop after natural or human disturbances are sometimes composed almost entirely of a single pioneer species, often a species of *Macaranga* in Southeast Asia, *Cecropia* in tropical America, and *Musanga cecropiodes* or *Trema orientalis* in Africa. Large areas dominated by a single tree species do sometimes occur, however, on normal soils and with no evidence of recent disturbance. These seem to be particularly common in Africa (see below) and particularly rare in Asia. In South America, dominance by one or a few tree species is most common in the Guiana Shield region, where – as in Africa – tree species diversity is relatively low even in "normal" forests (Henkel 2003).

Widespread plant families

Pantropical families

Except on the most remote oceanic islands, the tree flora of any area of tropical rain forest is likely to include members of the following families: Annonaceae,

Arecaceae, Burseraceae, Clusiaceae (mangosteen family), Ebenaceae (ebony family), Euphorbiaceae (spurge family), Fabaceae, Lauraceae, Meliaceae, Moraceae, Myristicaceae, Myrsinaceae, Myrtaceae, Olacaceae, Phyllanthaceae (the phyllanthus family), Rubiaceae, Rutaceae, Sapindaceae (litchi family), and Sapotaceae (Heywood et al. 2007). There are also many genera that are shared by two or more continents, but very few species. One unusual example is the rain forest tree *Symphonia globulifera*, which is widely distributed in both African and American rain forests. The detailed fossil record of *Symphonia* has permitted a DNA-based reconstruction of its African origin, some 45 million years ago, and its trans-oceanic colonization of both Central and South America 15–18 million years ago, before they were joined by the Panama land bridge (Dick and Heuertz 2008).

Legumes

Among the shared families, the legumes (Fabaceae) (Fig. 2.5) stand out because they typically dominate the rain forests of both Africa and the Neotropics, in terms of basal area and overall biomass. In Madagascar, legumes dominate the dry forest types but are less important in the rain forest, although the endemic rain forest rosewoods (*Dalbergia* spp.) are extremely valuable timber trees that attract illegal loggers (Patel 2007). The legumes are generally less important

Fig. 2.5 Members of the legume family are easily recognized by their pod-like fruits, pea-like flowers, and compound leaves. Leaves and flowers of the legume *Fordia hemsleyana* in Chukai, Malaysia. (Courtesy of K.M. Wong.)

in New Guinea and Australia, while in tropical Asia the dipterocarps dominate (see below), but there are many prominent rain forest legume species in all rain forest regions. As well as producing many of the biggest rain forest trees, the legumes are also an important family of climbers. In contrast to most of the widespread plant families, the legume family does not appear to be particularly old, suggesting that its current pantropical distribution may be a result of dispersal across northern land routes during the early Eocene thermal maximum or transoceanic dispersal (see Chapter 1) rather than an ancient Cretaceous origin on Gondwanaland.

A special feature of the legume family is that symbiotic, nitrogen-fixing bacteria inhabit "nodules" on the root systems of many species. These legumes can absorb nitrogenous compounds from these bacteria, giving them a potential competitive advantage; the nitrogen absorbed from the bacteria allows legumes to build extra enzymes and proteins that can be used for rapid growth. It has also been suggested that nitrogen fixation is particularly advantageous on phosphorus-limited tropical soils because it allows legumes to secrete nitrogen-rich enzymes (or to stimulate enzyme production by microbes) that can release inorganic phosphate from organic compounds in the soil (Houlton et al. 2008). These nitrogen supplies can also be used to produce poisonous chemicals, such as alkaloids, that deter herbivorous animals from eating the leaves, bark, young fruits, and other plant parts.

Despite the potential importance of symbiotic nitrogen fixation by rain forest legumes, few species have been studied and it is not yet clear how widespread this capability is among leguminous trees and climbers, or how important it is to tropical forest ecosystems (Hedin et al. 2009). On a global scale, most genera in the subfamily Caesalpinioideae – the most important subfamily of legume trees in rain forests – cannot fix nitrogen, while most genera in the other two subfamilies, Mimosoideae and Papilionoideae, can. Even among those species that can produce nitrogen-fixing nodules, nodulation appears to be rare in undisturbed forests, with individual trees apparently increasing their intake of nitrogen only in response to nitrogen shortages. Given the huge range of legume abundance across tropical rain forests, from less than 2% of tree basal area in some dipterocarp-dominated Asian forests to near 100% at some African sites, a better understanding of the role of symbiotic nitrogen fixation should be a high priority for research.

Figs

The Moraceae (the fig family) is another important pantropical rain forest family, recognized by the milky latex that is found throughout the leaves, bark, and even immature fruit. In many tropical Moraceae genera, such as the breadfruit and its relatives (*Artocarpus*), individual fruits are embedded in a large, fleshy receptacle to form a complex "multiple fruit." By far the most important genus in the family is the fig genus, *Ficus*, in which both the flowers and fruits are completely enclosed inside a fleshy receptacle – the fig (Fig. 2.6). Fig plants have a wider range of growth forms than any other genus of plants: trees, shrubs, climbers, and epiphytes, as well as hemiepiphytes, which start as epiphytes but then send roots down to the ground, and the bizarre "stranglers," in which the aerial roots eventually enclose and kill the host, leaving a free-standing fig tree.

Fig. 2.6 Everybody loves a fig. A Bioko Allen's bushbaby (*Galago alleni*) feeding on the fruits of *Ficus sur* in the coastal forest of Bioko Island, Equatorial Guinea. (Courtesy of Christian Ziegler.)

Although the fig genus is pantropical – indeed, it is the only genus found in all the CTFS tropical forest plots – the diversity of species and growth forms differs greatly between the major rain forest regions. This may be a reflection of the apparently Tertiary age of the main fig radiation, with the spread of genus dependent on rare long-distance dispersal events (Lopez-Vaamonde et al. 2009). The more than 750 fig species are currently classified into six subgenera and 19 sections (www.figweb.org). Only two, endemic, fig sections occur in the Neotropics: around 110 species of epiphytes, hemiepiphytes, and stranglers in the section *Americana*, subgenus *Urostigma*, and 21 species of free-standing trees plus one strangler in the section *Pharmacosycea*, subgenus *Pharmacosycea*. In contrast, the islands of Borneo and New Guinea each support at least 150 species, representing all six subgenera and 10 or more sections, with the full range of growth forms found in the genus. Moreover, the local diversity of fig species in rain forests in Asia and New Guinea is extremely high, with 75 species (including 27 hemiepiphytes) recorded in the Lambir Hills National Park in Borneo (Harrison et al. 2003). As a continent, Africa is intermediate in fig diversity, with 81 species in seven sections, but only around 50 species occur in the rain forest, many in the near-endemic section *Galoglychia* in the subgenus *Urostigma*. Madagascar has only 25 species (16 endemic), of which around two-thirds occur in the rain forest. Local fig diversities in Madagascan rain forest also appear to be low, with only 12 species known from the Ranomafana National Park (Goodman & Ganzhorn 1997).

These differences in fig diversity between rain forest regions do not necessarily translate into differences in the role that figs play in the ecology of the forest. In all the rain forest regions, the largest crops of fig fruits are produced by the giant hemiepiphytic and strangler species, and there is little evidence to suggest that the density of these plants – which is low everywhere – varies significantly between regions. Fig plants are very important in supporting the vertebrate communities of tropical rain forests because ripe fig fruits are available throughout the year (Shanahan et al. 2001). This contrasts with most other tropical forest plants, where all individuals of a species fruit together at one time. At some times of the year, figs may be the only fruit available and are eaten by many birds and mammals. A single hemiepiphytic tree of *Ficus pertusa* at Cocha Cashu in the Manu National Park had its figs eaten by 44 species of diurnal vertebrates, ranging in size from 10 g (0.5 oz) manakins to 10 kg (22 lb) spider monkeys, over a 21-day period (Tello 2003). For this reason, figs have been termed "keystone species" – species with a disproportionate influence on other species in their ecosystem. Some animals depend on figs for most of their diet year round. In areas of Sumatra and Sulawesi, the densities of hornbills and primates in the forest appear to be directly related to the abundance of fig fruits (Kinnaird & O'Brien 2007). On the rare occasions when no figs are available, hornbills will leave the area in search of other food sources.

Fig trees produce these successive crops of fruits year round as part of a coevolved relationship with tiny (c. 2 mm) fig wasps (Agaonidae). Fig flowers are pollinated by these highly specialized fig wasp pollinators, which must enter the figs and lay their eggs in fig flowers to complete their life cycle. In general, each fig species is pollinated by a different wasp species, although an increasing number of exceptions to this strict one-to-one relationship have been found, including a case of at least three different wasp species pollinating the figs in a single crop on one African rain forest fig tree (Compton et al. 2009). Each wasp species recognizes its own fig species by the volatile chemicals the figs release when ready for pollination. If a population of fig plants stopped producing crops of fig flowers, the short-lived pollinators would die out, and the fig plants would not be pollinated. This was dramatically illustrated with several fig species in northern Borneo after exceptionally severe droughts during the 1998 ENSO (El Niño–Southern Oscillation) event caused a break in fig production (Harrison 2001). It took up to 2 years for the fig wasps to recolonize from unaffected regions.

Palms

The palms are one of the most characteristic families of tropical plants: indeed, in the popular imagination palms symbolize the tropics. There are more than 190 genera and 2000 species, most of which occur in rain forests. Palms are immediately recognizable by their unbranched woody trunks, either solitary or in clusters, a rosette of large, divided, feather- or fan-like leaves at the top of the trunk, and often complicated, hanging clusters of flowers followed by single-seeded fruits (Fig. 2.7). The most notable feature of the distribution of palms, like that of many other tropical families, is their relative impoverishment in Africa. Africa has only 16 genera and 116 palm species, many outside the rain forest, in contrast with 64 genera and 857 species in the Neotropics. New Guinea, with more than 300 species, and Madagascar, with more than 170 species, both have more palm species than Africa, and even the tiny island of Singapore at the tip of the Malay Peninsula has more genera (18). Both fossil evidence and molecular

Fig. 2.7 A rattan – a spiny, climbing palm – with characteristic scaly fruits, in Bukit Timah Nature Reserve, Singapore. (Courtesy of Fam Shun Deng.)

studies suggest that the palm family is old enough for its pantropical distribution to reflect the break-up of Gondwanaland, although subsequent dispersal events have contributed to modern distributions.

Palms are not only diverse in tropical rain forests, but also abundant, particularly in the understorey. Relatively few palm species reach the forest canopy, but where they do their elegant stems and feathery leaves give the rain forest a distinctive appearance. Shrub and tree palms are ubiquitous, but climbing palms have a more restricted distribution. By far the most diverse group of climbing palms is the spiny rattan palm (*Calamus* and related genera) (Fig. 2.7). Rattans are found in rain forests from Africa to New Guinea, Australia, and Fiji, but they attain their greatest abundance and diversity in Southeast Asia, where up to 30 species can coexist in the same area. Rattan palms climb into the forest canopy, aided by spines on the young stem and the underside of the leaf blade and leaf stalk. Climbing species also produce specialized climbing organs covered in hook-like spines, either as a whip-like extension of the leaf tip or, in some species of *Calamus*, in the form of a modified sterile inflorescence. The result of all of these thorns and spines is that rattans are highly effective at grappling onto surrounding vegetation and using these supports to grow towards canopy openings. Rattans are particularly abundant in Borneo, although the great demands for rattan in furniture manufacture means that they are being ever more intensely harvested from the forest, with plantations being developed to grow them. In much of Southeast Asia, rattans are the most important forest product after timber.

Spiny climbing palms evolved independently in Neotropical rain forests, in the unrelated genus *Desmoncus*, but these are smaller, far less diverse – only seven species are currently recognized – and usually less abundant than the Old World rattans. The Neotropical understorey palm genus *Chamaedorea* (*c.* 100 species) also has a single, nonspiny, climbing species, *C. elatior*, found in Central American rain forest. Madagascar was thought to have no climbing palms until the discovery of *Dypsis scandens*, which looks very like *C. elatior*, although it is in a different subfamily (Dransfield & Beentje 1995).

In addition to the economic importance of many rain forest palms, they play an important role in supporting communities of large vertebrates by producing large crops of medium to large, single-seeded fruits, often with a fleshy layer inside the fruit covering. Many animals rely on eating the fleshy palm fruit or the seeds inside when there are no other fruits available.

Pandans and cyclanths

Both pandans (Pandanaceae) and cyclanths (Cyclanthaceae, the Panama hat family) are sometimes mistaken for palms, to which they are not closely related, because of their tufts of large leaves. Neither family is pantropical on its own, but they have complementary distributions and both morphological and molecular evidence suggest that they are "sister groups," i.e. they are more closely related to each other than to any other family. The modern distribution of these families is consistent with a common ancestor originating on Gondwana in the Cretaceous, with their descendents becoming separated as the supercontinent broke up, but the occurrence of abundant fossils of *Cyclanthus* in Eocene Europe implies a more complex history (Smith et al. 2008). Today, the Cyclanthaceae are exclusively Neotropical, with at least 225 species of herbs, shrubs, climbers, and epiphytes. The leaves of cyclanths can look very like those of palms, but the plants are much less woody and the climbing species are attached by means of short roots, not grappling spines. The Pandanaceae occur in rain forests from West Africa and Madagascar throughout Asia to New Guinea, Australia, and the Pacific, i.e. in every rain forest outside the Neotropics. The family contains around 1000 species of small trees, shrubs, and climbers, with a few epiphytes. The 700 or so tree and shrub species in the genus *Pandanus* are often called "screw pines" because the long, spiny-toothed leaves are arranged in a spiral (see Fig. 2.8). This genus is most diverse and abundant in the rain forests of Madagascar, Southeast Asia, and New Guinea. The other major genus, *Freycinetia*, with more than 200 species, ranges from Sri Lanka to the Pacific, and consists of woody root-climbers that are rather similar in appearance to some climbing cyclanths.

The fruits of pandans and cyclanths are common in rain forests, but little work has been done on what eats them and disperses the seeds. One of the few studies involved a hemiepiphytic cyclanth, *Asplundia peruviana*, in Peru (Knogge et al. 1998). For epiphytes, hemiepiphytes, and branch parasites, such as mistletoes, seed dispersal is a particular problem because the seeds must be dispersed to a suitable branch or stem, or else be wasted. Hemiepiphytic figs seem to depend on massive seed production, so the wastage can be tolerated; mistletoes make use of specialist birds (see Chapter 5), and many epiphytes rely on ants (see Chapter 7). *Asplundia peruviana*, in contrast, exploits a small generalist primate, the saddle-back tamarin (*Saguinus fuscicollis*), which normally deposits most seeds

Fig. 2.8 Close-up of the stem of a rain forest *Pandanus* treelet, showing the spiral arrangement of the long, spiny-toothed leaves. (Courtesy of Fam Shun Deng.)

on the ground. However, after consuming *Aspludia* fruits, the monkeys suffer severe diarrhea. The liquid feces run down the trunk, leaving *Aspludia* seeds stuck to the bark. No information is available on the chemistry of the fruits, but these observations strongly suggest that this cyclanth has evolved a strong laxative that increases the chance of its seeds being deposited on a suitable site for germination and growth.

Climbers and epiphytes

Big woody climbers and epiphytic vascular plants are two of the most characteristic features of lowland tropical rain forests. Woody climbers – lianas – often form tangles in the tops of trees and are an important element of tropical rain forests (Paul & Yavitt 2010). Climbers are particularly abundant where forests are affected by natural disturbances, such as windstorms that knock down trees, or by selective logging. Dense vine tangles also occur along rivers where flooding scours the banks. In all rain forest regions, many of the vines belong to such pantropical families as the Apocynaceae (the oleander family, which now includes the family Asclepiadaceae, the milkweed family), Convolvulaceae (the morning glory family), Fabaceae, Araceae (aroids), Cucurbitaceae (the squash family), and Rubiaceae, but there are also families with climbers only or predominantly in one region, such as the Bignoniaceae, Malpighiaceae, and Sapindaceae in the Neotropics, the Dichapetalaceae in Africa, and the palms in Africa, Asia, and New

Guinea (Turner 2001a). The rain forests of Southeast Asia appear to be relatively poor in climber species, apart from palms, though the species present are still sufficient to form vine tangles in disturbed areas.

Around 10% of all vascular plants are epiphytes and the great majority of these are restricted to tropical forests (Nieder et al. 2001). Although their diversity and abundance is greatest in montane forests, epiphytes are also an important component of all lowland rain forests. Frequent rain and mist supplies the water that they need, while their position in the tree canopy allows them to gain access to higher light levels than occur on the forest floor. Epiphyte communities play an important role in the food chains of the forest, providing nectar, fruit, leaves, and shelter for many of the insects and vertebrates that inhabit the canopy. Epiphyte communities are particularly well developed in Neotropical forests and partly account for the high overall number of plant species in these forests as compared to rain forests elsewhere. A 1 ha (2.5 acre) plot at Cuyabeno, Ecuador, in western Amazonia, contained 172 species of epiphytes (Balslev et al. 1998). Orchids (see below) and ferns are important as epiphytes throughout the tropics, but several large families with many epiphytes, including the Bromeliaceae (see below), Cactaceae, and Cyclanthaceae, are exclusively or almost exclusively Neotropical, while several others, such as epiphytic members of the Araceae, Gesneriaceae (the African violet family), and Piperaceae (the black pepper family), are concentrated in the Neotropics. Neotropical rain forests are also notable for the abundance of "ant gardens" (see Chapter 7), which have allowed a small number of specialist epiphyte species to become very common.

There are other epiphyte families that are primarily found in tropical Asia and New Guinea, but these families have fewer species. The most important of these families is the Apocynaceae, with more than 200 epiphytic species in the subfamily Asclepiadoideae, only two of which are found in the Neotropics. The two largest genera, *Dischidia* (around 80 species) and *Hoya* (more than 200 species), are closely related and largely confined to the rain forests of Asia and New Guinea (Wanntorp & Kunze 2009). Both genera consist largely of epiphytic climbers with succulent, drought-resistant leaves. Many epiphytic species of *Hoya* and *Dischidia* have associations with ants, including some that produce specialized leaves that house the ants and others that grow on arboreal ant nests (see Chapter 7).

Another notable feature of rain forests from East Africa through Southeast Asia to northern Australia and the western Pacific is the abundance of litter-trapping bird's nest ferns (*Asplenium* species) (Fig. 2.9). These ferns can attain both massive individual sizes (with a fresh weight of up to 200 kg) and a high density (30 large ferns per hectare) in lowland dipterocarp forest (Ellwood & Foster 2004). The basket-shaped rosette of long, broad leaves traps dead leaves and other organic matter falling from above, resulting in an estimated 3.5 tons per hectare of suspended soil and plant material at one site in Sabah. This accumulation of nutrient-rich, water-absorbing organic matter is a very important habitat for invertebrates as well as many other species of epiphytes. Recent DNA studies have shown that ferns previously identified as "*Asplenium nidus*" form a complex of morphologically similar species with different distributions and habitat requirements (Fayle et al. 2009; Yatabe et al. 2009).

Rain forest epiphyte communities are impoverished in Africa, probably as a result of repeated past episodes of aridity, a relative lack of suitable forest refuges during glacial periods, and lower current rainfall in most forest areas. The relationship between rainfall and epiphytes is illustrated by a series of rain

Fig. 2.9 Epiphytic bird's nest fern (*Asplenium nidus*) in lowland dipterocarp forest at Danum Valley, Sabah. (Courtesy of Kalsum Mohd. Yusah.)

forest research plots; the diversity of epiphytes declines by half when annual rainfall drops below 2500 mm per year (Turner 2001a). Australian rain forests are similarly poor in epiphytes.

Orchids

The orchids (Orchidaceae) (Fig. 2.10) are the largest angiosperm family, with approximately 20,000 species, and are also the largest family of rain forest epiphytes. The family appears to have originated in the late Cretaceous, but the major clades of epiphytes diversified in the Tertiary (Ramirez et al. 2008), when the major rain forest areas were separated by distances that would have been hard for even tiny-seeded orchids to cross. Of the world's orchids, about 41% occur in tropical America and 34% are found in tropical Asia and New Guinea, but only 15% occur in Africa and 3% in Australia. The highest orchid diversities are found in the northern Andes of South America and in New Guinea (Schuiteman & de Vogel 2007). Orchids reach their greatest abundance at mid-elevations where fog and clouds provide ideal growing conditions for epiphytes. In the Neotropics, the genus *Pleurothallus* and related genera have 4000–5000 species, mostly tiny epiphytic plants with whitish to yellowish to reddish flowers pollinated by small flies; the greatest diversity is found in mid-elevation cloud

(a)

(b)

(c)

Fig. 2.10 Three rain forest orchids from Borneo. (a) A slipper orchid (*Paphiopedilum sanderianum*) with long pendant petal tips, growing on limestone cliffs. (Courtesy of Hans Hazebroek.) (b) The panther-like *Phalaenopsis* (*Phalaenopsis pantherina*). (Courtesy of Tim Laman.) (c) Close-up of a slipper orchid flower (*Paphiopedilum* sp.). (Courtesy of Tim Laman.)

forests. In the Old World, the genus *Bulbophyllum* takes on a comparable role with around 2000 species. Species of *Bulbophyllum* occur throughout the world, but the genus reaches its peak of diversity in New Guinea and Southeast Asia, with 215 species in Borneo and around 600 in New Guinea (Schuiteman & de Vogel 2007). The flowers are usually tiny, and are often purple-colored and foul-smelling – the odor from flowers of *B. beccarii* has been likened to a herd of dead elephants (Pridgeon 1994)! Not surprisingly, these species are pollinated by carrion flies (Pemberton 2010). The flowers are often delicately shaped with elaborate fringes. *Bulbophyllum* species have thickened stems called pseudobulbs, which can aid in water and food storage.

In addition to these large genera of mostly tiny orchids, there are also examples of large but geographically restricted genera with larger flowers. In the Neotropics there are over 1000 species of *Epidendrum* orchids, species often with cane-like leaves, and the lip petals forming a nectar tube. Notable among the *Epidendrum* species are those, such as *E. ramosum*, with red and orange flowers that are hummingbird-pollinated. Overall, there is a bewildering array of flower colors, sizes, and shapes in the genus, with flowers ranging from white to yellow to pink to green to red, and with very delicate and elongate petals to other species with compact and thick flowers. The diversity of *Epidendrum* orchids in the Neotropics is paralleled by the genus *Dendrobium* in Asia and New Guinea, with over 1000 species with pseudobulb stems, with every possible color and shape of flower imaginable except for pure black. The flowers of *Dendrobium sinense* on the tropical Chinese island of Hainan are white with a red center and are pollinated by prey-hunting hornets (*Vespa bicolor*) attracted by a scent that mimics the alarm pheromone of *Apis cerana* honeybees, a frequent prey (Brodmann et al. 2009). Deceit is an orchid specialty, with around a third of all species attracting pollinators without offering any reward (Scopece et al. 2010). Allied to these giant orchid genera are several other large genera with hundreds of species restricted to particular places, such as *Lepanthes*, *Maxillaria*, and *Oncidium* in the Americas and *Eria* in Asia. The genus *Oncidium* (c. 600 species) includes species that offer no reward and others that provide floral oils: both appear to mimic Neotropical members of the family Malpighiaceae that offer an oil reward and they attract the same oil-foraging bees (see Chapter 7) (Pemberton 2010).

Africa is impoverished in epiphytic orchids, probably due to past episodes of drying and the relatively small area of upland forest. Madagascar is richer, presumably because it has been wetter for longer. One interesting – and historically important – example is the largely epiphytic genus *Angraecum*. Two-thirds of the 218 species in this genus occur in Madagascar, with the remainder in Africa or on islands in the Indian Ocean. Many species have night-scented white flowers with a very long spur, which secretes nectar at the base, and are apparently pollinated by nocturnal hawkmoths with very long tongues. In 1862, Charles Darwin predicted that the Madagascan star orchid, *A. sesquipedale*, which has a floral spur an amazing 29 cm long, would be pollinated by a giant hawkmoth with a tongue long enough to reach the nectar (Darwin 1862; Fig. 2.11). Such hawkmoths were subsequently found (Nilsson 1998). Flowers adapted for pollination by long-tongued moths are seen in other Madagascan genera as well, such as *Aerangis*. The abundance of white, long-spurred orchids in Madagascar is striking and appears to reflect a similar diversity and abundance of long-tongued species in the hawkmoth fauna. Amazingly, the genus *Angraecum* also includes, in the rain forests of Reunion, two species that are pollinated by birds (*Zosterops*)

Fig. 2.11 An illustration of Charles Darwin's 1862 prediction that the Madagascan star orchid, *Angraecum sesquipedale*, which has a 29-cm long floral spur, would be pollinated by a giant hawkmoth with a tongue longer than any then known. This drawing by Thomas Wood appeared in an 1867 article by Alfred Russell Wallace in the *Quarterly Journal of Science*, in which he defends Darwin's theory of natural selection against attacks by the Duke of Argyll.

and the first well-supported case of pollination by a cricket (Gryllacrididae, Orthoptera) (Micheneau et al. 2010)!

Ground herbs

In comparison with the deciduous forests of the temperate zone, ground herbs are a relatively inconspicuous part of the rain forest flora. In the interior of closed-canopy rain forests, most of the small plants at ground level are seedlings and saplings of trees, shrubs, and climbers. Herbaceous plants have a patchy distribution, but they are most prominent in canopy gaps, on steep slopes, and in the montane forest. In contrast to the uniformity of most rain forest leaves, those of forest herbs display a wide range of distinctive textures and colors, including some species with variegated patterns and others that are iridescent (Lee 2001). Some of these peculiarities are probably adaptations to the low light levels in the forest understorey, although the precise mechanisms by which they work are not always clear. It is also possible that leaf variegation is a defense against herbivores, disrupting the leaf outline for vertebrates and/or discouraging egg-laying by insects (Konoplyova et al. 2008). The dominant herbaceous plants of all rain forests include ferns and the fern-like spike mosses (*Selaginella* spp.), plus several families of flowering plants: the Zingiberaceae (gingers), Cyperaceae (sedges), Araceae, Rubiaceae, Gesneriaceae (African violet family), and Orchidaceae.

Neotropical rain forests

Trees and shrubs

In addition to the rain forest families found throughout the tropics, there are also certain families that are characteristic of one or two regions, giving them a distinctive quality. Four large, woody families form important components of the Neotropical rain forests and are much less common or absent elsewhere: the Vochysiaceae, with more than 200 species in the Neotropics and three in West Africa, and the Bignoniaceae, Lecythidaceae, and Chrysobalanaceae, which are pantropical but most diverse in Neotropical rain forests (Table 2.2).

The Vochysiaceae are mostly trees, plus a few shrubs and climbers, which produce abundant sprays of distinctive and delicate flowers often attracting large numbers of pollinators. The Bignoniaceae (the jacaranda family) includes trees, shrubs, and numerous woody climbers, of which trumpet creepers and the *Catalpa* tree will be familiar to temperate gardeners. The family is usually readily recognized by its opposite, compound leaves, large tubular flowers, and winged seeds. The tendency of many family members to lose leaves during the dry season and then to produce an entire crown of large, brilliantly colored flowers, often yellow, pink, or orange, is one of the most spectacular sights of Neotropical forests. When these bignoniaceous trees flower, they attract large numbers of pollinators, including hummingbirds, butterflies, and bees. The Lecythidaceae (the Brazil nut family) is a family of small to large trees recognized by their large, showy flowers, with 100 or more stamens. The sweet-smelling flowers of Neotropical species are typically pollinated by bees. The fruits are distinctive;

Table 2.2
Some important plant families in different rain forests. Families found only (or almost only) in one rain forest region are in bold, while families in regular print are especially abundant in that region. Underlined families are absent or relatively rare in a region, but present in others. The families consist mostly of trees and other woody plants unless otherwise indicated: families especially common as epiphytes are indicated with an "**E**"; vines and climbers with a "**V**"; and large herbs with an "**H**."

Neotropics	Africa and Madagascar	Asia and New Guinea
Vochysiaceae (Vochysia)	Dichapetalaceae	**Dipterocarpaceae** (dipterocarp)
Bignoniaceae	Olacaceae (African walnut)	**Fagaceae** (oak and chestnut)
Chrysobalanaceae	Lauraceae (laurel)	Myrtaceae (myrtle)
Lecythidaceae (Brazil nut)	Moraceae (fig)	Arecaceae (palm)
Arecaceae (palm)	Arecaceae (palm)	Apocynaceae E
Pandanaceae (screw pine)	Orchidaceae (orchid) E	
Bromeliaceae (pineapple) E		
Cactaceae (cactus) E		
Cyclanthaceae (Panama hat) V		
Passifloraceae (passionflower) V		
Heliconiaceae (Heliconia) H		

Modified from Turner (2001a).

they are often woody pots ("monkey pots") with lids that must be gnawed off by rodents to get to the large, flesh-covered, angular seeds within. Brazil nuts are an example of this type of seed.

Neotropical rain forests are also notable for the abundance of small trees and shrubs that flower and fruit in the understorey, including many Melastomataceae, Piperaceae, and Malvaceae, providing a food source for flower visitors and fruit-eaters (LaFrankie et al. 2006). This is in particular contrast with the dipterocarp forests of Southeast Asia, in which the forest understorey is dominated by the non-flowering saplings of canopy trees.

Hummingbird flowers

In the New World tropics, a number of large genera and families of plants are pollinated predominantly or exclusively by hummingbirds (see also Chapter 5). These plants present a diversity of typically red-flowered species and generate hummingbird activity throughout the rain forest. Hummingbird-pollinated species are usually red because this is a color that stands out against a green background, making it easier for birds to locate the flowers. In contrast, most bees and other insects see light in a spectrum from orange to ultraviolet, and cannot distinguish red from green. One genus of note is *Heliconia* (Heliconiaceae), which is centered in the Neotropics, with outlying species on Pacific islands as far west as New Guinea and the Moluccas (Fig. 2.12). The 200 or so hummingbird-pollinated *Heliconia* species are giant herbs with banana-like leaves, found in tree fall gaps and other forest openings. Hovering hummingbirds are attracted

Fig. 2.12 *Heliconia* species with female green-crowned brilliant hummingbird in Costa Rica. Unlike most hummingbirds, this large species almost always perches when feeding. (Courtesy of Dale Morris.)

to the pendulous or erect red inflorescences and visit the succession of small flowers produced inside the red bracts. In contrast, the two Polynesian members of the genus have erect, relatively inconspicuous inflorescences and are pollinated by perching honeyeaters, while the four species in New Guinea and Indonesia have green inflorescences, open at night, and are pollinated by nectar-feeding bats (Pedersen & Kress 1999).

An important climbing family, often pollinated by hummingbirds, is the Passifloraceae (the passionflower family). While the family is widely distributed in the tropics, the most important genus, *Passiflora*, is centered in the Americas; the genus has around 500 species in the Americas, fewer than 20 species in Southeast Asia and New Guinea, one species in Madagascar, and none in Africa. *Passiflora* species are generally recognized by their palmately lobed leaves and tendrils, making them similar in appearance to climbing squashes. Their most distinctive feature is their elaborate flowers, with a radiating star of five sepals and five petals, numerous filaments forming a colorful necklace on the surface of the petals, and then a fused structure of five large stamens and three large, knobbed stigma lobes (Fig. 2.13). This complex flower structure was seen by the Spanish conquistadors as a symbol of the crucifixion of Christ and thus a sign that the New World would be converted to Christianity. Christians call the crucifixion story the Passion, hence the name *Passiflora*.

Another noteworthy Neotropical family, the Marcgraviaceae, includes 135 species of thick-leafed woody climbers and shrubs, often epiphytic. A unique feature of the family is the production of pendulous whorls of flowers below which are

Fig. 2.13 Passion flower (*Passiflora* sp.) with two butterfly pollinators in Bolivia. (Courtesy of Louise Emmons.)

large, cup-shaped nectaries. These nectaries, which are highly modified sterile flowers, fill with nectar which, in different species, attracts predominantly hummingbirds, other nectar-feeding birds, nectar-feeding bats, or insects.

Bromeliads

Just as Asian rain forests are commonly called "dipterocarp forests" because of the prominence of this tree family, American rain forests could reasonably be called "bromeliad forests." Species of the family Bromeliaceae are the preeminent group of Neotropical epiphytes, with more than 3100 species (not all epiphytes) in the Neotropics and one outlying species, *Pitcairnia feliciana*, growing on rocky outcrops in West Africa, where it presumably arrived by long-distance dispersal (Givnish et al. 2007). The bromeliads give a special "look" to the forest that is unmistakable. This is an easily recognized family, known for the cultivated pineapple, which is native to arid northeastern Brazil, the Spanish moss that drapes trees in subtropical residential areas, and the numerous cultivated plants in the family. Bromeliads have adaptations for living in dry conditions, which have made them ideally suited for living perched on tree branches as epiphytes. For example, most epiphytic species have a special form of photosynthesis (crassulacean acid metabolism, CAM) in which carbon dioxide is fixed during the cooler, more humid, night and then used for photosynthesis during the day, when the stomata remain closed. Many bromeliad plants produce a distinctive rosette of elongated stiff leaves, often with sharp spines along the edges, sometimes with reddish, brown, or white coloring on the leaves, and with overlapping leaf bases, creating a basket-like appearance to the plant (Fig. 2.14). The overlapping leaf bases can hold water for days or even weeks or months following a rainstorm. In some bromeliads, the leaf bases are enlarged to form tanks, and specialized leaf hairs, called trichomes, absorb water and minerals from the enclosed reservoir. The reservoir in some species can reputedly hold up to 45 liters of water (12 gallons) and all the bromeliad species together can hold up to 50,000 liters per hectare of forest (Williams 2006).

Mature bromeliads produce an elongated inflorescence. The flowers and inflorescence structures are often brightly colored, in many cases bright red, and in many species are pollinated by hummingbirds. The flowers are typically tubular, with three petals and three sepals. The fruit is either a small berry eaten by birds, or a capsule, which opens to release the wind-dispersed seeds. Bromeliads are well known as ornamentals, because of their attractive variegated foliage, and often brilliantly colored inflorescences. Notable genera include *Aechmea*, *Vriesea*, *Bilbergia*, and *Guzmania*. Bromeliads are easy to grow as houseplants as they can grow with minimal soil, they can store water in their tanks so they can survive without being watered often, and they can grow in relatively dry conditions.

Bromeliads provide a wide range of benefits to canopy animals in the Neotropics. They provide pollen, nectar, and fruits to many birds and mammals, while other animals eat the leaves and young inflorescences. Many species of birds forage among bromeliads for the numerous arthropods and small vertebrates that live there. Mammals, such as the coati (Beisiegel 2001), also take advantage of this foraging opportunity. The water tanks of bromeliads provide sources of drinking water for monkeys and other canopy animals during periods of drought (Bennett

Fig. 2.14 Members of the family Bromeliaceae, an important group of New World epiphytes. (a) Epiphytic bromeliad (*Aechmea* sp.) in flower in French Guyana. (Courtesy of David Lee.) (b) Tank bromeliad (*Neoregelia pimeliana*) in flower and filled with water. (Courtesy of David Lee, taken in a garden.)

(a)

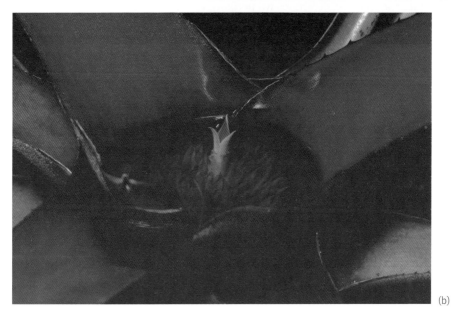

(b)

2000). Using these water tanks, animals do not have to descend to the ground to obtain drinking water.

Many species of invertebrates, especially aquatic insects, as well as frogs and salamanders, have taken this a step further by completing the aquatic phase of their life cycle in these canopy water tanks. These bromeliad tanks form self-contained aquarium-like communities, with herbivores, detritivores, and predators feeding, breeding, and hiding in the tank and water-filled spaces between the leaf bases. Among the insects that breed in bromeliad tanks are many species of mosquitoes, contributing to the relatively high diversity of this group in the Neotropics. A recent molecular phylogenetic analysis of a group of diving beetles (Dytiscidae) found that one lineage of strict bromeliad specialists was around 12–23 million years old, which is similar to the estimated timing for the major diversification of the Bromeliaceae (Balke et al. 2008).

Other Neotropical epiphytes

A second noticeable family of Neotropical epiphytes is the Cactaceae (the cactus family). While cactus species are most common in arid desert environments, around 150 species grow as epiphytes, many of them in drier Amazonian rain forests. In many ways, growing in the canopy of trees has certain similarities to deserts; the plants there are exposed to full sun and often high temperatures and plants are cut off from the soil water supply and nutrient pool. Eighteen genera of cacti have epiphytic members, with 58 species in the genus *Rhipsalis*, 18 in the genus *Hylocereus*, and 13 in the genus *Selenicereus*. With their thick waxy stems and absence of leaves, cacti are eminently suited to conserving water and surviving long periods without rain. A single species, the mistletoe cactus (*Rhipsalis baccifera*), is also found in Africa and on several islands in the Indian Ocean, but this is an exception to the general pattern, perhaps resulting from long-distance dispersal by birds. The third important Neotropical epiphyte family is the Piperaceae (the black pepper family), a large group of succulent herbs or small shrubs, often with thick, rounded, beautiful leaves, and tiny flowers in long, greenish or white spikes. The largest genus is *Peperomia*, which is pantropical, but with species concentrated and often extremely abundant in the Neotropics. All of the epiphytic members are apparently found in the Neotropics, where they are readily recognized by their distinctive flower spikes.

Asian rain forests

Dipterocarps

The tree family Dipterocarpaceae (literally, "two-winged fruits") plays a dominant role in the ecology and economics of Asian forests in a way that no comparable family plays in the other rain forest regions (see Table 2.2). Dipterocarps dominate forests in Borneo, Sumatra, Java, and the Malay Peninsula, as well as the wetter parts of the Philippines, with the majority of the large trees being members of this one family, which accounts for the majority of the biomass. First-time visitors to these forests will be amazed to slowly turn around and realize that virtually every giant tree is a dipterocarp, and yet that they belong to several separate genera and dozens of distinct species. Outside this core everwet area, dipterocarps gradually decline in diversity and abundance (Fig. 2.15). A secondary center of dipterocarp diversity exists in Sri Lanka. A few species of dipterocarps are found in the African tropics and Madagascar, though not in the rain forests,

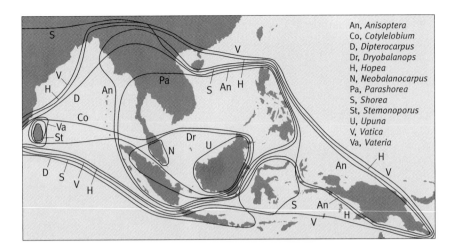

An, *Anisoptera*
Co, *Cotylelobium*
D, *Dipterocarpus*
Dr, *Dryobalanops*
H, *Hopea*
N, *Neobalanocarpus*
Pa, *Parashorea*
S, *Shorea*
St, *Stemonoporus*
U, *Upuna*
V, *Vatica*
Va, *Vateria*

Fig. 2.15 The ranges of dipterocarp genera showing the center of diversity in the Sundaland region, with a secondary center in Sri Lanka. (From Whitmore 1998.)

and in the highlands of South America, providing a testament to the family's southern Gondwana origin. Dipterocarps were already present in equatorial rain forest in India by 50–52 million years ago (Rust et al. 2010), but did not reach Southeast Asia until later, when a moist corridor between India and Southeast Asia resulted in a major influx of plants with Gondwanan affinities (Morley 2003). Despite this relatively late arrival, the dipterocarps underwent a massive evolutionary radiation in Southeast Asia. Enumerations of research plots in Southeast Asian forests show a conspicuous proliferation of tree species within the dipterocarp genera *Shorea*, *Hopea*, *Dipterocarpus*, and *Vatica*. In any one forest in the Malay Peninsula, Sumatra, or Borneo it would be common to find 25 species or more of *Shorea*, and six or more species of the other three genera. More recently, in geological terms, a few species have managed to disperse from island to island across the narrowing water gap to New Guinea, where they dominate in scattered patches.

Why should the dipterocarps be so dominant in Asian rain forests? There is no obvious single answer, but certain common features hint at the reasons behind their success. Dipterocarps tend to have smooth, straight trunks rising to great heights, without any side branches or forks until the canopy is reached (Fig. 2.16a). Dipterocarp forests often have canopy heights at and above 50 m (150 feet), which is higher than rain forests elsewhere (de Gouvenain & Silander 2003). The base of the tree is often buttressed. These growth characteristics emphasize the strength and stability of individual dipterocarp trees; trees do not typically fall over or get blown over as is often seen in Neotropical trees. Rather, dipterocarps often die standing, gradually losing their branches until only the trunk remains. As a result, the dipterocarp forest tends to be darker and more stable than forests in Africa (where the trees are shorter) and the Amazon (where trees may have a greater tendency to fall over and create large canopy gaps soon occupied by sun-loving trees and vines). Once dipterocarp trees reach the canopy and emerge from it, they produce a characteristic crown that is shaped like a cauliflower,

(a)

(b)

(c)

Fig. 2.16 The dipterocarp family is extremely important in Asian forests. (a) The trunk of a giant dipterocarp tree in Borneo. (Courtesy of Richard Primack.)
(b) Flowers of the dipterocarp *Hopea ponga* in India. (Courtesy of N.A. Aravind.)
(c) Winged seeds of a dipterocarp tree (*Shorea* sp.) in the rain forest canopy in Borneo. (Courtesy of Tim Laman.)

with clusters of leafy branches evenly spaced around a dome. A tendency toward lower wind speeds in Southeast Asian rain forests than in the other regions may favor this growth habit.

Another possible key to the dipterocarps' success and the long lives of individual trees is the presence in all plant parts of an oily, aromatic resin that presumably aids the plant in defense against attack by bacteria, fungi, and animals. This resin often accumulates where the bark is bruised and is encountered as hard, crusty, glass-like pieces on the trunk or on the ground. This resin, called dammar, is collected by the local people and used in varnishes or as boat caulking. The value of this resin is illustrated by the kapur tree, also known as the Bornean camphor tree (*Dryobalanops aromatica*). Historically, this species was one of the main commercial sources of camphor, an essential oil of importance for its use in medicine and as a preservative. The crushed leaves have a distinctive camphor or kerosene-like smell. Dipterocarps also contain bitter-tasting tannins as a further deterrent to attack. Although non-dipterocarp trees also have chemical defenses in their foliage, dipterocarp leaves do seem to be peculiarly inedible, at least to vertebrates. This is illustrated by the fact that the colugo, a leaf-eating gliding mammal, lives in dipterocarp forests, and forages widely in the tree canopy for new leaves, but does not eat dipterocarp leaves (see Chapter 6). The orangutan and proboscis monkey, which also eat young leaves, again do not eat dipterocarp leaves.

The flowers of dipterocarps vary in size, some being small (Fig. 2.16b) and others being relatively large and showy with five white, yellow, or pink petals and often with numerous stamens. The flowers are often scented and are adapted for pollination by a variety of insects – thrips, beetles, bees, or moths – depending on the species (Corlett 2004). Following flowering, a fruit is produced consisting of a single-seeded nut with a membranous wing-like calyx, looking like a badminton shuttlecock (Fig. 2.16c). The ratio of the fruit weight to the total wing area – known as the wing loading – is much higher in dipterocarps than in most other winged fruits, so they spin to the ground within a few meters of the parent tree (Osada et al. 2001). At least this is what usually happens. The fact that certain dipterocarp species have crossed major water barriers to reach New Guinea and the Philippines suggests that they can sometimes travel long distances. The key to their success must lie in occasional windstorms plucking the winged fruits off the tall trees and transporting them across rivers and seas.

A further reason for the success of the dipterocarps in the everwet areas of Southeast Asia may be the way most of the dipterocarp species over a wide region flower and fruit together only once every 2–7 years (Sakai et al. 2006). In an entire forest, only a few dipterocarp trees will flower in an ordinary year, but during a so-called "mast" year, most large trees reproduce (Fig. 2.17). Individual plant species have mast years in all rain forests, but only in Southeast Asia do the mast years of so many species coincide over such a large area. The trigger for the initiation of flower development appears to be an irregular period of drought, but the magnitude of the event depends on the time for accumulation of resources since the last such episode.

Several hypotheses have been put forward to explain the evolution of mass flowering and fruiting, of which two seem most plausible. First, in the Asian everwet climate, with no distinct wet and dry seasons, plants need some cue to trigger the onset of reproduction. In this way, all the individuals of a species can

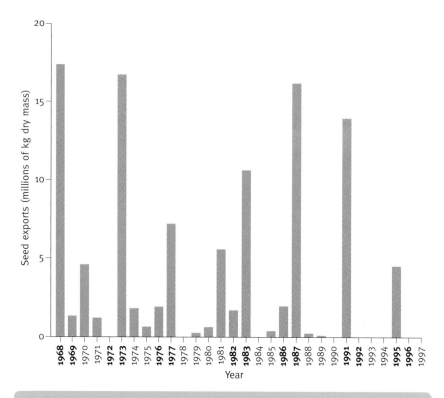

Fig. 2.17 In Southeast Asian rain forests, dipterocarp trees mass flower once every 2–7 years, with little or no flowering in intervening years. This is illustrated by the export figures of illipe nuts (*Shorea* spp., section Pachycarpae), common dipterocarp species, from West Kalimantan, Indonesian Borneo, from 1968 to 1997. Strong flowering years are associated with El Niño–Southern Oscillation (ENSO) events; years with such events are shown in bold. (From Curran & Leighton 2000.)

flower at the same time and cross-pollination can occur. The multiyear seasonality of the El Niño cycle (see Chapter 1) provides the distinct set of conditions needed to coordinate reproduction. The problem with this hypothesis is that the use of the same cue means that numerous different species flower at the same time, increasing competition for pollinators. Second, and perhaps more important, mass fruiting at long intervals may prevent the build-up of populations of insects, birds, and mammals that would destroy the large and highly nutritious, oil-rich fruits (Janzen 1974). Synchronization of fruiting by many dipterocarp species across large areas is necessary for this to work, otherwise nomadic seed eaters, such as wild pigs, could simply move to wherever the trees were fruiting and destroy the whole crop (Curran & Leighton 2000). Thus, it is only in the mast years that any seeds survive long enough to germinate and grow into seedlings. Dipterocarps invest so much energy in reproduction during flowering years that they stop growing; in practice, they often have 5 or so years of growing without reproducing, followed by a heavy flowering year with no growth (Primack et al. 1989). This dipterocarp pattern is similar to the growth patterns seen in apple trees in temperate fruit orchards, with a year of heavy fruiting and little woody growth followed by a year of less fruit production and strong woody growth.

Producing successful crops of seedlings only every 2–7 years could be a major disadvantage in responding to the short-term recruitment opportunities that occur following the death of adult trees. Another important element in the success of dipterocarps is therefore the ability of dipterocarp seedlings of some species to survive for many years under the dense, shady canopy of established trees. The resources provided by a large seed are an obvious advantage here. This effectively creates a "seedling bank" that can respond to opportunities created by an opening in the canopy overhead. In forests in Borneo, the seedlings of some dipterocarps species last for more than 15 years on the forest floor after a single fruiting event (Delissio et al. 2002). The variation among dipterocarp species in how well the seedlings can survive in deep shade and in how rapidly they can increase their growth rate in response to an increase in light levels allows the family as a whole to take advantage of a wide range of conditions (Brown et al. 1999).

Dipterocarp seedlings may also have an increased chance of survival as a result of a special form of the mutualistic relationship between roots and fungi called mycorrhiza (literally, "fungus root"), in which the plant receives mineral nutrients and water from the fungus in exchange for carbohydrates. Almost all plants form mycorrhizae, but unlike most other rain forest trees, dipterocarps are ectomycorrhizal – that is, the fungus forms a sheath over the outside of the roots. Ectomycorrhizal trees and their seedlings are linked by an extensive network of fungal hyphae that transfer nutrients from decaying organic matter to the plants. As soon as it germinates, a dipterocarp seedling can plug into the existing network and may obtain resources from its nearby parent, although this has not yet been convincingly demonstrated in dipterocarps. Whether this suggested ectomycorrhizal advantage exists or not, it is very striking that the same fungal association occurs in the oak family, which often dominates in Southeast Asian montane forests, and in legume trees in the subfamily Caesalpinioideae, which form extensive stands dominated by single species in parts of Africa (see below) and South America (McGuire 2008). Ectomycorrhizal associations are also the norm in low-diversity temperate zone forests. It is striking too that many of these ectomycorrhizal tree species have large, poorly dispersed seeds and a pattern of heavy fruiting at multiyear intervals – mast fruiting – which is similar to that shown by the Southeast Asian dipterocarps (Henkel 2003; Newbery et al. 2006). Note, however, that the community of ectomycorrhizal fungi in a lowland dipterocarp forest is very diverse and it is unlikely that all species function in the same way (Peay et al. 2010).

Many of these elements of the dipterocarp strategy for rain forest success seem to fit together. Wind-dispersed fruits are only practical in the rain forest for very tall trees that emerge from the forest canopy: there is too little air movement inside the forest (Slik et al. 2010). Large seeds produce seedlings that can establish and survive in deep shade. What is food for a seedling is also food for a beetle, rat, pheasant, or pig, but mast fruiting at long time intervals can satiate these seed predators so that many seeds escape consumption to grow into seedlings. Even in the rain forest, however, there are exceptions to some of these generalizations, including dipterocarp trees too small to emerge from the canopy, fruits without wings, and species that flower every year. Their strategies for survival must be different, and require further research. Moreover, in areas outside the everwet zone, such as in Thailand and the Western Ghats of India, dipterocarps flower and fruit on annual cycles in response to seasonal weather changes. In the rain forest at Sinharaja, in southwest Sri Lanka, which has a brief annual dry period, some dipterocarps have annual cycles while others show synchronized masting at multiyear intervals.

Oaks and chestnuts

Walking in Southeast Asian rain forests, one may be surprised to find acorns from the genera *Lithocarpus* and *Quercus* (Fig. 2.18) and chestnuts from the genus *Castanopsis* on the forest floor. The acorns come in a wide variety of nut and cup shapes, but they are all immediately recognizable as acorns. Acorns are associated in most people's minds with temperate forests, but the Fagaceae (the oak family) is surprisingly well represented in lowland, tropical forests in Southeast Asia, and is one of the most dominant families at mid-elevations where the dipterocarps and other lowland rain forest families do not grow as well. Although rarely as conspicuous as in the northern temperate forests, the family attains its greatest diversity in tropical and subtropical Asia. The small equatorial island of Singapore, whose highest point is only 164 m above sea level, has 21 species (8 *Castanopsis*, 12 *Lithocarpus*, and 1 *Quercus*). A few species of *Lithocarpus* and one of *Castanopsis* extend to New Guinea, where they are prominent in montane forests, along with the Gondwanic, and largely southern temperate genus, *Nothofagus*, the southern beeches, in the related family Nothofagaceae. Oaks (*Quercus* spp.) are also prominent in the montane forests of Central America, but only

Fig. 2.18 Rain forest acorns (*Lithocarpus* sp., Fagaceae) in Borneo. (Courtesy of K.M. Wong.)

a single species, *Q. humboldtii*, has reached the northwestern margins of South America, apparently within the last 2 million years (Graham 1999), where it occurs at altitudes above 1500 m. Lowland oaks and chestnuts are almost entirely a Southeast Asian phenomenon, and there are no members of the oak family at all in southern India and Sri Lanka, in the whole of sub-Saharan Africa, or in most of tropical South America.

In the lowland forests of Southeast Asia, the acorns and nuts of the oak family represent food for squirrels, monkeys, pigs, and other vertebrates. The strong nutshell may help to protect the large seeds inside against the powerful bites of these animals. However, unlike the similar large nuts of the dipterocarps, acorns and chestnuts have no wings and depend for their dispersal on animals – probably rats and ground squirrels – that cache them in the ground for later consumption.

Speciose genera

The term "speciose" does not appear in most dictionaries, but it is used by biologists to mean "with many species." It is widely believed that Southeast Asian rain forests are distinguished by the presence of plant genera with numerous coexisting species. There has been no systematic pantropical study of this question, but data from the CTFS rain forest plots shows that Lambir, Sarawak, has more species per genus than any other site (Lee et al. 2002; Ashton et al. 2004). The greater number of species per genus in Southeast Asia, if confirmed, could reflect relatively recent diversification of the rain forest flora there, but the pattern is also consistent with the alternative hypothesis that extinction has reduced diversity elsewhere.

The coexistence of multiple similar species of *Shorea* and other dipterocarps has already been mentioned. Another spectacular example is the genus *Syzygium*, in the Myrtaceae (the myrtle family). This huge genus of around 1100 small to medium-sized trees occurs throughout the Old World tropics, from Africa to Australia, but is particularly diverse in Southeast Asia and New Guinea. In the 52 ha (130 acre) forest plot at Lambir, Sarawak, 49 species occur together and there are 45 species in a 50 ha plot at Pasoh, Peninsula Malaysia. *Syzygium* species are readily recognized by their opposite, leathery leaves, fluffy flowers with numerous stamens, and round crunchy fruit with a characteristic cup of sepals at the top. These fruits are a major food source for birds, fruit bats, and other mammals. The Myrtaceae family is notable for its many cultivated fruit trees, among them the Neotropical guava tree, and a number of Asian *Syzygium* species have been brought into cultivation for their fruit known as jambus. Other genera with many coexisting species in Southeast Asian rain forests include *Aglaia* (Meliaceae), *Diospyros* (Ebenaceae), *Ficus* (figs), and *Garcinia* (Clusiaceae).

Rain forests in New Guinea and Australia

The lowland rain forests of New Guinea present an apparent paradox. The vertebrate fauna is very different from that of Southeast Asian rain forests, yet the composition of the flora is generally similar at the family and genus levels, although relatively few species are shared. The major floristic difference – one of considerable ecological importance – is that the dipterocarps are much less important, with the large Asian genera *Dipterocarpus* and *Shorea* entirely absent. Dipterocarps (in the genera *Anisoptera*, *Hopea*, and *Vatica*) are widespread in New Guinea and dominate over quite extensive areas, but they are far less diverse than in Southeast Asia and are entirely absent from many areas. The dominant families in the few rain forest plots that have been enumerated in Papua New Guinea included the Lauraceae, Meliaceae, Moraceae, and Myristicaceae (Wright et al. 1997).

Many of the floristic similarities between New Guinea and Southeast Asia appear to reflect an influx of Asian rain forest plants into lowland New Guinea after the mid-Miocene collision between the Australian and Asian plates. The absence of a dry land connection between the two regions limited the exchange of vertebrates, but water gaps of a few tens of kilometers are much less of a barrier to many plants. The predominance of Asian plants in New Guinea, rather than the other way round, seems to be a result of the relatively recent arrival of the Australian tectonic plate at tropical latitudes, which has not given time for a diverse indigenous tropical lowland flora to evolve. New Guinea's montane forests, in contrast, have a major component of southern origin, including the southern beech, *Nothofagus*.

In contrast to New Guinea, Gondwanic and Australian plants are more prominent in the small area of tropical rain forest in Australia. This forest occupies less than 1% of the island continent, but supports 1060 vascular plant genera, of which two-thirds have only one species, and over 2800 species, more than half of which are endemic (Gross 2005). Here, the fossil record shows little evidence for a post-Miocene influx of Asian plants, suggesting that the shared elements of the flora may reflect older connections and dispersal routes. The Australian rain forest is also noticeable for its concentration of primitive (basal) flowering plant families, including two woody climbers in the endemic family Austrobaileyaceae (Fig. 2.19). These differences from New Guinea presumably reflect the very different history of the Australian rain forest, which, as discussed in Chapter 1, covered much of northern Australia during the early to middle Miocene, before becoming restricted to the northeast because of the drying of the continent. The lower representation of Asian lowland rain forest genera in Australia may have enabled the surviving southern genera to hold their own, while in New Guinea they were excluded from the lowlands by the Asian influx (Ashton 2003).

African rain forests

One of the most notable of rain forest patterns is that African rain forests are relatively poor in plant species (Turner 2001a; Plana 2004), particularly in palms, orchids, and the tree families Lauraceae, Myrtaceae, and Myristicaceae, as well as in epiphytes and woody vines in general. In many prominent pantropical plant families, Africa only has 10–20% of the numbers of species in Asian and American forests. An extreme case is the pantropical genus *Piper* (Piperaceae), which includes the black pepper, *P. nigrum*, and around 2000 other species of herbs, shrubs, treelets, and climbers. There are more than 1000 species in the Neotropics and several hundred in tropical Asia through to Australia, but only

Fig. 2.19 Flower of *Austrobaileya scandens*, a woody climber found only in the rain forest of northeast Queensland, Australia. This species belongs to one of the most ancient groups of flowering plants and shows many features that are believed to be primitive. (Courtesy of Dennis Stevenson and PlantSystematics.org.)

three in Africa (Jaramillo et al. 2008). In contrast with these absences, two less well-known families of rain forest trees without common names in English, the Dichapetalaceae and the Olacaceae, are more abundant in African forests than in rain forests elsewhere (see Table 2.2).

This relative poverty of species in African forests is probably due to multiple factors, but fossil evidence suggests that it is largely a result of extinctions in the 30 million years since a species-rich rain forest spanned the entire continent (see Chapter 1; Plana 2004). Today, most African rain forests are drier and more seasonal than rain forests elsewhere, leading to the absence of drought-sensitive species, particularly among the epiphytes, most of which need abundant moisture to survive because their roots are not in the ground. Africa was also more strongly affected by drying during past glacial periods than other rain forest areas, resulting in forest loss and fragmentation (Maley 2001). In South America and Asia, species survived in the remaining wet areas. African forests have also been affected by human activity over a longer period. This contrasts with the relatively recent arrival of people in South America and the very low human population density in much of Southeast Asia's everwet forests until recently. It is important to note, however, that African rain forests are species-poor only in comparison with other rain forests: a 50 ha plot in southwest Cameroon contained nearly 500 tree species (Kenfack et al. 2007), which far exceeds the total for any other forest type.

Monodominance

Although African rain forests do not have a major distinctive element in their flora, they are notable for the large expanses of forest dominated by a single tree species. Single-species dominance occurs in rain forests elsewhere, notably on the Guiana Shield in South America, but not over such large areas. The most striking examples involve leguminous trees in the subfamily Caesalpinioideae, all apparently lacking the nitrogen-fixing symbiosis. The mbau, *Gilbertiodendron dewevrei*, dominates the canopy of large patches from southeastern Nigeria and Cameroon across the Congo River basin. In the northeast Congo basin, some of these patches extend over hundreds of square kilometers, with mbau contributing 80–90% or more of the large trees. Mixed-species rain forests, with a diversity of trees more typical of rain forests elsewhere, grow adjacent to mbau forests. Monodominance does not mean that other tree species are absent, however, just that they are rare; the total number of tree species in 20 ha plots of mixed and monodominant forest in the Congo was very similar (Ashton et al. 2004).

Several factors seem to contribute to the persistent dominance of *Gilbertiodendron* (Hart 1995). The dense, continuous tree canopy casts a deep and uniform shade, in which only the most shade-tolerant plants can survive. Moreover, the litter layer is much deeper than in adjacent mixed forests, making it very difficult for the roots of germinating seeds to penetrate through to the soil below. These conditions favor *Gilbertiodendron*, which has large seeds that supply food reserves for the young seedling, as well as shade-tolerant saplings that can survive for years under an intact canopy (Fig. 2.20). It is also likely that *Gilbertiodendron*

Fig. 2.20 Shade-tolerant seedling of mbau, *Gilbertiodendron dewevrei*, a legume that dominates large areas of rain forest in the Congo River basin. (Courtesy of Fiona Maisels, Wildlife Conservation Society.)

seedlings benefit nutritionally from the ability to "plug into" a common ecto-mycorrhizal network, as demonstrated for monodominant forests in South America (McGuire 2008) and suspected for the dipterocarps (see above). This mechanism may enhance shade tolerance, because seedlings would not have to depend on their own photosynthesis.

The dominance of the forest by a single tree species probably reduces the diversity of insects, as there are fewer types of plants on which to feed. Conversely, it may increase the risk of damaging outbreaks of a single insect species, as observed in Gabon where moth caterpillars defoliated another monodominant caesalipinioid legume, *Paraberlinea bifoliolata* (Maisels 2004). Mbau forests also seem to support a lower density and diversity of primates and other vertebrates than nearby mixed-species forests, although they may be an important habitat for some individual species. Colobus monkeys eat the new leaves and elephants uproot saplings to eat the roots, but the real bonanza occurs when the mbau trees fruit. As in Asian dipterocarp forests, fruiting is not an annual event, but occurs at intervals of several years. In mast years, these forests produce as much as 5 tons per hectare of large, edible, nutritious seeds (Blake & Fay 1997). Beetles destroy much of the crop, but enough is left for almost the entire forest mammal fauna to switch from their normal diets. Rodents, duikers, pigs, buffalo, elephants, gorillas, chimpanzees, and humans all find them irresistible. In non-fruiting years, the animals in the monodominant forests must either survive by eating figs, leaves, and fungi, or else migrate out of the area. In these periods, animals lose weight and have a greater chance of dying.

Elephants in African rain forests

Forest elephants appear to play a uniquely important role in African rain forests, as a result of their size, movements, and feeding habits (see Chapter 4). No equivalent large mammals currently exist in the Neotropics and there is no direct evidence that any of the elephant-sized species that inhabited the Neotropics until the arrival of people lived in the rain forests. Neither Madagascar nor New Guinea ever had any animals this big. Forest elephants and rhinoceroses occur in Southeast Asian forests, but they appear to have much less influence on the vegetation. However, this impression may partly reflect the greatly reduced densities of elephants throughout most of Asia in historical times and it is possible that the importance of forest elephants in Asian rain forests has been underestimated.

African elephants damage or kill plants both accidentally, by trampling them in the course of other activities, and deliberately when feeding. During their feeding, they uproot or knock over small trees and debark or tear branches off larger ones. This damage can create gaps in the forest, as well as help maintain existing openings. The movements of elephants while foraging and over longer distances between favored sites create more or less permanent paths, with no vegetation at ground level. The importance of these activities to other forest herbivores is discussed in Chapter 4.

Fruit is a major item in the diet of African forest elephants and they are known to eat a huge range of fruit species (Blake et al. 2009). Many of these fruits are eaten by other forest mammals, such as gorillas, chimpanzees, or duikers, but African rain forests are also notable for the presence of very large, hard fruits that fall to the ground when ripe. One of the most striking is the large round yellow fruit of *Strychnos aculeata*, which looks like a large billiard ball and is almost as hard. These fruits are formed by a vine in the tree canopy and come crashing down to the ground when ripe. The fruits are so smooth and tough that it is hard to imagine any animal being able to open them. However, elephants are capable of breaking through the 7 mm (0.25 inch) thick shell to get at the soapy slime inside, which surrounds the dozen or more round seeds. This slime is poisonous to many animals and is used both as a medicine and as a fish poison by local people. It has been speculated that elephants eat these fruits for the intoxicating effects of the slime rather than for nutrition (Martin 1991).

Elephants seem to be the primary dispersers of many large-seeded fruit trees in the African rain forest, such as the Guinea plum *Parinari excelsa*, the mango-like *Irvingia gabonensis*, and others such as *Balanites wilsoniana* and the maskore tree *Tieghemella heckelii* (White et al. 1993; Yumoto et al. 1995). Elephant paths lead to fruiting individuals of favored species. Fruits favored by elephants are typically at least 4–5 cm in diameter, dull-colored, and hard, with a shell protecting the strong-smelling flesh. Many of the fruits contain large, nut-like pits, which protect the seeds as they pass through the intestines of the elephant. Fruit consumption by an elephant has been shown to provide multiple benefits in various plant species: reliable consumption of the entire fruit crop, long-distance seed dispersal over tens of kilometers, enhanced seed germination as a result of passage through the gut, and more vigorous seedling growth as a result of the fertilizing effect of the elephant dung (Blake et al. 2009; Fig. 2.21). Numerous forest trees and climbers benefit from seed dispersal by elephants and several, including most of those named above, seem to be entirely dependent

Fig. 2.21 Seedlings of several plant species sprouting from elephant dung on an abandoned logging road in Waka National Park, west-central Gabon, West Africa. (Courtesy of David Wilkie.)

on elephants. The elimination of elephant populations will eliminate these dispersal services and, in the long term, lead to the decline and eventual extinction of the plants on which they depend.

The hard shells of these very large, elephant-dispersed seeds not only shield them on their slow passage through the elephant, but also protect the nutritious interior from rodents and other seed-eaters. This protection is important because elephant dung attracts seed-eaters. The kernels of *Panda oleosa* are reputedly the toughest in Africa. One animal, however, has learned to overcome this barrier. Rain forest chimpanzees in several parts of West Africa use stone tools to crack open the nuts of *Panda* and other elephant-dispersed species (Mercader et al. 2002). The nuts are placed on stone or rock anvils and hit repeatedly with stone hammers. A skilled chimpanzee can crack up to 100 nuts in a day.

Madagascan rain forests

Despite their small total area, the rain forests of Madagascar are surprisingly species-rich, probably because they were not subject to the severe drying during glacial episodes that occurred in Africa. As a result of the island's long history of isolation from other regions, these forests also have a very high proportion of endemic plant species – that is, species found nowhere else. An estimated 96% of the 4220 tree and large shrub species found on the island are endemic (Schatz 2001). In contrast to the vertebrate fauna considered in later chapters, this floristic distinctiveness does not run very deep and, although there are seven endemic families and many endemic genera, most species are in genera that are shared with Africa, showing that there have been multiple colonization events. There are some conspicuous exceptions, however. Notably, Madagascar has a diverse palm flora that includes genera with Asian affinities not represented in Africa (Dransfield & Beentje 1995). The largest endemic family is the Sarcolaenaceae, with around 60 species of trees and shrubs. The Sarcolaenaceae is the sister group of the dipterocarps and shares the ectomycorrhizal symbiosis of this family (Ducousso et al. 2004). There have also been spectacular endemic radiations in widespread genera, such as *Diospyros* (ebony; > 100 spp.) and *Dalbergia* (rosewood; 43 spp.).

Madagascan rain forests are distinct in their generally low canopy heights, high tree densities, and the rarity of large-diameter trees (de Gouvenain & Silander 2003; Grubb 2003). Exposure to frequent cyclones is the simplest explanation for this structure, but low soil fertility may also contribute. Cyclone damage creates gaps in the canopy that benefit the famous "traveler's palm," *Ravenala madagascariensis* (Strelitziaceae), a giant relative of the bananas. Other distinctive features of Madagascan lowland rain forests include an abundance of palms, pandans, and bamboos.

One feature of the rain forest fauna also requires a botanical explanation. This is the striking poverty of the frugivore (fruit-eater) community in Madagascan rain forests (Ganzhorn et al. 2009). There are very few frugivorous birds compared with other rain forest regions (see Chapter 5), only three species of fruit bat, and, although many lemurs eat fruits when they are available, none depend on them year round. This seems to reflect a generally low density of plants that produce edible, fleshy fruits, relatively small fruit crops on these plants, a strong seasonality in fruit production, with near zero fruit availability at certain times of the year, and, possibly, relatively low protein content compared with fruits in other tropical forests (Ganzhorn et al. 2009). In many other rain forests, gaps in fruit production are filled by figs, which, as discussed earlier, produce ripe fruits throughout the year. Madagascan rain forests have relatively few rain forest fig species (fewer than 20) in comparison with other regions, but there is no published information on their abundance.

Conclusions and future research directions

In contrast to the rain forest vertebrates, where very few families are pantropical, all the major rain forest regions share many plant families and a number of genera, but each region also has some distinctive botanical characteristics. Asian rain forests are notable for the dominance of dipterocarp trees and the presence of oaks and chestnuts in the lowlands. The tendency of dipterocarps and numerous other unrelated plant species to reproduce over a wide area at multiyear intervals results in long periods of flower and fruit poverty interspersed with brief and unpredictable periods of superabundance, with major consequences for the animal community. American rain forests have the greatest diversity and abundance of epiphytic species, most notably the water-filled bromeliads, providing canopy resources to many other species, as well as a number of distinctive tree, vine, and herb families. African rain forests are noted for their relative impoverishment of plant species due to past episodes of drought and forest contraction, the large areas of forest dominated by a single legume species, and the presence of plants with elephant-dispersed fruits. Madagascan rain forests are richer than might be expected from their small area and isolation. The lowland rain forests of New Guinea have a predominantly Asian flora – minus the dominance of dipterocarps – which, as we will see in subsequent chapters, interacts with a decidedly non-Asian fauna. Finally, the tiny area of tropical rain forest in Australia cannot be considered as simply an outlier of the more extensive rain forests of New Guinea, but shows evidence of its very different and much longer history. In each of these regions, a combination of biogeographical differences, past climates, and modern climate determine the types of plant communities that we see today. In later chapters, the impact of these plant communities on animal community structure and abundance will be developed.

The great diversity of rain forest floras makes rigorous comparisons between continents very difficult. In addition, the huge differences between years in plant reproductive activity – most strikingly in Southeast Asia but also, to some extent, in all rain forests – coupled with the very long life span of individual canopy trees, means that comparative studies must extend over many years or even decades to be meaningful. Understanding the causes and consequences of the observed differences will undoubtedly need an experimental approach, but experiments on the spatial and time scales required are beyond the capabilities of all but the largest and richest research organizations. Even impractical ideas for experiments can have great value, however, in stimulating thought and the design of more realistic experiments.

Although the rain forests of the Neotropics are the most distinctive in terms of the families and genera of plants that occur there, the lowland dipterocarp forests of Southeast Asia are undoubtedly the most distinctive in terms of their ecology. Both the family-level dominance by numerous species of dipterocarps and the supra-annual pattern of mass flowering and mast fruiting are unique to the region and have major consequences for the animals that live there. In comparisons between regions, dipterocarp dominance and supra-annual reproduction tend to be lumped together in the "food desert" view of dipterocarp forests. However, they are very different phenomena with different – and separate – consequences for other organisms in the forest. Dipterocarp dominance, for instance, is largely irrelevant to an understorey frugivore, but supra-annual fruiting cycles are highly relevant. Comparative studies are needed that separate the two phenomena.

There are no masting non-dipterocarp rain forests in Southeast Asia, but there are extensive areas of non-masting dipterocarp forests. Dipterocarp trees mass flower and fruit every 2–7 years in the everwet zones of Southeast Asia, but change over to annual reproduction in the more seasonal areas to the north. It would be very valuable to investigate a series of forest sites along a transect from the southern tip of the Malay Peninsula up into northern Thailand to determine how the changeover from supra-annual to annual flowering affects the entire animal community. One hypothesis would be that the forest stands with an annual cycle of dipterocarp reproduction would support greater densities of flower-, fruit-, and seed-dependent insects and insect-eating frogs, reptiles, and birds. These seasonal forests would be predicted to have annual cycles of abundance in animal density and reproduction, reaching peaks of abundance when dipterocarp trees are reproducing and declining when trees are not reproducing. In contrast, everwet forests might have animal populations that are both lower in density and less variable over the course of a non-flowering normal year. Unfortunately, this comparison is unavoidably compounded by the direct effects of climatic differences on animal abundance.

In order to investigate the effects of masting independently from dipterocarp dominance, comparisons need to be made with masting non-dipterocarp forests. Although mast fruiting at multiyear intervals involves many more species and much larger areas in Southeast Asian dipterocarp forests than elsewhere, masting also occurs on a smaller scale in some legume-dominated forests in Africa and the Neotropics. These mast-fruiting legumes show a number of other parallels with the dipterocarps, including large, poorly dispersed seeds and ectomycorrhizal associations with the roots. Studies of these various mast-fruiting tropical forests have so far used different techniques to test different hypotheses in each area. It would be instructive to do a pantropical comparison of mast-fruiting trees and forests, using standard methods, to look for common mechanisms behind their similarities. Parallel studies could contrast communities of vertebrates and invertebrates to determine how they respond to the masting cycle of the dominant plants.

An alternative, experimental, approach to understanding the role of dipterocarps in Asian rain forests would be to remove all the dipterocarps from a particular locality and observe the effects. This might seem hopelessly impractical, but selective logging already removes all the big dipterocarp trees from large areas of forest. This process could be completed by removing the remaining dipterocarp trees (perhaps by poison-girdling) and letting the residual non-dipterocarp trees dominate the forest. The hypothesis would be that the non-dipterocarp forest, because of its greater abundance of more edible plant families, would support higher densities of insects, leaf-eating vertebrates, and fruit-eating vertebrates than nearby dipterocarp forests. Such a project would need to be monitored for a period of decades as the trees grew to full size and the numerous pioneer trees gradually died off. Also, to be effective, the dipterocarps would have to be removed over a large enough area to encompass the dispersal distance of relevant animal species. The monodominant forests of Africa or South America could perhaps be investigated in a comparable manner, with all of the dominant species removed by repeated cutting over several years. However, the observation that adjacent monodominant and mixed forests have very similar floras – it is just the proportions of each species that are different – suggests that comparisons between the existing forest types in the same area would serve the same purpose. Southeast Asia is exceptional in that dipterocarp dominance is universal, so that non-dipterocarp forest can only be created artificially.

Bromeliads are the key distinguishing family of Neotropical forests. Experimental approaches might assist in understanding their role in forest ecology. An obvious experiment, that we think still has not been done, would be to remove all the bromeliads from an area of forest and compare this experimental forest with a nearby control forest with the bromeliads untouched. The hypothesis would be that the forest where the bromeliads had been removed would suffer a substantial loss of animal life, as insects, birds, lizards, and other animals of the canopy would no longer have a source of food, water, and breeding sites in the canopy. This decline in animal life would presumably be most severe during dry periods. It would also be valuable to check on microclimate conditions to determine if the removal of bromeliads had an impact on the humidity and temperature of the tree canopy.

Another method of investigating the importance of tank-forming bromeliads would be to place "artificial" bromeliads in the canopy of both Neotropical and African or Asian rain forests. "Artificial bromeliads" could be made by folding together strips of stiff, green plastic to create a water tank similar in structure and water-holding capacity to a tank bromeliad. When these artificial bromeliads are placed in the canopy of Neotropical trees would they be colonized by animals in the same way as living bromeliads? Is it simply the physical structure of the bromeliads, holding water in their overlapping leaf bases, that makes bromeliads ecologically important, or do the bromeliad leaves have some active role in the process, such as the leaves releasing oxygen or chemicals into the water? When they are placed in Asian or African canopies would they be utilized by the canopy fauna? Would the overall density of insects and other canopy animals increase as a result of the artificial bromeliads?

Elephants are keystone species in African forests. Where elephants have been removed from the forest by hunting, forest structure is expected to change and the fruits of many plant species to remain undispersed. A great deal of largely anecdotal evidence suggests that this is the case, but there is a need for more systematic studies. Africa now offers a range of rain forest sites from which elephants have been removed at different times in the past, providing an opportunity to investigate the long-term impact of elephant loss. Hopefully, the conclusions of these studies can be tested in future by the reintroduction of forest elephants to areas from which they had previously been removed. Comparative studies of other rain forests are needed to determine how the loss of large herbivores has affected the forest ecology in terms of tree density, the level of disturbance on the forest floor, and the types of plants growing on the ground. Reintroduction may be the only possible approach in Asian rain forests, where both rhinoceroses and elephants have been eliminated or greatly reduced in density in most areas. Indeed, large herbivores have been eliminated from most forests throughout the world; are there groups of large-seeded species that are now no longer being dispersed because their fruits are too big to be eaten by the remaining animals? The introduction of elephants to the Neotropics as a substitute for the extinct gomphotheres is almost certainly a bad idea, but exposing a domesticated elephant to "the fruits the gomphotheres ate" (Janzen & Martin 1982) in the forests of Costa Rica would be an alternative. Would the seeds survive passage through an elephant gut? Would germination be enhanced by gut passage and seedling growth by elephant dung?

   This chapter has reviewed the topic of plant diversity and community structure. Certain differences among rain forests have been described, and in this conclusion some ideas for comparative and experimental investigation have been developed. In the next chapter, we will consider patterns in the primates, probably the most completely investigated group of animals.

Chapter 3

Primate Communities: A Key to Understanding Biogeography and Ecology

No animal group illustrates the differences among the major rain forest regions better than the primates do. The unique biogeographical relationship that primates have with each of the five major rain forest areas sheds light on the forests' distinctive histories and features. These relationships depend on the evolutionary history of primates in each area of the world, past and current environmental conditions, and human impacts both past and present. Each continent represents a separate set of evolutionary radiations and ecological adaptations to the rain forest environment. Thus, New World primates are evolutionarily distinct from the Old World primates of Africa and Asia, and although African and Asian primate communities have some similarities at the family level, the species in each region belong to separate radiations within each family and represent adaptations to the distinctive environments of their region. In particular, many African species represent adaptations to a drier, more open forest, while most Asian species are adapted to the tall, everwet dipterocarp forests. Neotropical species are adapted to an almost purely arboreal existence in the dense forests of South and Central America. The primates of Madagascar represent an entirely novel radiation of an early primate group, the lemurs, which expanded and persisted in the absence of later primates and without the constraints of competition with many of the more recent mammals of the large continents. Finally, the New Guinea–Australian region lacks native primates altogether, so it is possible to see how other rain forest animal groups have responded evolutionarily to this lack of a key mammalian group.

What are primates?

Our closest living relative in the animal kingdom is a rain forest primate, the chimpanzee, so it is not surprising that these animals hold a great fascination for both biologists and the general public. But rain forest primates are far more than just a source of insight into human behavior and evolution. In most rain forests, primates represent a large proportion of the vertebrate biomass, are among the most abundant of large mammals, and play a key role in the ecology of the forest, particularly as important dispersers of seeds and as major consumers of leaves, fruits, and insects. Moreover, 90% of all primates are associated with tropical forests and the majority live in rain forests. Unfortunately, primates are also the most endangered group of rain forest animals, with many species threatened by hunting and habitat destruction. Primates as a group are valuable as indicator species for conservation purposes because they are easy to census, sensitive to hunting intensities, and typically are strongly affected by forest fragmentation.

Modern primates show an astonishing diversity of shapes, sizes (30–200,000 grams), and behaviors, but at the same time, they are one of the most easily identified mammalian orders (Fig. 3.1). Most primates can be identified by various anatomical features, including a large head associated with an enlarged brain, a flattened face with shortened nose, and grasping hands and feet, often with opposable thumbs. Associated with a well-developed brain is a complex set of behaviors, often involving elaborate social interactions.

The principal divisions of the primates at the family level and below are associated with geographical areas, as primates are not able to easily cross oceanic barriers. Further, most species are primarily arboreal; even ground-dwelling species climb trees to obtain food or to take shelter when threatened, so that primates are only able to disperse across landscapes with at least some trees. The distribution at higher taxonomic levels, in contrast, is less easily explained

(a)

Fig. 3.1 The diversity of primates. (a) Common brown lemur (*Eulemur fulvus*). (Courtesy of Dale Morris.)

(b)

(c)

Fig. 3.1 (*cont'd*) (b) Slender loris (*Loris tardigradus*) from Sri Lanka. (Courtesy of N. Rowe, taken in captivity.) (c) Spectral tarsier (*Tarsium spectrum*) from Sulawesi, Indonesia, holding a cockroach. (Courtesy of Tim Laman.)

(d)

(e)

Fig. 3.1 (*cont'd*) (d) Emperor tamarin (*Saguinus imperator*), a small New World monkey. (Courtesy of Tim Laman, taken in captivity.) (e) Red-eared guenon, an Old World monkey, in Central Africa. (Courtesy of Tim Laman.)

by present-day geography. In modern systems of classification (Fig. 3.2), the order Primates is divided into two suborders: the Strepsirrhini (literally "inward-turned nose," the wet-nosed primates) and the Haplorrhini ("simple nose," the dry-nosed primates). The Strepsirrhini includes the lemurs (Figs. 3.1a, 3.10, 3.11), which are found exclusively in Madagascar; and the lorises (Fig. 3.1b), pottos,

(f)

Fig. 3.1 (*cont'd*) (f) Adult male Bornean orangutan (*Pongo pygmaeus*), a great ape, in Gunung Palung National Park, Indonesian Borneo. (Courtesy of Tim Laman.)

and galagoes (also known as bushbabies; Fig. 2.6), which are found in Africa, India, Sri Lanka, and Southeast Asia. The Haplorrhini includes the tarsiers (Fig. 3.1c), which are confined to Southeast Asia; the New World monkeys (Platyrrhini, literally, "flat-nosed") (Figs. 3.1d, 3.7), found in South and Central America; and the Old World apes (Figs. 3.1f, 3.4, 3.9) and monkeys (Catarrhini, "down-nosed") (Figs. 3.1e, 3.3, 3.5, 3.6), widely distributed in Africa and Asia. Humans fall in the latter category, as *Homo sapiens* is an ape species of African origin. Note that the primate communities of African and Asian rain forests include species from several separate radiations, while those of the Neotropics and Madagascar are each composed of a single radiation.

The monkeys, apes, and humans – known collectively as anthropoids – differ from other primates in several respects: they are generally larger in body size but have a proportionally shorter torso, they have a larger brain in proportion to their body size, and they have nails rather than grooming claws. There are also many differences in their associated activity patterns. Anthropoids tend to be active during the day, while tarsiers and many strepsirrhines are active at night. Anthropoids also have short snouts with less developed nasal regions, relying more on vision than smell.

Old World versus New World primates

The main division of the anthropoids is between the Old World Catarrhini and the New World Platyrrhini. The timing and place of anthropoid origins is still

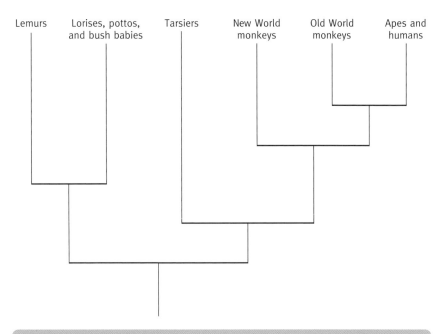

Lemurs   Lorises, pottos,   Tarsiers   New World   Old World   Apes and
          and bush babies   monkeys   monkeys   humans

Fig. 3.2 Our current understanding of the relationships between the major groups of primates mentioned in the text and illustrated in Fig. 3.1. Note that the position of the tarsiers is still uncertain.

unclear, but they were diverse in both Asia and Africa by around 40 million years ago (Jaeger et al. 2010). During this time, South America was an island continent, far enough separated from Africa that primates could not easily cross the water barrier of the Atlantic Ocean. However, approximately 30–35 million years ago, primates did somehow manage to cross from Africa to South America, as evidenced by their appearance at that time in the fossil record (de Oliveira et al. 2009).

The fact that all South American monkeys have similar characteristics to each other, which distinguish them from Old World monkeys, suggests that this dispersal across the water took place only once. The Atlantic was younger and narrower then, and there were islands along the now submerged Rio Grande Rise and Walvis Ridge between 20 and 30° South until around 40 million years ago, but it was still a formidable obstacle. Oceanic currents sweeping northward along the western coast of Africa would be deflected from the bump of West Africa directly westward toward the South American coast. A small group of African monkeys exploring a floating clump of trees in a river, or perhaps feeding on a tangle of trees and vines when the entire bank slumped into the river, could have been swept out into the Atlantic Ocean on a river current. The river current then joined the ocean current, which would have carried the tree raft and the monkeys perched on top directly to the South American coast within a week or two. Alternatively, dispersal could have taken place over several generations as the oceanic islands in the South Atlantic were occupied by monkeys rafted from the African coast, and then served as dispersal points for further westward migrations. While both scenarios – "floating island" and "island-hopping" – may seem highly improbable, they only had to occur once to explain the arrival of monkeys in South America. It is also striking that the ancestors of the Neotropical caviomorph rodents arrived around the same time (see Chapter 4) – perhaps even on the same floating island?

The primates that arrived in South America underwent an episode of adaptive radiation into a diversity of forms that resemble various Old World monkeys, strepsirrhines, apes, and even non-primates, such as squirrels. Despite this radiation, they are still a relatively uniform group and share several unique characteristics that distinguish them as New World monkeys. One main difference is the shape of the nostrils. New World primates have broad nostrils, pointing to the sides, whereas Old World primates have narrow nostrils, directed forward. Also, the New World primates have a number of distinctive anatomical characteristics of the head, such as three premolars (the teeth between the molars and the canines), whereas the Old World primates have two premolars. (Check your own mouth to confirm that you are an Old World primate by ancestry, if not in modern residence.) In general, New World primates have relatively short forearms and lack an opposable thumb. Most obviously, all New World primates have a tail, whereas many Old World primates lack a tail. In five of the New World genera, the muscular tail is used like a prehensile fifth limb, which lets them hold on to branches while using their front limbs for picking up food. The prehensile tail allows for spectacular acrobatics in species such as the spider monkey.

Another interesting contrast is in the area of color vision (Jacobs & Nathans 2009). Color vision is clearly very important for primates, with only two nocturnal groups – bushbabies (*Galago*) in Africa and owl monkeys (*Aotus*) in the Neotropics – known to lack it. Unlike other mammals, Old World monkeys, apes, and humans are uniformly trichromatic – that is, all normal individuals have light receptors in the retina of the eye that are most sensitive to three different ranges of wavelengths. In practice this means that they see what we see. In contrast, the New World monkeys are very variable in their color vision: only howler monkeys (*Aloutta*) have uniform trichromacy while in other species some females are trichromatic and other females and all males are dichromatic, i.e. red-green color-blind. The situation in some lemurs may be similar.

Trichromacy is expected to be advantageous in any activity that involves detecting reddish objects against a background of green foliage, such as finding ripe fruits or young leaves to eat. It is noteworthy that the New World monkey that consumes most leaves, the howler monkey, is the only species currently known in which all individuals are trichromatic. Another potential advantage of trichromacy is in detecting variations in skin color used as social signals, threats, or indications of emotional states (Changizi et al. 2006). Reddish signals are widespread in trichromatic catarrhines, such as the naked pink or red rumps of the drill and mandrill (Fig. 3.3), the reddish facial patterns of some guenons and macaques, and the pink faces of embarrassed humans. Red signals are less striking in New World primates, but both trichromatic and partly trichromatic species tend to have bare skin on the face, while dichromats have furry faces. Theory suggests that for dichromacy to persist in mixed populations it must also have an advantage, and captive dichromat marmosets outperform trichromats when foraging at low light intensities (Caine et al. 2010). The varied visual capabilities within a group of New World monkeys may allow different individuals to

Fig. 3.3 Male and female mandrills (*Mandrillus sphinx*) showing the striking coloration whose impact depends on the viewer having trichromatic color vision. (Courtesy of Fam Shun Deng, taken at the Singapore Zoo.)

specialize on finding particular types of foods or detecting certain kinds of predators, although there is, as yet, no strong evidence for this.

The New World monkeys are relatively uniform in comparison with the older radiation of anthropoids in the Old World. The primary division among the Old World anthropoids is between the apes and the Old World monkeys, which are distinguished by many skeletal and dental characteristics. In addition, the apes have broad palates, broad nasal areas, and even larger brains. They are also extremely intelligent, and some species show complex tool use and the transmission of cultural behavior across generations (Fig. 3.4). However, this intelligence does not seem to have done them much good: the 20 ape species that survive today are the remnants of a much more diverse radiation during the Miocene that not only dominated the rain forest primate fauna, but also occupied many other habitats. The Old World monkeys, by contrast, were less diverse during the Miocene and seem to have occupied mainly non-rain forest habitats, but have radiated dramatically since and now greatly outnumber the apes. The living apes are divided into the lesser apes, composed of 16–18 species of gibbons from Asia (Fig. 3.9), and the great apes, including orangutans (Figs. 3.1f, 3.4) in Asia, and gorillas, chimpanzees, and bonobos in Africa. The great apes have a longer time to first reproduction and a longer gestation period than Old World monkeys. Humans are classified with the great apes and originated in Africa, but now are the most widely dispersed primate.

Fig. 3.4 A subadult male Sumatran orangutan using a carefully prepared twig to extract the nutritious seeds of *Neesia* from among the irritant hairs that fill the woody capsule. (Courtesy of Perry Van Duijnhoven, who painted this picture.) The prevalence of this behavior in some populations in northwestern Sumatra but not others is most easily explained as the result of differences in local culture (socially transmitted innovations), rather than ecology or genetics (van Schaik 2009).

Primate diets

Primates are intelligent and adaptable animals, and many species are more or less omnivorous, taking whatever food is most available and nutritious at any one time. This tends to obscure neat categories of dietary specialization, although most species depend heavily on one or two types of food for a major part of each year. On each continent, there are species specializing on insects, fruits, leaves, and even plant saps and gums, although the proportion of primate species with each diet differs greatly between continents. Many small primates feed on mixtures of fruit and insects, and many large primates feed on mixtures of leaves and fruits.

Dietary differences are sometimes reflected in the classification of primates, indicating the conservative nature of diet, linked in turn to a wide range of associated behavioral, morphological, and physiological adaptations. This is particularly obvious in the Old World monkey family, Cercopithecidae, which is divided into two very different subfamilies. The Cercopithecinae, including baboons, mandrills, mangabeys, guenons, and macaques, have cheek pouches (Fig. 3.6) and a simple stomach, and are omnivorous feeders, with a strong preference for ripe fruits when they are available. The Colobinae, including colobus monkeys, langurs, and the proboscis monkey, have no cheek pouches and a complex stomach, and are primarily leaf- and seed-eaters (Fig. 3.5). Members of both subfamilies coexist in most African and Asian rain forests, but the Cercopithecinae are most diverse in Africa while the Colobinae are most diverse in Asia.

Intelligence and foraging adaptability are taken to the extreme in our primate cousins, the chimpanzees (*Pan troglodytes*). There is evidence that chimpanzees are able to remember the precise locations of thousands of fruiting trees in their forests (Normand et al. 2009). Although primarily fruit-eaters, there are big differences in diet between different groups, some of which seem to reflect differences in culture rather than resource availability or genetics (Lycett et al. 2007; Schöning et al. 2008). In some areas, cooperative hunting of red colobus monkeys or other mammals by adult chimpanzee males has a significant impact on the populations of their prey. There are also striking differences between chimpanzee groups in the amount and type of tool use, including the use of sticks to "fish" for ants and termites, and of stones to break open hard nuts. Some chimpanzee populations in the Congo use separate tools sequentially to first puncture the nests of army ants (*Dorylus* spp.) and then harvest the ants (Sanz et al. 2010). Similar cultural differences have also been observed between groups of orangutans (*Pongo pygmaeus*), which are relatives of both humans and chimpanzees (van Schaik 2009; Jaeggi et al. 2010). As with the chimpanzees, some of these differences involve methods of foraging, such as the use of sticks to extract the highly nutritious seeds of *Neesia* trees from the irritant hairs that surround them (Fig. 3.4), and the capture and eating of slow lorises (*Nycticebus coucang*) hiding in dense vegetation.

Leaf-eaters

Some primates are predominantly leaf-eaters. The great advantage of a leaf diet is that leaves are readily available; the disadvantage is that leaves are of relatively low food value and often contain toxic chemicals. Leaves take longer to digest than other foods, leading to a larger gut, slower processing time, a reduced metabolic rate, and lower activity levels. As a result, most leaf-eaters are relatively large.

Each rain forest area that has primates includes some species that eat leaves, but the extent to which these leaf-eating species rely exclusively on leaves differs from region to region, as do the physical adaptations that allow these species to survive on such a low-quality diet. The leaf specialists of African and Asian rain forests are all Old World monkeys in the subfamily Colobinae, which includes the colobus monkeys of Africa and the leaf monkeys (including the langurs, lutongs, surilis, doucs, snub-nosed monkeys, and the proboscis monkey) of Asia (Fig. 3.5) (Kirkpatrick 2007). These colobines vary in their dependence on leaves and are perhaps better categorized as feeders on "difficult" plant materials, since different species also take varying amounts of seeds and unripe fruits. The key colobine adaptation is their complex, ruminant-like stomach, which can hold a third of their body weight in food while bacteria detoxify plant defensive chemicals and digest the cellulose.

The leaf-eating lemur species of Madagascar, such as the 600–900 g (20–30 oz) sportive lemurs (*Lepilemur*), are very different animals. These lemurs are exceptions

Fig. 3.5 Dusky leaf monkey (*Trachypithecus obscurus*) in southern Thailand. This species feeds predominantly on leaves, but also consumes fruits (mostly unripe) and flowers. (Courtesy of Robert Pollai, www.the-ninth.com.)

to the rule that leaf-eaters are large and, unlike colobines, they have a relatively simple stomach. Instead, they are "hindgut fermenters," digesting leaves in the colon and an expanded cecum, excreting large particles rapidly while selectively retaining smaller, more digestible particles.

The New World lacks lemurs and colobine monkeys, and no species has developed such extreme adaptations for eating leaves as these Old World primates have. The howler monkeys (*Alouatta*) and woolly spider monkey (*Brachyteles*) are the closest to being leaf specialists in the Neotropics, but, unlike most colobines, they also eat substantial amounts of ripe fruit when it is available. Howlers and woolly spider monkeys are hindgut fermenters, with a long hindgut and expanded cecum. The nearest ecological equivalent to the Old World colobines are not primates but sloths, a group that is confined to the New World. Like colobines, sloths have a complex, ruminant-like forestomach for cellulose digestion and can reach high densities in Neotropical rain forests. The low metabolic rate, slow movements, long periods of inactivity, and general slothfulness of sloths are an extreme example of a trend seen in other arboreal leaf-eaters, none of which are particularly lively animals. This is presumably an energy-saving mechanism, made possible by the abundance of their food supply and necessitated by the low nutritional quality of their diet.

Insectivores

Other primates specialize on eating insects and other invertebrates. Insects are an important source of protein for many primates, including such large species as chimpanzees. However, the insect specialists are mostly small, active primates, such as the tarsiers (*Tarsius*) of Asia (Fig. 3.1c), the marmosets (*Callithrix*) of the Americas, the mouse lemurs (*Microcebus*) of Madagascar (Fig. 3.10), and the dwarf galagos (*Galagoides*) of Africa. Small size is no disadvantage in overcoming insects, whereas a larger size would make it impossible to find enough insects to eat, since a large animal cannot catch insects significantly faster than a small one. The earliest ancestors of the modern primates may have been insectivores and such key primate adaptations as grasping hands and binocular vision may have evolved first as an aid to catching such small, active prey. Most insectivores seek other food, such as fruits, nectar, or plant exudates, when insects are scarce, but the tiny (60–140 g) tarsiers appear to feed exclusively on invertebrates – mostly large insects.

Frugivores

The third major diet type is frugivory – fruit eating. Fruit is conspicuous and often readily available, larger in size than insects, and more easily digestible than leaves, although it is typically low in protein. Frugivorous primates usually consume only the fruit pulp so the seeds, which are hard and often poisonous, are an unwanted waste product. Fruit-eating species form a major part of all rain forest primate communities, but there are striking differences in the ways that the different groups of primates deal with the seed problem. The fruit-eating Old World monkeys in the subfamily Cercopithecinae, which includes the African guenons (Fig. 3.1e) and Asian macaques (Fig. 3.6), all have well-developed cheek pouches. These are used to hold excess fruits, which are then returned one at a time to the mouth for processing, with the larger seeds being spat out. This allows these monkeys to harvest many fruits quickly and process them more slowly as they move between fruiting trees. Fruit-eating apes, lemurs, and New World monkeys, in contrast, lack cheek pouches and often swallow fruits whole. This also permits rapid harvesting, but has the disadvantage that indigestible seeds can make up a large proportion of the material in the gut at any one time. Both seed-processing strategies seem to work equally well for the primates, but may have very different consequences for the plants whose fruits they consume (see below). Note also that there are species in all four regions that destroy the seeds in most fruits that they eat: colobine monkeys in Africa and Asia, pitheciine monkeys in the Neotropics, and the sifakas (*Propithecus*) in Madagascar.

Consumers of tree exudates

Tree exudates – sap, resin, and gum – are consumed opportunistically by many species of primates, but only a few species are known to cause the exudates to flow by actively gouging with their front teeth (Vinyard et al. 2003). Gouging holes through bark requires dental and skull adaptations that have evolved independently in small primates in each rain forest region. Moreover, the complex carbohydrates (β-linked polysaccharides) in plant gums are resistant to mammalian digestive enzymes and need to be fermented by gut microbes for efficient digestion (Power & Myers 2009). In tropical Asia, slow lorises (*Nycticebus* spp.) use their lower front teeth to perforate the cambium before lapping up the exposed phloem sap (Wiens et al. 2006; Swapna et al. 2010). In

Fig. 3.6 Rhesus macaque (*Macaca mulatta*) with cheek pouches full of food. (Courtesy of C.T. Shek.)

Madagascar, all the dwarf and mouse lemurs (family Cheirogaleidae) feed on tree exudates and they are a major food source for the fork-crowned lemur (*Phaner furcifer*) and the hairy-eared dwarf lemur (*Allocebus trichotis*) (Biebouw 2009). In Africa, the needle-clawed galago (*Euoticus elegantulus*) feeds on exudates in the forest canopy. Finally, in the Neotropics, the pygmy marmoset (*Cebuella pygmaea*) (Fig. 3.7) and, to a varying extent, other marmoset species (*Callithrix* spp.) specialize on exudates (Power & Myers 2009). These Neotropical exudate specialists are all diurnal, while those from the other regions are all nocturnal. Since most exudates consist largely of carbohydrates, these primates usually meet their protein needs by consuming insects.

Primate communities

African primates

One of the most noticeable differences among primate communities in different rain forest regions is in the range of body sizes (Fig. 3.8) (Fleagle 1999). Africa has the widest range of sizes among living rain forest primates, from the largely insectivorous dwarf galagos (also known as bushbabies), weighing less than 100 g (3.5 oz), to the gorillas, which can attain 200 kg (440 lb) or more. The large size of gorillas confers a number of advantages, including safety from predators and dietary flexibility. Lowland gorillas (*Gorilla gorilla*) feed largely on fleshy fruits

Fig. 3.7 Pygmy marmosets (*Cebuella pygmaea*) feeding on exudates from holes they have gouged in a tree. (Redrawn from Moyhihan [1976] by Helga Schulze, www.loris-conservation.org.)

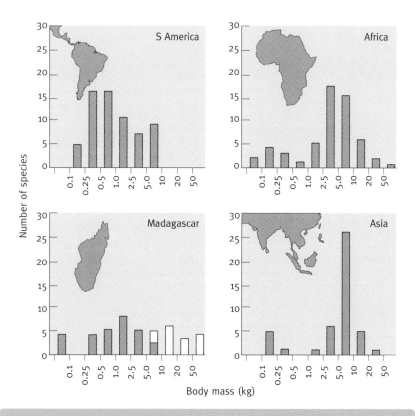

Fig. 3.8 The range of primate body weights in four land areas. Recently extinct Madagascan species are indicated as white columns. Data include both rain forest and non-rain forest species. (From Fleagle 1999; courtesy of Peter Kappeler.)

when they are available, but can also cope with fibrous alternative foods when fruit is scarce. Mountain gorillas (*Gorilla beringei*) have an almost entirely folivorous (leaf-eating) diet.

A typical African rain forest community will have several species of galagos, pottos, and their relatives – small (0.1–2 kg), nocturnal strepsirrhines that feed on various combinations of insects, fruits, and gums – and one or two species of apes, of which the chimpanzee (up to 60 kg) is the most widespread. The other members of the community are all medium to large Old World monkeys. Typically, one to three species of leaf- and seed-eating colobus monkeys (3–14 kg) are joined by up to eight species of frugivorous or omnivorous cercopithecine "cheek pouch" monkeys (1.5–50 kg), ranging from primarily terrestrial to entirely arboreal and active during the day. These include the approximately 24 species of guenons (Fig. 3.1e) – up to 6 species at one site – in the genus *Cercopithecus*, a genus that includes the Diana's, Mona, and de Brazza's monkeys (Glenn & Cords 2002). These are medium-sized monkeys, weighing from 4 to 12 kg, depending on the species, with a diverse diet. Most notable is their sleek, beautiful fur, often with contrasting white and darker patches on the face, throat, and body.

The mandrill (*Mandrillus sphinx*) is a very large – up to 35 kg in males and 11 kg in females – African cercopithecine monkey, found on the ground and at mid-level in the canopy in dense rain forests in Cameroon, Equatorial Guinea, Gabon, and the Congo. Mandrills are the most colorful of primates, at least to a viewer with trichromatic vision (Fig. 3.3), and there is striking variation between individuals in relation to sex, age, and social factors (Setchell et al. 2009). This bright signaling may an adaptation for life in huge foraging groups (< 1300 individuals) in which strangers must be able to evaluate each other quickly.

American primates

The New World monkeys are the youngest and most uniform primate radiation, although local species diversities are similar to those in the phylogenetically diverse primate communities of Africa (Fleagle 1999). American primate communities are unique in being dominated by arboreal, small to medium-sized insect- and fruit-eating monkeys, with no living species weighing more than 14 kg (30 lb) (Fig. 3.1d). The Americas lack the ecological equivalents to many of the primate species found elsewhere, with no terrestrial species, only one nocturnal genus, *Aotus* – with 10 species known as the night monkeys or douroucoulis – and few leaf-eating or suspensory species. There is, however, a distinct subfamily (Pitheciinae), containing the sakis (*Pithecia*), bearded sakis (*Chiropotes*), and uacaris (*Cacajao*), which are specialized feeders on fruits and seeds with hard coverings. The majority of New World primates weigh less than 2.5 kg, with the species-rich, squirrel-like marmosets and tamarins typically weighing less than 1 kg.

None of these small New World species readily descends to the ground, where their small size would make them highly vulnerable to predators. This is in marked contrast with African primates, such as gorillas, chimpanzees, bonobos, drills, mandrills, and mangabeys, that are larger in size and are found more often on the ground. Perhaps because of their small size and arboreal habitats, marmosets are apparently unable to cross rivers, and many species have distributions restricted to one side of a river system.

Both the absence of large primates from tropical American forests and the low diversity of leaf-eaters may be explained by the prior presence and high density of sloths as arboreal leaf-eaters when the primates first arrived. There were, however, several, considerably larger, 20–25 kg species of primates in the Americas in the recent past, but they have since died out, perhaps because of hunting by early human inhabitants of the continent (MacPhee & Horovitz 2002; Guedes & Salles 2005). The low diversity of nocturnal primates may be similarly explained by the prior presence and diversity of night-foraging marsupials. Thus, in striking contrast to the situation in Madagascar, considered below, the primate radiation in the New World was constrained by the diversity of preexisting mammal species that already occupied several potential primate niches.

Asian primates

Although the New World rain forests have an abundance of small, active insect-eating primates, tropical Asia has only the big-eyed, acrobatic tarsiers (nine or more *Tarsius* spp.) (Fig. 3.1c), found on the larger Southeast Asian islands but absent from the mainland, and the slow-moving lorises (two species of *Loris*, three of *Nycticebus*). There are also two very large (40–100 kg) species of arboreal, frugivorous ape, the orangutans, *Pongo pygmaeus* in Borneo (Fig. 3.1e) and *P. abelii* in Sumatra (Fig. 3.4). All the other primates in Asian rain forests, including the cercopithecine macaques (Fig. 3.6), the colobine leaf monkeys (or langurs) (Fig. 3.5), and the gibbons (Fig. 3.9), which are small apes, have adult weights in the 5–12 kg (10–25 lb) range. Gibbons have amazingly long forearms that allow them to swing from branch to branch through the canopy, with an agility matched only by the Neotropical spider monkeys.

One possible explanation for both the relatively low diversity of most Asian primate communities and the deficiency in small species is that Asian rain forests have a high diversity of potential competitors. Most striking is the diversity and abundance of squirrels (more than 30 species) that have been present since Miocene times, including ground squirrels, flying squirrels, and giant squirrels. As many as a dozen species can coexist in the same area. Asian rain forests also have the squirrel-like tree shrews (Tupaiidae, 20 species), which vary in weight from 20 to 400 g and feed largely on insects and small fruits (Table 3.1) (Emmons 2000). Different species range from largely terrestrial to largely arboreal. African forests are intermediate in both the numbers of small primates (14 species) and the numbers of squirrels (14 species). Africa also has the anomalures (Anomaluridae), a group of seven species of gliding rodents ecologically similar to flying squirrels, but belonging to an unrelated endemic African clade. The earlier diversification of squirrels and other squirrel-like mammals in Asian forests prior to the arrival of the primates probably put small primates at a competitive disadvantage in Asian forests, and to a lesser extent in African forests. The paucity of squirrels in tropical America (only seven species), and their recent arrival in the Pleistocene period, allowed American primates to undergo a radiation of small-sized species that was closed to them elsewhere.

Another, perhaps complementary, explanation for the low primate diversity in Asian rain forests is the dominance of the forest canopy by the inedible leaves and branches of dipterocarps, coupled with the tendency of these and many other species to flower and fruit at intervals of 2–7 years. As described in Chapter 2,

Fig. 3.9 A glimpse of the Hainan gibbon (*Nomascus hainanus*), which may be the most endangered primate species in the world. Fewer than 25 individuals survive in a single forest block on Hainan Island, China. (Courtesy of Lee Kwok Shing, Kadoorie Farm & Botanic Garden.)

Table 3.1
Species richness of some small canopy vertebrates in rain forests.

	Small primates*	Squirrels	Tree shrews	Parrots	Tree kangaroos	Other†
Australia	–	–	–	13	2	–
New Guinea	–	–	–	46	8	–
Java, Borneo, Sumatra	2	37	10	7	–	1
Southeast Asian mainland	2	31	5	6	–	1
Philippines	1	2	1	12	–	1
India	–	7	1	6	–	–
Sri Lanka	1	6	–	5	–	–
African rain forests	14	14	–	8	–	7
Madagascar	35	–	–	3	–	–
Amazon basin	22	7	–	50	–	2
Central America	4	7	–	32	–	4

* Includes lorises, galagos, bushbabies, pottos, tarsiers, small lemurs, marmosets, and spider monkeys.
† Includes colugos (Dermoptera) in Asia, anomalurids in Africa, and kinkajous and olingos (Carnivora, Procyonidae) in the Americas.
Data from multiple sources. Due to an incomplete knowledge of ranges and habitat types, these values should be regarded as approximate.

the everwet forests of tropical Asia are truly "fruit deserts" most of the time, and also appear to be impoverished in the insects that feed on flowers, fruits, and seeds, as well as the amphibians, reptiles, and birds that depend on those insects as basic food items (Duellman & Pianka 1990). For Asian primates, dietary specialization would be fatal, since fruits and insects are unreliable sources of food, even if the squirrels were not already there as competitors. Thus, the majority of Asian primates are either primarily leaf-eaters, such as some leaf monkeys, or are at least capable of subsisting largely on leaves when fruit is rare. As discussed earlier, leaf-eating favors a large size. Presumably it is because good-quality food is so often scarce and in such small patches that orangutans are wide-ranging, largely solitary, foragers, contrasting with the social feeding of the African great apes. The orangutans are frugivores by preference, but will eat leaves, bark, and even insects when fruits cannot be found. Reproduction in orangutans and other Asian primates is enhanced during dipterocarp fruiting years, when a majority of rain forest trees follow the lead of the dipterocarp species and fruit at the same time, allowing the animals to feast on the rich abundance of fruit.

Madagascan primates

The island of Madagascar provides a useful natural experiment, in that there was an independent radiation of the lemurs (see Box 3.1), filling the ecological niches occupied by monkeys, apes, and other primates in the rain forest areas of other regions. Molecular evidence suggests that all the Madagascan lemurs are descended from a single species that arrived around 65 million years ago, soon after the catastrophic global extinctions that marked the Cretaceous/Tertiary boundary (Horvath et al. 2008). Madagascar has been separated from the African mainland for 100 million years, so they must have crossed a formidable water barrier to get there. However, paleogeographical models suggest that strong surface currents flowed eastwards from Africa to Madagascar in the early Tertiary, before switching direction by the early Miocene, 20 million years ago, as Madagascar drifted slowly north (Ali & Huber 2010). No apes or monkeys have managed to cross the Mozambique Channel, allowing the lemurs to diversify in their own unique way. The channel has also kept out squirrels, terrestrial herbivores (see Chapter 4), and many important groups of rain forest birds (see Chapter 5), and mammalian carnivores did not arrive until much later (see Chapter 4). In striking contrast to the Americas, therefore, the lemurs radiated into an environment with no mammalian competitors. The Madagascan primate radiation is also much older than the American radiation. As a result of both these factors, they now dominate the rain forest fauna in a way not seen in any other rain forest region.

Using living species to explore the evolutionary radiation of the lemurs presents some problems. Many of the larger species and genera were driven to extinction when people first arrived on the island 2000 years ago (Burney et al. 2003). Most of the at least 13 known extinct species were larger in size than any surviving species. Among the most notable species that have gone extinct is *Palaeopropithecus ingrens*, a 50 kg leaf-eating sloth lemur. Its forelimbs were much longer than the hindlimbs, suggesting it was able to hang in a manner comparable to a modern sloth or orangutan. Another large, extinct sloth lemur is *Archaeoindris*, a 160 kg leaf-eating ground-dweller, probably comparable to

Box 3.1 The lemur radiation

The lemurs have undergone a broad diversification, evolving at least 95 living species, in 15 genera, in five families, widely separated from each other and only distantly related to the other strepsirrhines (Orlando et al. 2008). At least 17 species in 9 genera and three additional families have become extinct within the last thousand years. The eight families are:

1 The Cheirogaleidae is a family of nocturnal, nest-building, solitary animals, with a weight of less than 500 g (20 oz). Most notable are the mouse lemurs, which feed on a diverse diet of insects, small vertebrates, fruits, flowers, nectar, and gums. A recently discovered dry forest species, *Microcebus berthae*, has the distinction of being the world's smallest primate at a weight of around 30 g with a total length of around 20 cm, and the slightly larger brown mouse lemur (*M. rufus*) (Fig. 3.10) is the smallest rain forest primate. Mouse lemurs have soft fur, long limbs and tail, large eyes, and a relatively short snout. They are probably the most numerous of all lemur species and the number of species recognized has increased more than seven-fold in the last decade, largely as a result of studies based on mitochondrial DNA (Weisrock et al. 2010).

2 The Lemuridae are the typical Madagascar lemurs of medium size (1–4 kg, 2–9 lb), organized in groups, active during the day or intermittently during the day and night, and feeding on leaves and fruit. A well-known species is the ring-tailed lemur. Also in this group are the bamboo lemurs, which feed almost entirely on bamboo shoots, and the common brown lemur (Fig. 3.1a).

3 The Indriidae are specialized leapers with an enlarged intestine and cecum, which aid in the digestion of leaves. A well-known group of species is the sifakas, which leap actively between trees, hang by their limbs when feeding, and hop on their back legs when moving on the ground. The indri (*Indri indri*) is the largest of all living lemurs (weighing up to 7.5 kg).

4 The Lepilemuridae consist of the sportive lemurs, small drab lemurs that are nocturnal and eat leaves.

5 The Daubentoniidae diverged very early on from the other lemurs (Horvath et al. 2008) and now consist of a single living species, the aye-aye (*Daubentonia madagascariensis*), which is certainly the strangest looking primate. The aye-aye is a medium-sized black lemur with coarse fur, huge ears, a bushy tail, two large, rodent-like incisors, and long clawed digits, particularly the third digit on the hand. The aye-aye forages for insects at night, tapping branches with an elongated finger, and gnawing on bark with their teeth, sort of a primate equivalent of woodpeckers, which did not reach Madagascar. There was also a second, five times larger, species, the giant aye-aye (*D. robusta*), until its recent extinction.

6 The Archaeolemuridae is an extinct family of rather monkey-like lemurs with estimated body weights in the 14–27 kg range and believed to be semi-terrestrial.

7 The Megaladapidae is an extinct family of large-bodied (38–75 kg) lemurs that were probably slow-moving leaf-eaters.

8 The Palaeopropithecidae is an extinct family of "sloth lemurs" with sloth-like hanging skills, which included the largest known lemur, the gorilla-sized *Archaeoindris*.

a modern gorilla. Two thousand years ago, therefore, the Madagascan lemurs spanned much the same size range as the apes, monkeys, and strepsirrhines do together in Africa. However, there are no fossil records from eastern Madagascar, so we can only speculate on whether any of these extinct species lived in the rain forest. As with other recent "megafaunal extinctions," it is tempting to look for natural environmental changes that can at least share the blame with people (Virak-Sawmy et al. 2010), but for a 65-million-year-old radiation to suffer such massive extinctions in the brief period they have coexisted with human populations is surely more than a coincidence.

There are several characteristics that distinguish the ecology of lemur communities from other primate communities. First, Madagascar is unique in the diversity and

Fig. 3.10 The smallest rain forest primate, the brown mouse lemur (*Microcebus rufus*), in Ranomafarma National Park, Madagascar. (Courtesy of Jon Atkinson.)

abundance of relatively small, leaf-eating species. Folivorous primates make up most of the biomass of primate communities elsewhere, but in Asia, Africa, and the Americas this biomass is made up of a few species of large (more than 5 kg) primates. In Madagascan rain forests, in contrast, most of the biomass comes from a very high density of 1–5 kg folivores (although the recently extinct, very large lemur species were probably also folivorous). Second, some species of lemur often descend to the ground to search for food, in the same way as many unrelated African species and a few species in Asia do; although none do in the Neotropics. This is probably partly due to the relatively open nature and short height of most Madagascan forests and the lack of competition for food on the ground. Also, being on the ground does not add substantially to the risk of attack from predators, since the greatest danger to lemurs is from hawks and the fossa, an endemic species of carnivore (see Chapter 4), both of which can attack lemurs in trees. Third, lemurs use their sense of smell more than primates do elsewhere. They mark their bodies and their territories extensively with scents produced from scent glands on their face, wrists, and urogenital region, depending on the species (Fig. 3.11). In the relatively dry forests where most lemurs live, these scents may be more lasting than in wetter rain forests elsewhere. Fourth, many lemurs are active at night or during both day and night, in contrast to primates in other regions that are more specialized for being active at just one time period, often during the day. There are no other large Madagascar mammals active at night, so this niche is available to lemurs. Nocturnal lemurs use acoustic signaling more than diurnal primates, which may provide a mechanism for mating isolation and thus help explain the large numbers of morphologically similar species found in several nocturnal genera (Weisrock et al. 2010).

Fig. 3.11 Red-bellied lemur (*Eulemur rubriventer*) scent-marking a branch in mid-elevation rainforest in Marojejy National Park in northeastern Madagascar. (Courtesy of Rachel Kramer.)

Despite the apparently low fruit supply in Madagascan forests (see Chapter 2), a number of lemur species, ranging in size from the 40 g brown mouse lemur (*Microcebus rufus*) (Fig. 3.10) to the 6.5 kg diadem sifaka (*Propithecus diadema*), eat mostly fruit when it is available. These frugivorous lemurs have various ways of surviving the long nonfruiting season. Sifakas rest more, have a low metabolic rate, and females stop lactating. Red-bellied lemurs (*Eulemur rubriventer*), in contrast, travel further in times of fruit shortage, eat a wide variety of alternative foods, and may continue feeding after dark. The most extreme adaptation to the strongly seasonal supply of high-quality food is found in the squirrel-sized greater dwarf lemur (*Cheirogaleus medius*), which accumulates large amounts of fat in its tail when sugar-rich fruits are available and then hibernates for several months during the nonfruiting season (Fietz & Ganzhorn 1999). Most female and some male brown mouse lemurs do the same. This ability to hibernate or go into torpor is just the sort of adaptation that may have enabled the ancestors of modern lemurs to survive the long water crossing to Madagascar 65 million years ago (Kappeler 2000).

In addition to the role of lemurs in seed dispersal, discussed later, there is also evidence that several plant species, including the endemic traveler's palm (*Ravenala madagascariensis*), may depend on lemurs for pollination (Kress et al. 1994). Thus the lemurs of Madagascar have not only occupied the niches filled by primates in other rain forest regions, but they have also taken over some of the roles fulfilled by fruit- and nectar-eating birds and bats elsewhere. The role of the aye-aye, as a sort of primate woodpecker (see Box 3.1), is another example of this phenomenon. The larger species of recently extinct lemurs, such as *Archaeoindris*, may have extended the dominance of the lemurs to ground level, making use of resources consumed by other groups of terrestrial mammals in other rain forests (see Chapter 4). It is no exaggeration to call the Madagascan rain forest the "lemur forest."

Primate equivalents in Australia and New Guinea

The rain forests of Australia and New Guinea are useful places to examine how animal communities respond in the absence of primates. The example is not clear-cut because these areas also lack other major groups of mammals, such as carnivores, ungulates, and squirrels (see Table 3.1). In these rain forests, some of the activities of primates have been taken over by marsupials, including species of possum and tree kangaroo. Marsupials are also abundant in Neotropical rain forests and the fossil record shows that they were once much more diverse than they are today. However, the presence of other mammalian groups, such as sloths and, later, primates, has restricted their diversity in the Neotropics. The rain forest marsupials in Australia and New Guinea, in contrast, vary greatly in size, diet, and time of activity in a manner somewhat similar to the divisions found in primate communities elsewhere, particularly in Madagascar (Smith & Ganzhorn 1996).

The tree kangaroos (*Dendrolagus* spp.) (Fig. 3.12) are the largest arboreal mammals in the rain forests of New Guinea and Australia, with adult weights of 6–15 kg (13–33 lb). In contrast to ground-dwelling kangaroos, the front and

Fig. 3.12 Two Huon tree kangaroos (*Dendrolagus matschei*) in the Yopno, Uruwa, and Som (YUS) Conservation Area, Papua New Guinea. (Courtesy of Tim Laman.)

hindlimbs of tree kangaroos have similar proportions. They are agile climbers, leaping from branch to branch or to the ground. Their diets are little known, but appear to consist largely of leaves, although fruits and flowers are also eaten. Tree kangaroos are foregut fermenters, like colobines and sloths, digesting cellulose in a modified forestomach. Interesting, although the ancestors of the order to which tree kangaroos belong, Diprotodontia, were probably arboreal, the tree kangaroos themselves evolved from terrestrial, wallaby-like ancestors (Meredith et al. 2009).

The possums are an extremely diverse group of marsupials, with at least 35 species in New Guinea, ranging in size from the 20 g long-tailed pygmy possum (*Cercartetus caudatus*) to the 6.5 kg black-spotted cuscus (*Spilocuscus rufoniger*) (Flannery 1995a). For reasons that are not fully understood, the possums are most diverse in the highlands and there are relatively few species in lowland rain forests. The possum radiation shows some parallels with that of the primates (Rasmussen & Sussman 2007). The smallest species seem to feed mostly on insects, while most of the larger species are apparently more or less omnivorous. The ringtail possums (Pseudocheiridae) are slow-moving arboreal leaf-eaters. They are much smaller animals (150–1500 g) than the tree kangaroos and are hindgut fermenters, digesting cellulose in a greatly expanded cecum like the sportive lemurs of Madagascar (Hume 1999). In the absence of squirrels, marsupials have also evolved squirrel-like forms, including the sugar glider (*Petaurus breviceps*) (Fig. 6.9), which has converged toward the form of a flying squirrel. There are even marsupial equivalents of the Madagascan aye-aye, the striped possums (*Dactylopsila*), with powerful jaws, specialized teeth, an elongated fourth finger, and a long tongue for extracting wood-boring insect larvae (Rawlins & Handasyde 2002). Another curious feature of these animals is their rather skunk-like smell, associated with their skunk-like, black-and-white striped, color patterns.

Old World primates and marsupials meet only on the large island of Sulawesi, in the center of the Indonesian Archipelago. Here it is a marsupial, the bear cuscus (*Ailurops ursinus*), which fills the arboreal leaf-eater niche, while the only primates are insectivorous tarsiers and omnivorous macaques. The bear cuscus is unusual for a cuscus, both in its large size (*c.* 7 kg) and the fact that it is active during the day. There is also a nocturnal frugivorous species on Sulawesi, the dwarf cuscus (*Strigocuscus celebensis*).

Primates as seed dispersal agents

Primates play multiple roles in the ecology of tropical rain forests, as predators of invertebrates and (in some cases) vertebrates, as leaf-eaters, as seed dispersal agents, and even occasionally as pollinators. Primates are also the most threatened of rain forest animals and there are few areas left with an intact primate community in which all species occur at their natural densities. It has been suggested that the decline or local extinction of primates will have major consequences for plant species that depend on them for seed dispersal (Chapman & Russo 2007). Yet rain forests have survived without primates for millennia in New Guinea and Australia, as well as on many tropical islands. What is so special about seed dispersal by primates?

Primates make up a substantial proportion of the total frugivore biomass in all rain forests in which they occur and eat huge amounts of fruit, but frugivory does not necessarily result in seed dispersal. What matters is the fate of the seed. Most primates kill some seeds by breaking them in the mouth, drop or spit out others, and swallow and defecate the rest. Seeds are usually hard and often toxic, so only a minority of specialized primate species destroy large numbers of seeds. The proportion of seeds that are dropped or spat out also varies greatly from species to species, with larger seeds and those that are easily separated from the flesh most likely to suffer this fate. Most of these seeds end up directly under the fruiting tree, which is not a good place for a seed to be, since it must compete with its siblings and parent, as well as survive the pests and diseases that are concentrated there. The cercopithecine monkeys of Africa and Asia, however, are seed-spitters with a difference. Many seeds are dropped while feeding, but these monkeys also have cheek pouches in which fruit is carried away from the fruiting tree. The seeds are then spat out one by one as the monkey moves through the forest.

Apes, most lemurs, and most New World monkeys swallow most of the seeds in the fruits they eat and defecate them later, unharmed. In most cases, these primates are probably high-quality dispersal agents, carrying the seeds far from the parent tree and scattering them through the forest. A study of gibbons in Borneo, for example, found that more than 90% of the seeds swallowed were dispersed more than 100 meters (McConkey & Chivers 2007) and one of bonobos in Central Africa found average dispersal distances of more than 700 meters (Tsuji et al. 2010). Larger primates, such as chimpanzees or howler monkeys, may provide a lower-quality dispersal service, particularly if they defecate on the ground or from low branches, because large numbers of seeds are deposited in a clump. This problem is increased in howler monkeys by the concentration of feces in group "latrines" associated with sleeping trees (Bravo 2009). In contrast, some smaller primates, such as the Neotropical tamarins, defecate large seeds singly and small seeds in small clumps, while the feces of canopy primates, such as gibbons, often shatter before they reach the ground. This is an over-simplification, however, and the postdispersal fate of the seeds depends on numerous factors, including the attractiveness of primate feces to seed-predating rodents and the abundance of feces-burying dung beetles.

Are primates providing a unique seed dispersal service in rain forests, or could other animals compensate for their loss, as they presumably have compensated for their natural absence in New Guinea? Birds and fruit bats are the other major seed dispersal agents in tropical rain forests (see Chapters 5 and 6, respectively), followed by mammalian carnivores and various ground-dwelling large herbivores (see Chapter 4). Studies that have compared the fruit diets of primates with those of other dispersal agents have invariably shown some degree of overlap, but there are also always fruits that are consumed largely or only by primates. Most of these "primate fruits" are relatively large, with large seeds, and many have a thick, inedible husk, which is hard to remove without the coordinated use of hands and teeth. If primates were lost from a rain forest area, these plant species would lose their major seed dispersal agents.

Overall, it appears that rain forest primates provide an important service for many plants species and an essential service for some. Primates and rain forests are truly co-dependent. The data currently available are insufficient for pantropical comparisons, but dependence on primates may be highest in Madagascar, where the primate radiation is particularly ancient and there are relatively few frugivorous birds, bats, and other mammals.

Conclusions and future research directions

Primates give valuable insight into differences among rain forest regions because of their diversity and the variety of their ecological roles. These differences are well known because primates have been intensely investigated for insights into human evolution. Humans are currently altering primate distribution and abundance in a way that might incidentally give additional insight into rain forest ecology.

Primates are being intensively hunted for meat in many parts of the world and there are now vast areas of rain forest without any primates or with a greatly reduced density. When primates are eliminated from the forest, how will that influence the ecology of the remaining species? As discussed above, the most important impact is likely to be on seed dispersal. Several studies have now looked at the consequences of primate removal for primate-dispersed plants, but there is an urgent need for broader, longer-term studies and for pantropical comparisons. Although primates feed on fruits in all areas of the world, the importance of primates as seed dispersers may be greater in some areas than others. It may also be possible to look at the roles of particular primate species, where only one or a few have been eliminated. For instance, macaques in Asia and howler monkeys in the Neotropics often persist in rain forest fragments from which other primate species have been lost.

Unfortunately, comparisons between forests from which primates have been eliminated by human activities and forests with intact primate communities are not as straightforward as might at first appear. Because all large vertebrates are hunted in addition to primates, it may be difficult to separate the special ecological role of primates from the impact of other large animals if all the large vertebrates have been removed. A better comparison would be between a forest where primates have been hunted out and a forest in which a new population of primates has been established. Alternatively, a before/after comparison could be made at a single site where primates are being reintroduced. Primate reintroductions are rare at present – one exception being the successful reintroduction of the charismatic golden lion tamarin (*Leontopithecus rosalia*) (Fig. 9.20) to forest fragments in southeast Brazil – but their number is certain to increase, providing a range of opportunities for studying their impact.

There are also lessons to be learned from the introduction of primate species *outside* their natural ranges, although we emphasize that this is not a reason to encourage such introductions, which can potentially cause massive ecological damage. In this case, the main interest is not in the primates' role in seed dispersal, but in their impact as predators, particularly on vertebrates. A surprising number of primate species have been released, deliberately or accidentally, in other parts of the world, usually on islands, but most of the "successes" have been with Asian or African cercopithecines (Long 2003). Macaques, in particular, are now established on previously primate-free islands throughout the tropics. The most worrying of these "successes" has been the establishment of a small wild population of long-tailed macaques (*Macaca fascicularis*) in the Indonesian province of Papua on the island of New Guinea (Kemp & Burnett 2007). These are the first wild primates in New Guinea and it is feared this opportunistic omnivore may adversely affect the endemic wildlife, through competition for fruits and other food items and by predation on small vertebrates and bird eggs. If ultimately "successful," this movement of macaques across one of the world's major biogeographical boundaries could be seen as analogous to the arrival of monkeys in South America 30 million years ago, or of lemurs in Madagascar even earlier.

The successful introductions have almost all been with widespread "weedy" species like the long-tailed macaques. Other possible movements of primates across natural biogeographical barriers should remain forever as thought experiments. What if the apes had reached Madagascar or South America? Is the Amazonian rain forest unsuitable for large primates or is their absence simply a biogeographical accident? Imagine releasing populations of chimpanzees, gorillas, or other large African primates into a suitable Amazonian rain forest. Would these primates survive or perhaps thrive? Would these primates alter the surrounding biological community? How would orangutans do in an African or Amazonian forest? Such experiments are certainly not appropriate to carry out because of the potential for introducing new diseases and other unforeseen changes. The best strategy for most species is to protect them where they are currently living.

Primates have great interest for humans because of their close relationship to people. But they are also useful for understanding rain forest differences. Other mammalian groups provide further valuable insights, and these will be considered in the next chapter.

Chapter 4

Carnivores and Plant-eaters

Primates are not the only group of mammals that show major differences among rain forest regions. In this chapter, we use two relatively well-known groups of mammals with very different diets and behaviors – the flesh-eating and the plant-eating mammals of the forest floor – to illustrate general differences between rain forests. These are both ecological, rather than taxonomic, groups, although a single taxonomic order, the Carnivora, accounts for most carnivorous mammals. Their inclusion in the same chapter is justified by their occupation of the same forest floor habitat – although a few carnivores are mainly arboreal – and the fact that the plant-eaters provide the major food for the carnivores. They have also had major reciprocal influences on each other over evolutionary time, so that many elements of the ecology and behavior of herbivores make sense only in relation to the carnivores that threaten them, while many features of carnivores are clearly adaptations to their preferred prey. In addition, most members of both groups are very poor at crossing even narrow ocean barriers, so their spread across the Earth's surface has been subject to similar constraints and opportunities. Finally, naturally low densities and susceptibility to hunting pressures make large carnivores and large terrestrial herbivores especially vulnerable to human impacts, so understanding their ecological roles and how these differ between rain forest regions is particularly urgent.

Carnivores

Large carnivores have become flagship species for conservation in many parts of the world. Much of this attention reflects their charismatic image, but there is also evidence that carnivores can play an important role as "top-down" controls on community structure and processes. Carnivores reduce the population numbers of their prey, including large terrestrial herbivores, primates, rodents, birds, reptiles, insects, and other, smaller, carnivores. Prey animals also avoid places and times of day associated with greater exposure to attack. By influencing the population sizes and behaviors of herbivorous animals, carnivores have

Tropical Rain Forests: An Ecological and Biogeographical Comparison, Second edition.
© Richard T. Corlett and Richard B. Primack. Published 2011 by Blackwell Publishing Ltd.

an indirect impact on the structure of plant communities. Equally important may be the influence of large carnivores on the abundance of the smaller carnivores, which are the main predators of rodents, birds, and other small vertebrates. The removal of large carnivores can therefore have major consequences for the rest of the rain forest community as the effects propagate from level to level down the food web, from top carnivores to plants, in what is sometimes called a "trophic cascade." At some isolated sites, the removal of carnivores has led to substantial increases in rodent populations, and a consequent decline in seedling abundance (Terborgh et al. 2001). Unfortunately, we currently know far too little about the complexities of rain forest food webs to predict what the general impact of the loss of particular carnivore species will be in larger areas.

Paradoxically, many rain forest carnivores are also important seed dispersal agents. Despite their obvious adaptations for killing and eating other animals, most carnivores eat at least some fruit, bears eat a lot, and some, including many civets, procyonids, and mustelids, are best described as opportunistic omnivores. The teeth and jaws of carnivores are not designed for chewing and the digestive system is unspecialized, so seeds pass through undamaged into the feces.

Although carnivores and Carnivora are practically synonyms today, the conquest of the rain forest by members of this mammalian order has been achieved in stages. By the beginning of the Miocene, modern families of Carnivora were already established in Asia, Africa, and North America. A single, mongoose-like species then crossed the substantial water gap from Africa to Madagascar around 20 million years ago, giving rise to a remarkable radiation of endemic carnivores (Yoder et al. 2003). This water-crossing feat was not repeated until around 7 million years ago, when a kind of raccoon (now extinct), from the North American family Procyonidae, traversed the narrowing water gap between Central and South America. However, it was then not until the first definite land connection arose, around 3 million years ago, that more procyonids and five additional families of Carnivora – the cats (Felidae), dogs (Canidae), bears (Ursidae), skunks (Mephitidae), and weasels (Mustelidae) – finally crossed from Central America into the southern continent. Among the major rain forest regions, only Australia and New Guinea remained free of Carnivora until recently, but this too is changing due to human influences. The Australian dingo is a descendent of domesticated dogs brought to Australia (and New Guinea) several thousand years ago, while the European fox and feral cats were brought by European settlers more recently. All three species have had a major impact on native wildlife but, so far, are not established within the rain forest.

The mammalian carnivore niche in the rain forests of both modern Australia–New Guinea and pre-Pliocene South America was occupied by marsupials. Various giant marsupial carnivores from several lineages appear in the Miocene fossil record in both regions. Descriptions of these animals as "marsupial lions," "marsupial wolves," and so on, give an idea of their size, but understate the bizarreness of their appearance. However, most of the larger species had disappeared before the end of the Tertiary. In Australia and New Guinea, marsupials still dominate, but there are no longer any really large species.

Asian rain forests are richest in mammalian carnivores, with 15–25 species found at sites where they are not hunted (Corlett 2007a). This compares with estimates for the Neotropics of up to 18 in western Amazonia and up to 15 in Central America. Diversities of mammalian carnivores are lower still in African forests, which have no dogs or bears and only two cats. It is also important to remember that mammals are not the only flesh-eating vertebrates in tropical forests. In all the major rain forest regions, all but the largest prey species are shared with snakes (see below) and, in many cases, birds of prey (see Chapter 5). Both the Amazon and African rain forests also support small (< 2.5 m; 7 feet), stream-dwelling crocodiles, and Old World rain forests, from Africa through Asia to Australia, also support monitor lizards (*Varanus* spp.) of various sizes.

Cats

Members of the cat family (Felidae) are the ultimate killing machines, with none of the compromise adaptations to an omnivorous diet shown in other rain forest carnivores. Cats include the largest carnivores hunting in the forest, some of which prey upon even the largest herbivores (Fig. 4.1). Cats are present in the Americas, Asia, and Africa, but their diversity and size range in the rain forest differs markedly among these three regions. Many of these rain forest cats also occupy a range of other vegetation types and, in these species, the rain forest animals are usually considerably smaller than their cousins in other habitats, perhaps as a result of the generally smaller size of the available prey in rain forests. Some species are among the most widespread of all vertebrates, with the cougar (*Puma concolor*) distributed from the Canadian Yukon in the north to the Straits of Magellan in the south, while the leopard (*Panthera pardus*) used to occupy most available habitats in both Africa and Asia.

Asia is the richest in rain forest cats, with 11 species in Southeast Asia, ranging in size from the 200 kg (440 lb) tiger (*Panthera tigris*) (Fig. 4.1a) and the 60 kg (130 lb) leopard (*P. pardus*), to the tiny flat-headed cat (*Felis planiceps*), at 2 kg (4.4 lb) weighing less than a housecat (Sunquist & Sunquist 2002). Up to six of these species can coexist at the same site. An even smaller species, the 1–1.5 kg rusty-spotted cat (*Prionailurus rubiginosus*), is found mostly in dry forests in southern India, but also occurs in rain forest on the island of Sri Lanka, which supports only three other cat species. Tigers are the largest of all rain forest carnivores and the only one that regularly takes prey weighing more than 100 kg. Tigers hunt on the ground, alone, often at night, preying upon even very large animals such as the gaur, a relative of domestic cattle weighing up to 900 kg, although pigs and deer are usually most important. Individual rain forest tigers can roam over tens of square kilometers. The leopard hunts mainly at night, taking a huge range of prey, including ungulates, monkeys, smaller vertebrates, and even insects. Where tigers and leopards coexist, the tiger takes bigger prey.

The clouded leopards (*Neofelis*) are intermediate in size (10–23 kg). Their relatively small size, long tail, short legs and broad paws have been considered adaptations to an arboreal life. Most sightings have been on the ground, but recorded prey includes arboreal squirrels, slow lorises, macaques, leaf monkeys and orangutans, as well as a range of medium-sized terrestrial species. Clouded leopards have the largest canine teeth in relation to body size of any living cat and the widest maximum gape angle (*c.* 85°), but the significance of this is unknown. On the island of Borneo, which lacks tigers and leopards, the clouded leopard is the largest carnivore and reaches a greater size than on the mainland. This Bornean form, also found on Sumatra, is now considered a separate species, *Neofelis diardi*, from *N. nebulosa* of the Asian mainland.

(a)

(b)

Fig. 4.1 Large cat species found in tropical rain forests. (a) Tiger (*Panthera tigris*) in India. (Courtesy of Harald Schuetz.) (b) Jaguar (*Panthera onca*). (Courtesy of Tim Laman, taken in captivity.)

Many of the smaller Asian cat species typically hunt near water, catching prey such as birds, rodents, other small vertebrates, and insects, as they come to drink or feed near riverbanks or wetland margins. One species is known as the fishing cat, *Felis viverrinus*, for its readiness to swim and its ability to catch fish and crustaceans. The flat-headed cat also lives along riverbanks where it catches fish and frogs. With its short legs and tail, long head, and tiny ears, this species looks, in some ways, more like a civet or mustelid than a cat.

Although Africa has many species of cats, including the familiar lion and cheetah, only two species live in the rain forest, the versatile leopard and the African golden cat (*Profelis aurata*), which is restricted to forest. Golden cats are large-pawed, lightly built, 8–16 kg predators that, like the leopard, mainly hunt on the ground, taking mostly small mammals and birds, but also duikers and monkeys. African rain forests have neither a tiger- or jaguar-sized cat nor any really small species. The ecological role of the small cats is perhaps taken over by the forest genets, the smallest members of the rather cat-like civet family (Viverridae), which is described in more detail later in this chapter.

Although cats did not enter South America until 3 million years ago, the American rain forest is intermediate in numbers of cat species, with six species that range in size from the 120 kg jaguar (*Panthera onca*) (Fig. 4.1b) and the 60 kg puma (*Puma concolor*, also known as the cougar or mountain lion), to four smaller species including the ocelot (*Leopardus pardalis*, 16 kg), the jaguarundi (*Herpailurus yaguarondi*, 7 kg), the margay (*L. weidii*, 4 kg), and the oncilla (*L. tigrinus*, 3 kg), also known as the tiger cat, tigrina, or little spotted cat (Sunquist & Sunquist 2002). All these species can coexist in the same area of rain forest, although the oncilla is less widespread than the other five species. Most, if not all, of these cat species appear to have originated in North or Central America before the interchange, and thus represent recent invaders, rather than a separate South American radiation of rain forest cats.

The jaguar is more closely related to the big cats of the Old World, from which it separated only about 2 million years ago, than to other American cats (Davis et al. 2010). This appears less surprising when it is remembered that the modern distribution of the tiger extends north to the Arctic Circle, suggesting that the big cats had no need for a tropical land route to the Americas. The jaguar hunts mainly on the ground, often near water, and primarily at dawn and dusk. Although jaguars can kill tapirs, the largest prey available in the rain forest, they usually specialize on smaller prey (< 25 kg), including peccaries, capybaras, agoutis, armadillos, deer, caimans, river and marine turtles, and even fish. The leopard-sized pumas have a diet similar to the jaguar, but, like leopards, also include smaller vertebrate prey and even large insects. Ocelots appear to hunt opportunistically, taking small terrestrial prey (< 4 kg) more or less in proportion to their availability. The rather weasel-like jaguarundi is a largely terrestrial feeder on small vertebrates and arthropods, but it also climbs well. The margay, in contrast, seems to forage mostly in the trees, catching small mammals, birds, lizards, and tree frogs. Margays have the ability to rotate their hindfeet through 180°, allowing them to run straight down a tree trunk like a squirrel (de la Rosa & Nocke 2000). These are the only cats that include fruit as a regular part of their diet. Finally, the oncilla is a largely terrestrial consumer of small vertebrates and insects, which is reported to adapt well to disturbed and even suburban habitats.

Asian linsangs

The two species of Asian linsangs (*Prionodon* spp., Prionodontidae) have traditionally been seen as cat-like civets, but molecular studies have shown that they are,
in fact, civet-like relatives of the cats (Gaubert & Veron 2003). This is consistent
with their cat-like appearance and behavior and retractile claws. Linsangs are
distinguished by their elongated bodies, long narrow heads, short legs, and long
tails. Both species are smaller than any cat, with the banded linsang (*P. linsang*)
weighing 600–800 g and the spotted linsang (*P. pardicolor*) around 600 g. Both
species appear to be equally at home in trees and on the ground and are reported
to feed on a wide range of small vertebrates. Like cats but unlike most civets,
they are almost exclusively meat-eaters and do not seem to eat fruit. Note that
the same molecular techniques that showed that Asian linsangs are not civets
confirmed, confusingly, that the rather similar-looking African linsangs (*Poiana*) are.

Dogs

Unlike the cats, members of the dog family (Canidae) are relatively uncommon
in rain forests. Dogs were confined to North America until the late Miocene, before
spreading to Eurasia (*c.* 9 million years ago), Africa (4.5 million years ago), and
finally South America (2–3 million years ago) (Hunt 1996). The Neotropics has
two small rain forest canids, both of which are little known and apparently rare
(Macdonald & Sillero-Zubiri 2004). The small-eared dog (*Atelocynus microtis*) is a
medium-sized (8–10 kg, 18–22 lb), rather cat-like, predator, with a long slender
muzzle, short ears, and bushy tail. It appears to be mostly solitary and at least
partly aquatic, with most sightings in or near water. Its diet includes fish, frogs,
small mammals, and fruits. Bush dogs (*Speothos venaticus*) are smaller (5–7 kg),
with a very short muzzle, legs, and tail, but they live in social groups that hunt
cooperatively. This allows them to prey on animals larger than themselves, up
to and including the largest of all Amazonian land mammals, the lowland tapir
(*Tapirus terrestris*) (Wallace et al. 2002). Most of their diet, however, seems to
consist of large rodents. The crab-eating fox (*Cerdocyon thous*) also occurs widely
in forests and savannas in South America, often near riverbanks, where it sub-
sists on small vertebrates, crabs, crayfish, insects, and fruit.
  African rain forests lack dogs, but Southeast Asian forests have the widely
distributed dhole (*Cuon alpinus*) (Fig. 4.2), a large-eared, 12–18 kg predator that
usually hunts in packs of 3–12 individuals. Cooperative hunting turns this
medium-sized wild dog into a fearsome predator, able to overcome prey as large
as deer, wild pigs, and wild cattle. There are even reports of large dhole packs
overcoming tigers. Another canid, the omnivorous raccoon dog (*Nyctereutes pro-
cyonoides*), is widespread in East Asia and enters the rain forest on the northern
margins of the tropics.

Weasels and their relatives

The weasel family (Mustelidae) has more species globally than any other group
of carnivores, but relatively few occur in tropical rain forests. Most are long,

Fig. 4.2 Dhole (*Cuon alpinus*) in India. (Courtesy of Sandesh Kadur.)

slender animals such as weasels, but there are also stocky, badger-like forms, and intermediate species such as martens. Rain forest mustelids are most diverse in Southeast Asia, where they include the mainly terrestrial Malayan weasel (*Mustela nudipes*), the larger, tree-climbing, yellow-throated marten (*Martes flavigula*), the very large (up to 14 kg, 31 lb) terrestrial, hog-badger (*Arctonyx collaris*), ferret-badgers (*Melogale* spp.), and the aquatic otters. African rain forests support only aquatic otters and the omnivorous honey badger or ratel (*Mellivora capensis*), which can kill animals up to the size of small forest duikers. Honey badgers get their common name from their habit of raiding the nests of wild bees. Mustelids are, again, relatively diverse in the Neotropics, where they include the large (< 7 kg), slender, arboreal tayra (*Eira barbata*) (Fig. 4.3), the predominantly terrestrial, short-legged and slender-bodied grison (*Galictis vittata*), the tropical weasel (*Mustela africana*), and several otters, including the giant river otter (*Pteronura brasiliensis*), which can exceed 2 m (6.5 feet) from head to tail and weighs 22–34 kg. As with cats, this Neotropical diversity is surprising in view of the late arrival of mustelids, which took place after the formation of the Panama land bridge 3 million years ago.

Mustelids in general are opportunistic carnivores, eating whatever invertebrates, small vertebrates, eggs, and fruit they can obtain. The slender, weasel-like species are fast, active hunters and can catch and kill vertebrates up to their own size or larger. For example, the Neotropical tayra takes prey on the ground or in the trees, and ranging in size from small rodents, birds and arthropods, up to small deer and juvenile primates (Bezerra et al. 2009). The stocky, badger-like species, in contrast, seem to concentrate on soil invertebrates, such as earthworms, although some species also take small vertebrates and plant material.

Fig. 4.3 Tayra (*Eira barbara*) in Costa Rica. (Courtesy of Matthias Dehling.)

Skunks

The skunks are notorious for the foul-smelling fluid produced by their anal scent glands. Skunks used to be included in the Mustelidae, but molecular evidence shows that they should be recognized as a distinct family, Mephitidae. One species, the hog-nosed skunk (*Conepatus semistriatus*), occurs in rain forests in Central America and adjacent parts of South America, while two species of stink badgers (*Mydaus* spp.) are found on the Southeast Asian islands of Sumatra, Borneo, Java, and Palawan.

Bears

An interesting contrast is found among the three species of tropical bears. The spectacled bear (*Tremarctos ornatus*) of South American montane forests is the most herbivorous of the three and feeds extensively on fruit, as well as the leaves of bromeliads and cacti. Its choice of food is biogeographically revealing, as both bromeliads and cacti are strictly New World families. In contrast, the Malayan sun bear (*Ursus malayanus*), which is the smallest of all bears, feeds extensively on the nests of termites and wild bees, palm hearts, fruits, and small vertebrates, using its strong curved claws to rip apart trees in search of food. Its legs turn inward, giving it expert tree climbing ability but an ungainly gait on the ground. The sloth bear (*U. ursinus*) of the Indian subcontinent and Sri Lanka has a diet also composed of fruits and social insects. Protrusible lips and the absence of the inner pair of upper incisors allow it to suck up termites like a vacuum cleaner. A third Asian bear species, the omnivorous Asiatic black bear (*U. thibetanus*), inhabits mostly mountainous and temperate regions of Asia, but also occurred in the northern range of Asian tropical rain forests. Africa has no bears south of the Sahara.

Civets, mongooses, and their relatives

A variety of other carnivores inhabits tropical forests in one or more regions. One of the most distinctive groups of small to medium-sized carnivores is the civets (Viverridae), including the genets, which are found from Africa through Asia, but are most diverse in the Sundaland region of Southeast Asia, where up to six species can coexist at the same site (Fig. 4.4). Civet species vary in weight from 600 g to 15 kg (1.5–35 lb). Somewhat cat-like in appearance, civets typically have an elongate body, short legs, a long, bushy tail, and a pointed muzzle. Many species have striped or spotted fur, perhaps in some species as warning coloration to predators that they can produce foul-smelling secretions from their anal glands. These secretions also function in territorial marking and social interactions, and the secretions from several species are used commercially in the production of perfumes. In other species, the patterns may aid in camouflage while hunting. Most civets are agile and skillful climbers, and some are also skilled swimmers. Civets eat small vertebrates, invertebrates, and fruits, and some species are important seed dispersal agents in both Asian and African forests. Several civet species, such as the common palm civet (*Paradoxurus hermaphroditus*) in Asia, have been able to adapt to human-dominated landscapes,

Fig. 4.4 Small-toothed palm civet (*Arctogalidia trivirgata*) in Borneo. (Courtesy of Hans Hazebroek.)

living in house rafters and drains. The commercial harvesting of the anal gland secretions for the perfume industry has resulted in the introduction of the Malay civet, *Viverra tangalunga*, to several islands between Southeast Asia and New Guinea that previously lacked mammalian carnivores (Nowak 1999).

Some civets, including the common palm civet, are largely arboreal, foraging in the canopy for small vertebrates, insects, and fruits. One of the most distinctive of the arboreal Asian civets is the binturong (*Arctictis binturong*), a very large species (up to 15 kg) with long coarse black hair and a prehensile tail that it uses during its nocturnal foraging. The only other carnivore with a truly prehensile tail is the Neotropical kinkajou (*Potos flavus*, Procyonidae), which is unrelated but ecologically rather similar (see below). The African palm civet (*Nandinia binotata*, Nandiniidae) is an arboreal omnivore with similar habits to the Asian palm civets, although molecular evidence shows that this animal is not a civet at all, but rather an early branch from the group that gave rise to cats, civets, and mongooses (Yoder et al. 2003). African rain forests support several species of partly arboreal genet: small, spotted or blotched, cat-like carnivores that are confined to Africa, except for one species that extends north into Europe. Other civet species in both regions, such as the Malay civet and the African civet (*Civettictis civetta*), live mostly on the ground. There are also civets in both regions that are semiaquatic, notably two Southeast Asian species in the genus *Cynogale* and the aquatic genet (*Osbornictis piscivora*) of the Democratic Republic of Congo.

The mongooses (Herpestidae) are related to the civets and have a similar Old World distribution. The rain forest mongooses have long, slender bodies, short legs with non-retractile claws, a tapering snout, small rounded ears, and a long tail. The smallest species weigh < 1 kg while the largest can weigh 4 kg. All the Asian species and some African species are solitary, opportunistic predators on large invertebrates and small vertebrates. These animals are known for their agility and quick movements; so fast and agile are some mongooses that they can attack and kill poisonous snakes without being bitten. Kusimanses (*Crossarchus* spp.) are dark, shaggy, social mongooses that inhabit the rain forests of West Africa, moving around in large, noisy groups.

Raccoons and their relatives

The Americas lack native civets and mongooses, though mongooses have been introduced into the region to control pest rats, most notably in islands of the West Indies, with disastrous consequences for the native wildlife. In the Neotropics, their ecological roles are, to some extent, played by a distinctive New World family of carnivores, the Procyonidae, which includes the familiar North American raccoon. The most closely related Old World mammal is the red panda (*Ailurus fulgens*) of the Himalayan Mountains, now placed in a separate family, Ailuridae. The procyonids are another of the many animal groups that have diversified in South America after the arrival of immigrants from the north when a land route became available. However, with the procyonids there is a curious additional twist to the story. A raccoon, *Cyonasua* ("dog-coati"), had reached South America by the late Miocene (7.5 million years ago), by rafting or island-hopping from Central America across the inter-American seaway, and gave rise to a small endemic radiation (Koepfli et al. 2007). This radiation included the bear-like *Chapalmalania*, which survived until the arrival of overland migrants,

Fig. 4.5 White-nosed coati (*Nasua narica*) in Costa Rica. (Courtesy of Matthias Dehling.)

including real bears, 5 million years later. The modern procyonids are considered to be descendants of this second invasion.

In tropical American forests, the coatis (*Nasua* spp.) (Fig. 4.5) are long-tailed and longer-snouted versions of the raccoon, weighing 3–6 kg (6.5–13 lb). Coatis are inquisitive diurnal omnivores, foraging on the ground but also going into trees. Males are solitary, except during the breeding season, but females and juveniles travel in stable groups. The long tail is often carried straight up while walking, giving a group of coatis an unmistakable appearance. The name "coatimundi" – often used in English for the species – is applied in South America only to the lone males. Other rain forest members of the family include the olingo (*Bassaricyon gabbii*), and, in Central America, the cacomistle (*Bassariscus sumichrasti*). The olingo weighs 1–1.5 kg and lives in trees. The cacomistle is a small (< 1 kg) arboreal procyonid that looks remarkably similar to the forest genets of tropical Africa. It has been little studied, but is said to be an opportunistic omnivore (de la Rosa & Nocke 2000). The crab-eating raccoon (*Procyon cancrivorus*) thrives in all Neotropical lowland habitats, including rain forest, but apparently always near water. It is largely nocturnal and forages largely on the ground, eating an unusually wide range of foods.

Until recently, the kinkajou (*Potos flavus*) was considered a close relative of the morphologically similar olingo, but molecular evidence suggests that it is either a very early offshoot from the procyonids or not a procyonid at all (Agnarsson et al. 2010). The kinkajou is a nocturnal, arboreal, monkey-like forager, weighing 1.5–4.5 kg, eating fruits, seeds, insects, and small vertebrates. Unlike other procyonids it has a fully prehensile tail and a long (13 cm, 5 inch), extrudable tongue, used to reach nectar in flowers.

Madagascan carnivores

Until recently, the seven living members of Madagascar's carnivore fauna (family Eupleridae) were assigned to two or three separate families, each believed to result from a separate colonization event. Molecular evidence has now shown that, despite their diversity of form, they are all descended from a single, mongoose-like ancestor that crossed the ocean from Africa, 18–24 million years ago (Yoder et al. 2003).

The largest Madagascan carnivore is the fossa (*Cryptoprocta ferox*) (Fig. 4.6), which presents a cat-like appearance similar to a small puma, weighing 7–12 kg (15–26 lb) with a compact body, short smooth hairs, a very long and slender tail, and short retractile claws (Garbutt 1999). Other distinctive features are its reddish brown color, long facial whiskers, and somewhat elongate muzzle. The fossa is widely distributed across Madagascar, primarily in woodlands and forests. It eats a range of vertebrates and invertebrates, but in rain forests the diet is dominated by lemurs, including adults of the largest species, such as the sifakas, for which they are the major predator (Wright 1998). Fossas are superb climbers and apparently take lemurs at night while they are sleeping. A 25% larger species of *Cryptoprocta*, *C. spelea*, which became extinct during the Holocene, may have predated the larger lemur species that were present at that time (see Chapter 3) (Goodman et al. 2004).

Two additional rain forest carnivores, the fanaloka (*Fossa fossana*) and falanouc (*Eupleres goudotii*), were previously considered to be civets. The fanaloka (or

Fig. 4.6 The largest Madagascan carnivore, the fossa (*Cryptoprocta ferox*). (Ran Kirlian, Wikimedia Commons, taken in captivity.)

Malagasy civet) is a shy, nocturnal animal the size of a domestic cat (< 2 kg), but more fox-like in appearance. It apparently prefers streams and marshy areas, and its diet includes insects, small mammals, and a variety of aquatic animals (Goodman et al. 2003). The falanouc is a larger animal (2–4 kg), with an elongated snout and tiny conical teeth, like an insectivore. It feeds almost exclusively on earthworms and other invertebrates. There are also five mongoose-like species, three of which occur in the rain forest. A true civet, the small Indian civet (*Viverricula indica*), has been introduced to Madagascar from tropical Asia some time in the last 2000 years, but is more common in degraded and agricultural habitats than in rain forest.

Australia and New Guinea

The rich fossil record suggests that the vast Miocene rain forests of northern Australia supported a mammal fauna as diverse as in rain forests elsewhere, including leopard-size marsupial lions and wolf-size carnivorous kangaroos (Long et al. 2002; Fig. 4.7). The subsequent drying of the Australian continent eliminated most of this fauna and the extensive rain forests of New Guinea are apparently too young to have evolved substitutes. As a result, the absence of large, or even medium-sized, mammalian carnivores is one of the most distinctive features of the rain forests of New Guinea and Australia today. The top carnivores in New Guinea and Australia today are reptiles (see below) and birds (see Chapter 5).

Australia and New Guinea lack native members of the order Carnivora and they have been only partly replaced by meat-eating marsupials in the order Dasyuromorphia. The largest species in New Guinea today is the New Guinea quoll (*Dasyurus albopunctatus*), which at around 700 g in weight (about the size of a squirrel) is no threat to anything larger than a rat. All other rain forest carnivores weigh less than 0.5 kg and feed mostly on insects. Australian rain

Fig. 4.7 Reconstruction of a leopard-sized marsupial carnivore, *Wakaleo vanderleuri*, from the middle Miocene rain forests of Australia. The tropical rain forests of Australia and New Guinea today have no large mammalian carnivores. (Courtesy of Anne Musser, the Australian Museum.)

forests have a larger (< 5 kg, 11 lb) species, the spotted-tailed or tiger quoll (*D. maculatus*), which is now the largest native mammalian carnivore on the Australian mainland, preying on animals as big as a small wallaby, as well as smaller mammals, birds, and insects.

Until recently, the largest mammalian predator in the region was the thylacine or Tasmanian wolf (*Thylacinus cynocephalus*), which weighed up to 35 kg. These marsupials bore a striking resemblance to wolves in their overall body shape, especially the shape of the head and the arrangement of teeth for stabbing and cutting meat. The marsupial ancestry, however, was shown by the presence in the female thylacine of a pouch for carrying her young. The thylacine occurred in New Guinea and throughout Australia, but underwent a dramatic decline following the introduction of the domestic dog by human settlers less than 5000 years ago, although it persisted in Tasmania until the 1930s. Records show that it could attack the largest native prey, including kangaroos and wallabies (Jones & Stoddart 1998). There is, however, no definite evidence that it occurred in tropical rain forests, and the fossil records from New Guinea are from alpine grassland.

The role of the thylacine as Australia's largest mammalian carnivore has now been assumed by the dingo, a large (up to 20 kg) wild dog descended from domesticated animals brought from Asia 3000–5000 years ago (Savolainen et al. 2004). Although generally not a forest animal, the dingo is an important predator in the margins of rain forest patches in the wet tropics.

Carnivorous reptiles

Reptiles are usually seen as ecologically inferior to mammals, particularly in the temperate zone, where their inability to stay active in cold conditions seems to be a fatal flaw. In the lowland tropics, however, year-round warmth means that "ectotherm" carnivores, which do not use food to maintain body temperature, are much more energy efficient than "endotherm" birds and mammals and thus have much lower food requirements, allowing them to live at higher densities. Moreover, while birds and mammals must minimize their surface to volume ratio to control heat loss, reptiles can be flattened, like geckoes, or elongated, like snakes. Reptiles are definitely not second-class citizens of the rain forest.

Reptilian carnivores include crocodilians (alligators, caimans, crocodiles, and gharials), lizards, and snakes. The largest living reptile is the saltwater crocodile, *C. porosus* (< 7 m, 1000 kg), but all modern crocodilian species are at least semi-aquatic and most are confined to more or less open habitats, presumably because direct sunlight is used for regulating body temperature. This limits their impact in rain forests. However, the smooth-fronted caiman (*Paleosuchus trogonatus*, < 2.6 m) in South America and the African dwarf crocodiles (2 or 3 species of *Osteolaemus*, < 2 m; Eaton et al. 2009) in Central and West Africa are true rain forest species, which can live in streams under a continuous forest canopy and forage on land at night. The smooth-fronted caiman has been estimated to have a higher biomass per unit area than any other large carnivore in the Amazonian rain forest (Magnusson & Lima 1991). This species often nests next to termite mounds, making use of the waste heat to incubate its eggs (Magnusson et al. 1985). Adult smooth-fronted caimans eat snakes and mammals up to the size of large rodents, but the diet of their African counterparts is poorly known. A relative of the African dwarf crocodiles, *Voay robustus*, survived into the Holocene on Madagascar and may have been part of the mass extinctions that followed human arrival on that island (Brochu 2007).

Large individuals of many rain forest lizards take some small vertebrate prey opportunistically, but really large lizards are confined to the Old World (except Madagascar), where the monitor lizard family (Varanidae) (Fig. 4.8) includes some formidable predators. It has recently been shown that some varanids produce a complex and potent venom from a gland on the lower jaw, although its role in subduing prey has not yet been investigated (Fry et al. 2006). Varanids are also noted for their relatively complex behaviors and "mammal-like" intelligence, and were described by Sweet and Pianka (2003) as "roughly speaking, dumber than civets and smarter than [marsupial] quolls." The largest living lizard, the Komodo dragon, *Varanus komodoensis* (< 3.1 m, 165 kg), does not occupy rain forest, but the second largest species, the widespread water monitor, *V. salvator* (< 2.5 m and > 20 kg, in some parts of its range), occurs in rain forests from Sri Lanka to Indonesia, along with several smaller species. Adult water monitors have been reported feeding on a huge range of prey items, from invertebrates to mammals as large as monkeys and small deer (Shine et al. 1996). The similar-sized crocodile monitor (*V. salvadorii*) in New Guinea is a specialized tree dweller (Horn 2004). Nothing definite is known about its diet in nature, but villagers say it can kill deer, pigs, and hunting dogs. The African rain forest species, the ornate monitor (*V. ornatus*) (Fig. 4.8), is somewhat smaller (< 2 m) and more closely associated with water. Most other varanids for which some dietary information is available consume a variety of invertebrates and small vertebrates, but three species are largely frugivorous (Welton et al. 2010).

Fig. 4.8 Ornate monitor (*Varanus ornatus*) in Gabon. (Courtesy of Stéphane Tridon, Precious Woods Gabon.)

Snakes are diverse and abundant in tropical rain forests, with more than 20 species coexisting at the richest sites. Unlike most other predatory vertebrates, all snakes swallow their prey whole. The maximum prey size is therefore set by the size of the mouth, although in most advanced snakes the left and right lower jaws are connected by an elastic ligament which allows them to separate when engulfing large prey. Changes in diet as a snake grows are the rule because size has such a large influence on the ability of an animal to capture, subdue, and swallow prey. Snakes catch and kill prey in three ways: by simply grabbing and swallowing an animal; by constriction; and with venom.

The largest of all snakes, pythons (Pythonidae) and boas (Boidae), kill by constriction. Pythons (*c.* 33 species, by no means all very large) occur from southern Africa, throughout tropical Asia, to New Guinea and Australia, but are absent from Madagascar and the Neotropics. They are not only the largest snakes in these regions, but are also among the largest predators. A big python has the body weight of a leopard and can capture and ingest all but the largest of leopard prey species. The largest python species, and probably the longest snake in the world, is the Southeast Asian reticulated python, *Broghammerus (Python) reticulatus* (Fig. 4.9), which can attain 10 m, although individuals longer than 7 m appear to be very rare. Large prey found inside pythons include deer, civets, monkeys, pigs, and a 23-kg adult female sun bear (Fredriksson 2005). Adult humans have been killed and at least one teenager swallowed. Another very large species, *Python molurus*, coexists with *B. reticulatus* in parts of Southeast Asia and also occupies most of the remainder of tropical Asia. Although shorter than *B. reticulatus*, it is heavily built and large individuals can attain over 70 kg and take similar-sized prey. The closely related African species, *Python sebae*, attains

Fig. 4.9 The world's longest snake species, the reticulated python (*Broghammerus reticulatus*), in the Danum Valley Conservation Area, Malaysian Borneo. (Courtesy of Tim Laman.)

a similar size and is also reputed to take large prey. Somewhat smaller (< 6 m) pythons in the genus *Morelia* inhabit rain forests in New Guinea and Australia, where they are the largest terrestrial carnivores and take prey up to the size of a wallaby.

Infrared-imaging pit organs on the lips of *Python molurus* (and, presumably, the similar structures in other pythons) are most sensitive in the 8–12 μm range, which matches the infrared body-heat emissions from mammals and birds (Grace et al. 1999). Infrared and visual information is overlain in the brain and these snakes can target prey accurately with their eyes covered. When a prey is captured there is a huge increase in the metabolic rate of the snake and large increases in the mass of most internal organs, which atrophy in the weeks or even months between meals (Secor 2008). No endothermic cat could wait in between meals for as much time as a snake.

Another group of large snakes are the boas, which occur throughout the tropics, except for Southeast Asia. In the Neotropics, this family includes the heaviest of all snakes, the green anaconda (*Eunectes murinus*, < 10 m, > 200 kg), but unlike pythons, which are equally at home in water and on land, anacondas are fully aquatic, although they may hunt at the water's edge. A recent video on YouTube shows one regurgitating a tapir. An even larger boid, *Titanoboa cerrejonensis*, with an estimated length of 13 m and mass of 1135 kg, inhabited Paleocene rain forest in Columbia 58 million years ago, and was probably also aquatic (Head et al. 2009). No terrestrial boid is as big, although the widespread Neotropical *Boa constrictor* (< 3–5 m) takes mostly mammalian prey, including ocelots, dogs, and young deer (O'Shea 2007). Two large boids in Madagascar, the Madagascan ground boa (*Acrantophis madagascariensis*, up to 3.2 m) and Madagascan tree boa (*Sanzinia madagascariensis*, up to 2.5 m), are reputed to take small lemurs. However, neither Madagascar nor the Neotropics have any real equivalent of the large python species, in terms of their ability to hunt for very large prey away from water.

Many of the advanced snakes (Colubroidea or Caenophidia) use venom to subdue their prey. The use of venom, rather than constriction as with the pythons and boas, may have allowed the evolution of more rapid locomotion, although many venomous species are slow-moving ambush foragers and some venomous species also use constriction. Venom delivery systems vary greatly among snakes, but the venom-secreting glands and some of the venoms apparently originated in the common ancestor of snakes and anguimorph and iguanian lizards, suggesting that they were subsequently lost in the ancestors of boas, pythons, and other constricting snake groups (Fry et al. 2006). In traditional systems of snake classification, only the 25% or so of the advanced snakes that have a front-fanged venom delivery system are considered "venomous," since only these species can poison animals too large to swallow, such as people. Such attacks on big prey appear to be usually defensive, although some may reflect underestimation of prey size. The majority of rear-fanged snakes, in contrast, can usually only deliver venom to prey that is already in the mouth and thus threaten only relatively small animals.

Front-fang venom systems have apparently evolved independently in the families Viperidae and Elapidae. Both families occur in the rain forests of Asia, Africa, and the Neotropics, but vipers are absent from New Guinea and Australia, and there are no front-fanged snakes at all in Madagascar. Vipers are stout-bodied, wide-gaped, snakes that inject venom through long, anterior fangs, which

Fig. 4.10 Gaboon viper (*Bitis gabonica*), a front-fanged snake from West Africa (Wikimedia Commons).

fold backwards when not in use. The longest fangs occur in the gaboon viper (*Bitis gabonica*) (Fig. 4.10), where they can reach 55 mm (2.1 inches). This species and its relatives are responsible for many human deaths in African rain forests (Chippaux 1998). Viperid venom not only immobilizes the prey but also contains powerful digestive enzymes that start the process of tissue break-down. Pit vipers have paired infrared (heat)-sensitive pits between the nostril and the eyes that, as with pythons, allow them to accurately target warm prey in the dark (Safer & Grace 2004). The largest of all pit vipers is the bushmaster (*Lachesis muta*), which can reach 3.6 m and, along with the lancehead pit vipers (*Bothrops* spp.), is responsible for many human deaths in Neotropical forests. Pit vipers are also common in Asian rain forests, but most are relatively small and human fatalities appear to be rare.

In comparison with the vipers, elapids generally have shorter, immobile fangs that remain erect when the mouth is shut. Their venoms generally have fewer components and are primarily neurotoxic. Elapids also have relatively narrow gapes and, usually, bodies, and take mostly elongate prey, such as other snakes. For example, the king cobra, *Ophiophagus hannah*, is a swift-moving diurnal snake that can attain a maximum of 6 m in length, making it the largest venomous snake in the world, yet it feeds mostly on other snakes, including large pythons. Forest elapids, including kraits (*Bungarus* spp.) and cobras (*Naja* spp.) in tropical Asia, and the death adders (*Acanthopis* spp.) in New Guinea, are responsible for many human deaths.

The characterization of the majority of the advanced snakes that lack front fangs as "rear-fanged" hides a great variety of venom delivery systems (Deufel & Cundall 2006). Recent work has also begun to reveal the diversity in the composition of venoms in these species, with evidence for prey-specific toxicity (Mackessy et al. 2006). However, the role of the venom in prey capture by rear-fanged species is still very poorly understood, since the prey is firmly held before venom is injected. Most prey seems to be small relative to snake body mass, but there are exceptions and the diet of most rain forest species is poorly known.

There are too many gaps in our knowledge of rain forest reptile biology for a robust synthesis at this stage, but some striking similarities and differences among rain forest regions are apparent. The most obvious similarity is that, except on the most remote oceanic islands, carnivorous reptiles are a major threat to the smaller vertebrates, including other reptiles, frogs, birds, and small mammals, such as rodents. In contrast, the threat to larger vertebrates, such as deer and primates, differs greatly between regions. In the rain forests of New Guinea and Australia, where large mammalian carnivores are absent, pythons and eagles (see Chapter 5) are the top predators, while in Asian and African rain forests they compete for prey with medium to large cats. Molecular phylogenies show that varanids, pythons, and elapids colonized Australia and New Guinea from Asia within the last 30 million years (Rawlings et al. 2008), suggesting that the role of reptiles as carnivores in this region has greatly increased over this period.

With the giant anacondas confined to water, the largest Neotropical mammals are relatively safe from snakes. The same is true in Madagascar, which lacks both giant constrictors and the front-fanged snakes that can bite large animals.

Herbivores of the forest floor

The tropical savannas and grasslands of Africa teem with highly visible herds of large mammals, yet a naturalist is considered lucky to glimpse a small duiker tip-toeing alone through the African rain forest. Similarly, it is a rare event to catch the eye-shine of a solitary mouse deer by torchlight in Asia or to see plant-eating mammals in rain forests elsewhere. Rain forests contain far more plant biomass than savannas, so why are the animals that eat it so hard to see? Part of the answer is that the large mammals of many rain forests have been almost hunted out, with the few survivors so wary of humans as to be invisible. In the few well-protected reserves where there has been no hunting for several decades, the greater visibility of forest floor mammals is striking. Even in pristine rain forests, however, the density of forest floor herbivores is limited by the low availability of edible plant matter that can be reached from the ground. Much of the forest biomass is in tree trunks and branches, and flushes of new leaf material, along with flowers and fruits, are produced mainly in the canopy, far out of reach of the animals below. The dense canopy of the rain forest greatly reduces the amount of light energy reaching the understorey, so grasses are rare and all plants grow slowly. Slow-growing understorey plants cannot easily replace lost tissues, so they protect themselves from herbivores with high concentrations of tough fiber and toxic chemicals, such as bitter tannins.

The limited availability of nutrients from plants near ground level has had a major role in shaping the terrestrial plant-eating faunas of rain forests, but the very different taxonomic compositions of these faunas show that historical processes have also had an overwhelming influence. There are some striking examples of convergence on an "ungulate-like" body form in unrelated animals, but also many equally striking differences in the form, size, and diversity of species in the different faunas. Although comparative ecological studies are still largely lacking, these differences are likely to be very significant for ecological processes because of the major role that forest floor herbivores play in shaping vegetation, through their patterns of foraging, their dispersal of seeds, and the way in which the larger species trample plants as they move.

No single order of mammal has managed to dominate the terrestrial herbivore niche to the extent that the Carnivora dominate the carnivore niche, but the even-toed ungulates (Artiodactyla, or Cetartiodactyla if the whales are included) have been by far the most successful. Familiar members of this order include the pigs, deer, cattle, antelopes, giraffes, and hippopotamus. The most successful artiodactyls are the ruminants, represented in the rain forest by the mouse deer and chevrotains (Tragulidae), a giraffe relative, the okapi (Giraffidae), the deer (Cervidae), and the bovids (Bovidae: cattle, antelopes, and their relatives). These four families are not evenly distributed, with the okapi confined to Africa, deer to Asia and the Neotropics, and both bovids and tragulids to Asia and Africa. Ruminants are distinguished by their specialized adaptations to herbivory, including their dentition and a compartmentalized stomach where symbiotic microorganisms process plant material. The tragulids are small, forest-adapted animals while the other ruminants appear to be primarily adapted to open, grassy habitats, with some lineages having become secondarily adapted to forest.

The artiodactyls are related to a second, much smaller, group of hoofed mammals, the odd-toed ungulates, or Perissodactyla. The perissodactyls flourished in the early Tertiary, when they included giants such as the 20-ton *Indricotherium*, the largest known land mammal, but are they now represented by only the horses, rhinoceroses, and tapirs. Although traditionally combined as the "true ungulates," the Artiodactyla and Perissodactyla are now known to be part of a larger group that includes not only the whales, but also the Carnivora.

Throughout the Tertiary, South America supported a diverse fauna of ungulates, assigned to several different orders (Janis 2008). The Notoungulata in particular radiated to fill many herbivore niches, from the rabbit-like typotheres to the rhinoceros-like *Toxodon*, while the Litopterna generally resembled horses or camels. Most of the fossil evidence for these animals comes from temperate and nonforest habitats, but a rich fossil fauna from La Venta, Colombia, dated at 12–14 million years ago (middle Miocene), appears to represent a lowland forest environment, along with more open riverine habitats (Kay & Madden 1997). These ungulates included species ranging in body size from less than 1 kg to more than 1000 kg (2–2200 lb), suggesting an analogy with the rich ungulate fauna of modern African rain forest, rather than modern South America. However, the relationship between these animals and the living ungulates is not clear and the similarities may be entirely a result of convergent evolution. They declined in both diversity and abundance during the Pliocene and most were extinct before the faunal interchange brought new predators and competitors, although the last survivors of the Notoungulata and Litopterna were seen and hunted by the first humans to reach the continent.

Ungulate-like forms also arose independently in other mammalian lineages, including relatives of the hyraxes that dominated the small to medium-sized herbivore niche in Oligocene Africa (around 25–30 million years ago) before the Miocene influx of true ungulates, and the giant caviomorph rodents that are still such a prominent part of the Neotropical rain forest community. The ancestral stock of caviomorphs arrived in South America by the end of the Eocene, around 31–36 million years ago (Sallam et al. 2009), at about the same time as the New World monkeys (see Chapter 3). A variety of alternative source areas and routes have been proposed, but the molecular evidence argues strongly for a single colonization event. The caviomorph invaders subsequently radiated into a wide range of ecological niches, apparently displacing marsupials and endemic ungulates from some of them.

There are also terrestrial plant-eaters with very different body forms, including the elephants in African and Asian rain forests, the now extinct ground sloths of the Neotropics, and the kangaroos and wallabies of Australia and New Guinea. Remember also that several of the Old World apes and monkeys, considered in the previous chapter, are either partly or largely terrestrial plant-eaters, as are some lemurs, whereas the New World monkeys are all arboreal.

Feeding strategies

Food scarcity is a problem for an animal that is limited to foraging on the forest floor. Although arboreal and flying animals may exploit fruits, flowers, and leaves in the forest canopy (see Chapters 3, 5, and 6), terrestrial animals must wait for these foods to fall from above (or to be dropped by monkeys, birds, and fruit bats) or make the most of what is available growing close to the ground. There are, however, some plant species in all the rain forest regions that make use of terrestrial animals to disperse their seeds; these species provide a more regular supply of fruits, which are often large, green in color, and fibrous in texture. In rain forests around the world, fallen fruits and seeds are important to forest herbivores. Many animals also seek out the newly sprouted leaves of forest herbs, tree seedlings, and shrubs in tree-fall gaps, at riverbanks, in human-made clearings, and at the forest edge – anywhere a break in the forest canopy lets in enough light to stimulate shoot growth at ground level.

Earning such a hard living has a number of consequences that unite the terrestrial herbivore faunas in different rain forest regions. Several feeding strategies are possible. Being small is one solution to the problem of resource scarcity. In general, smaller animals need less food, but it must be of higher quality, whereas larger animals can tolerate less nutritious foods but need larger quantities. This is because, while metabolic rate declines with increasing body weight, gut capacity remains a constant proportion of body weight. Larger animals can therefore afford to retain low-quality food in their guts for a longer time in order to digest it, while small animals must extract nutrients rapidly. Terrestrial rain forest mammals are often small, less than 25 kg (55 lb), especially if they include a substantial component of fruit in the diet. Fruits, though relatively nutritious, are scarce, so larger species must either spend the day foraging just to get enough to eat or include other plant material besides fruit in the diet.

Leaves, shoots, and grasses are much lower in nutritive value than fruits, and large volumes need to be processed to provide sufficient energy and nutrients. Very large species like the Asian guar (*Bos gaurus*), which weighs up to 900 kg, can accommodate that extra bulk, and eat fruit only opportunistically. Browsing on the leaves and twigs of shrubs and small trees is possible in continuous forest, but species that graze on grasses and other herbaceous plants are usually associated with water, such as rivers, coastal areas, and marshes, or they depend on forest gaps and the forest edge. Gap and edge habitats have been extended greatly by recent human activities, such as farming and logging, but were created in the past by landslides, windstorms, floods, and natural fires. Sodium is an essential nutrient for animals but not for plants, so except near the coast plant foods tend to be deficient in this element. Both browsing and grazing herbivores visit natural salt licks and the availability of these mineral sources may influence herbivore population densities (Matsubayashi et al. 2007).

The resource base in the rain forest is too limited and widely dispersed to support the extensive herds of grazers seen in tropical savannas, so rain forest herbivores are mostly solitary or occur in pairs, and they frequently defend a territory against others of their species. The most notable exceptions to this are the pigs (Suidae) and their New World relatives, the peccaries (Tayassuidae), in which an omnivorous diet, including roots, shoots, fallen fruits, and small animals, permits larger group sizes, although these must then travel very large distances in search of food. White-lipped peccaries, *Tayassu pecari*, form wide-ranging herds of 50 to 300 or more individuals. The bearded pig of Malaysia, *Sus barbatus*, may move hundreds of kilometers in the course of a year to keep pace with fruiting patterns in the forest. This usually solitary feeder is often found in montane areas, foraging for fallen acorns and chestnuts from trees in the Fagaceae family (Fig. 4.11). In dipterocarp masting years, pigs come together in large, loosely structured aggregations and descend into the lowlands to feast on the oil-rich dipterocarp nuts (Curran & Leighton 2000). In such years, pigs breed more actively and put on extra fat. In Africa, the red river hog (*Potamochoerus porcus*) feeds in groups of 4 to 30 individuals. These pigs travel

Fig. 4.11 Bearded pigs in Malaysia (*Sus barbatus*) travel over long distances in search of food. (Courtesy of Hans Hazebroek.)

up to 4 km per day to reach feeding sites, where they eat fruits and use their snouts to plow up the ground in search of roots. They also eat small animals. The forest hog (*Hylochoerus meinertzhageni*) – the largest pig species – forms groups of up to 20 individuals, but little is known about their diet.

Thus, it seems that to survive on the limited amount of plant material available at ground level, mammals foraging on the forest floor have had to adopt one or several of the following strategies: small body size; solitary or small social groups; continuous foraging; feeding on fallen fruit; extensive travel; or dependence on the lush and renewable plant material in swamps, gaps, or the forest edge. The same constraints apply in each of the five major tropical rain forest regions, and there has been some convergence in the foraging strategies adopted to deal with them. However, the actual animals involved, as well as the patterns of species richness on a local scale, vary enormously.

Frugivorous ungulates

The most common foraging strategy, seen today in all the rain forests except Madagascar, involves a combination of small body size and feeding on fallen fruit. Other plant material, such as fresh fallen leaves or new leaves, shoots, and flowers are commonly eaten as well in most species, but fruit is a large component of the diet.

In African and Asian rain forests, the artiodactyls (even-toed ungulates) predominate as terrestrial frugivores. African assemblages typically comprise species of duikers (*Cephalophus* spp., Bovidae, the cattle family) and the water chevrotain (*Hyemoschus aquaticus*, Tragulidae). Duikers appear to feed mainly on fibrous, tannin-rich fruits that are low in sugar and starch (Kendrick et al. 2009). Although they destroy many seeds, they also regurgitate intact hard seeds brought up from the rumen. The duikers are small, shy forest antelopes that look rather like squat, short-legged miniature deer (Fig. 4.12). Most duikers forage at night, although a few species feed during the day, and they forage alone or in pairs. In addition to plants, duikers will sometimes supplement their diet with animals such as frogs, birds, insects, and carrion. The diversity of duiker species is greatest in West Africa and the Congo basin, and is lower in the isolated forests east of the Congo basin. In the most westerly African rain forest region, centered around Liberia, there may be as many as seven species of duiker occurring in a single forest, and there is some suggestion that they avoid excessive dietary overlap by including different fruits in their diet (Happold 1996). Duikers occur at a lower density in forests dominated by one species of tree (monodominant forests, see Chapter 2) in comparison with mixed species forests, where their diet is more diverse and consistent.

Throughout most of the African rain forest, duikers are joined by the water chevrotain (*Hyemoschus aquaticus*, Tragulidae), a solidly built ungulate with a short, thick neck and a brown coat, with rows of white spots running along the body and forming lines along the flanks. Although, as its name suggests, it lives within a few hundred meters of water, it only enters water to avoid danger and its diet is largely fallen fruit. Two species of chevrotains, or mouse deer, also make up the small, fruit-eating guild of species in the forests of Asia, where there are no small species of bovid. They rival the dwarf antelope as the smallest artiodactyls; the lesser mouse deer, *Tragulus javanicus*, weighs only 2.0–2.5 kg (4.5–5.5 lb) and

Fig. 4.12 A forest duiker (*Cephalophus* sp.) in Gabon. (Courtesy of Stéphane Tridon, Precious Woods Gabon.)

the greater mouse deer, *T. napu*, is not much larger at 3.5–4.5 kg (Fig. 4.13). Like their African relative, the water chevrotain, mouse deer have no antlers or horns, and they are usually solitary. Although usually referred to as "deer," these tragulids represent an ancient, forest-adapted group of ungulates, which split early from the other ruminants, before the origin and diversification of the deer and bovids (Marcot 2007).

Frugivorous rodents

Although Neotropical rain forests appear to have supported small, fruit-eating ungulates in the past, the terrestrial frugivore niche is today occupied by a very different group of mammals: large caviomorph rodents in the families Dasyproctidae and Agoutidae. Old World rain forests also have frugivorous rodents, but these are mostly small, rat-like animals. The larger caviomorphs, in contrast, in some ways resemble their Old World ungulate counterparts more than rats (Fig. 4.14). Adapted for a more mobile lifestyle than most rodents, they have long legs and the hindfoot has a reduced number of weight-bearing toes. The high hindquarters give an arched appearance to the back that rounds into a much-reduced tail. Their heads are large and retain a rodent-like appearance, with strong jaw muscles and blunt noses. There are seven species of rain forest agoutis (*Dasyprocta*), and although they are very large for rodents (1.3–4 kg, 3–9 lb), they are much the same size as the small forest ungulates of Africa and

Fig. 4.13 Greater mouse deer (*Tragulus napu*) from Malaysia. (Courtesy of Tim Laman, taken in captivity.)

Fig. 4.14 Many of the terrestrial herbivores in New World rain forests are large rodents. Here a Central American agouti (*Dasyprocta punctata*) eats a seed. (Courtesy of P.M. Forget.)

Asia. In contrast to the local species richness observed in African duikers, it is rare to find more than one species of agouti in any one locality in Neotropical forests. Their ranges overlap very little, and the similarity in their habits suggest that the different species are ecological replacements for one another that may have arisen as populations became separated by geographical barriers such as major river systems.

In most places, the agoutis are joined by the larger paca (*Agouti paca*), which is most common near water, and one of the two species of acouchi (*Myoprocta*). The acouchies are smaller than the agoutis (0.6–3.1 kg) but look very much like them and have similar diets. Like the agoutis, the two acouchi species overlap very little in their distribution. The less-well known pacarana (*Dinomys branickii*) is added to the group in the Amazon highlands. The pacarana looks like an overgrown guinea pig at 10–15 kg in weight, and is the last survivor of a once diverse caviomorph family, the Dinomyidae. The genus name, *Dinomys*, means "terrible mouse," and reflects its reputation as a fighter when cornered. Among the many extinct species in this family is the largest rodent known, the rhinoceros-sized *Josephoartigasia monesi* (Rinderknecht & Blanco 2008).

All the Neotropical terrestrial rodent frugivore species tend to be rather similar in their habits, being fruit- and seed-eaters that take succulent plants when available. Unlike ungulates, but like many other rodents, they usually sit on their haunches to eat and manipulate the food with their forefeet (Fig. 4.14). This dexterity also enables them to bury seeds and nuts for use when fruit is not in season, and they are important seed dispersers as a result. Large rain forest seeds tend to be non-dormant, making them far from ideal for long-term storage, but the red acouchi (*Myoprocta acouchy*), and probably other species, produces storable "zombie seeds" by removing any sprouts (Jansen et al. 2006). Although no other rain forest region has anything like an agouti or acouchi, frugivorous ground squirrels and murid rodents, such as the long-tailed giant rat (*Leopoldamys sabanus*) in Malaysia and the white-tailed rat (*Uromys caudimaculatus*) in Australia, also scatter-hoard seeds for future consumption and may make a similar contribution to seed dispersal if they fail to retrieve them all (Forget & Vander Wall 2001).

African browsers

Most fruit-eating species are small. Larger species must either include more browse in their diet or – as in the case of the large African duiker (*Cephalophus sylvicultor*) and the predominantly frugivorous red river hog (*Potomochoerus porcus*) – they must broaden their diet and forage almost continuously. We can identify a loose foraging strategy of medium- to large-size browsers. Most of the medium-sized species will also tend to take seeds of woody legumes, fungi, or other accessible plant matter, and many of the larger species also graze, but the principal tendency, often associated with morphological adaptations, is toward browsing – that is, eating the shoots and leaves from woody plants. The largest browsers are the largest animals in the forest and can have a major impact on the structure and species composition of the forest, not only by what they eat but also by what they push over and tread on (Owen-Smith 1992).

The browsers are a variable group, both morphologically and taxonomically. In Africa, the bushbuck (*Tragelaphus scriptus*) and bongo (*T. eurycerus*) (family

Bovidae) have varied diets but are predominantly forest-edge browsers of leaves, twigs, shoot tips, and vines (Kingdon 1997). These species look like crosses between an antelope and a small cow, with large eyes and ears, spiraled horns, and a red-brown coat marked with white stripes. The bushbuck is much the smaller of the two, just over 1 m (3 feet) at the shoulder and weighing 24–42 kg (53–93 lb), and it tends to be associated with water and the edges of swamps. The bongo is much bigger, standing 1.4 m at the shoulder and weighing up to 220 kg. A third species, the sitatunga (*T. spekei*), is intermediate in size and predominantly a grazer. These species avoid large areas of dense, closed-canopy forest, where little fresh greenery is available within reach, and instead make use of areas disturbed by landslides, floods, tree falls, and the activities of elephants. Even then, they must range over huge areas to obtain enough nutritious browse.

One of the most unusual animals in the African rain forests is the okapi (*Okapi johnstoni*), confined to the equatorial forests of northern and east-central Democratic Republic of Congo. The okapi is not a bovid, but a deep-forest relative of the familiar giraffe of African savannas. Like its larger relative, its body is short and compact and the neck and legs are long, though not as pronounced as the giraffe. It can reach up to 1.8 m at the shoulder and weighs 250 kg, and has a beautiful short, shiny coat that is the purplish-maroon color of an eggplant, with white horizontal stripes at the top of the legs. The long tongue is prehensile and is used to pluck leaves and buds and even small branches. Adequate food is hard to come by, and the okapi is usually solitary as a result.

The suite of African browsers is completed by the African elephant (Fig. 4.15). Forest elephants are smaller, darker, and more solitary than their familiar savanna relatives, and are now recognized as a separate species, *Loxodonta cyclotis*. The

Fig. 4.15 African forest elephants in Gabon. (Courtesy of Stéphane Tridon, Precious Woods Gabon.)

forest elephant feeds largely on shoots, leaves, and bark, but also takes fruit whenever available. Some trees with very large fruits depend entirely on elephants for seed dispersal, as described in Chapter 2. In the process of feeding, forest elephants knock down many small trees and bushes, opening up the forest and creating feeding opportunities for many other animals, both immediately on the fallen plants and later when plants begin to grow up in the gaps.

Asian browsers

Although several species of bovids are found in Asian rain forests, the main browsers are deer (Cervidae) and two groups of odd-toed ungulates (Perissodactyla), the tapir and the rhinoceroses. The smaller deer species (at 14–50 kg, 30–110 lb) are known as muntjacs (*Muntiacus* spp.) (Fig. 4.16) or barking deer. Their diets include a fair amount of fruit as well as browse. Barking deer are named for their habit of emitting deep, fox-like barks, particularly in dense vegetation and when predators are detected. The larger species of sambar or red deer (*Cervus* spp.) are larger browsers weighing up to 260 kg and as tall as 160 cm (5 feet) at the shoulder. They feed on leaves, shoots, and fallen fruits in tree-fall gaps, riverside forests, and other areas of lush growth.

A browsing habit is also typical of the Malay tapir, *Tapirus indicus* (Fig. 4.17), which has a prehensile proboscis comprising the snout and upper lips with which it reaches out and pulls leaves and shoots into the mouth. Tapirs also

Fig. 4.16 Muntjac or barking deer (*Muntiacus muntjak*) from Bhutan. (Courtesy of Harald Schuetz.)

Fig. 4.17 Malay tapir (*Tapirus indicus*). (Courtesy of Jessie Cohen, taken at the Smithsonian's National Zoo.)

include fallen fruit in their diet, when available. In Asia, two species of rhinoceros browse in the forests, the Javan one-horned rhinoceros (*Rhinoceros sondaicus*) and the Sumatran rhinoceros (*Dicerorhinus sumatrensis*) (Fig. 4.18). These large Asian animals may have to travel long distances to fulfill their daily dietary requirements. In fact, the Javan rhino can travel 15–20 km (9–12.5 miles) in 24 h in search of adequate shoots, twigs, young foliage, and fallen fruit, and also needs a plentiful supply of water (Nowak 1999). Diet diversity is taken to extremes in the Asian elephant (*Elephus maximus*), which consumes a huge range of plant species (Campos-Arceiz et al. 2008). Despite their size, however, their trunks allow these elephants to be highly selective in the plant parts consumed, with many tree species eaten exclusively for their bark or fruits.

When the okapi was first described from the forests of the Congo in 1901, there were suggestions that this would be the last new large mammal to be found (MacKinnon 2000). Of course this turned out not to be true, but the discovery of completely new mammal species, as opposed to the splitting of known forms (such as the African elephants, mentioned above), is a very rare event. The description of three new artiodactyls species within four years from the rain forests of the Annamite Mountains that border Laos and Vietnam is therefore particularly exciting. The strangest of these new species is the saola or Vu Quang ox (*Pseudoryx nghetinhensis*), found first in 1992 in Vietnam and then in 1993 in adjacent regions of Laos (Groves & Schaller 2000). Although clearly a bovid, the long, thin, backward-curving horns and conspicuous white markings on the

Fig. 4.18 Sumatran rhinoceros (*Dicerorhinus sumatrensis*) browsing on leaves. (Courtesy of Nico van Strien, International Rhino Foundation.)

face give the saola a strikingly distinct appearance and make its late scientific recognition even more surprising. Less than 1 m at the shoulder, and weighing about 100 kg, the incisors suggest a browsing habit, but this animal probably also eats grasses and herbs. In the same region as the saola, at least two new muntjac species (*Muntiacus*) have been found, a very large species (*M. vuquangensis*) in 1994 and a small species (*M. truongsonensis*) in 1995 (Groves & Schaller 2000). Yet another new muntjac (*M. putaoensis*) was discovered in northern Myanmar in 1998 (Rabinowitz et al. 1999). The new discoveries emphasize how little we know about these shy and elusive rain forest browsers.

South American browsers

The Neotropics lacks bovids, rhinos, and elephants, and – probably as a result – the rain forests support a relatively low total biomass of browsers and grazers (Mandujano & Naranjo 2010). Deer and tapirs are the primary browsers in South America, although in contrast to Asia both groups are relatively recent (less than 3 million years ago) arrivals from the north over the Panama land bridge. There are two rain forest species of tapir, the South American tapir (*Tapirus terrestris*) and Baird's tapir (*T. bairdii*), found in Central America and west of the Andes in South America. Baird's tapir, at up to 300 kg (660 lb), is the largest land mammal in the rain forests of the Neotropics. Tapirs are found near water and often wear paths to the water's edge. They browse mainly on shoots of plants (especially aquatic species), as well as twigs, leaves, and fruit. Brocket deer (*Mazama* spp.) are shy, solitary animals of the forest weighing up to 25 kg. Their diet includes leaves, shoots, and grass, but they also feed heavily on fallen fruits when available (Gayot et al. 2004). They appear to lack the endurance of other deer and can be overtaken and killed by domestic dogs (Nowak 1999).

Although the Neotropics today has no equivalent of the elephants, rhinoceroses, and large bovids of Africa and Asia, this was not always true. The middle Miocene tropical forests of La Venta, Columbia, supported a variety of very large (> 500 kg) terrestrial herbivores, including ground sloths and a variety of endemic "ungulates" which appear to have been the ecological equivalent of the living Old World megaherbivores. As recently as 10,000–12,000 years ago, at the end of the Pleistocene, Central and South America supported mastodons and gomphotheres (both relatives of elephants), as well as giant sloths (Koch & Barnosky 2006). These animals may have played a similar role in maintaining open habitats within the forest and dispersing large seeds. However, there is no direct evidence that these Pleistocene species occurred in rain forests.

Grazers

Grazers cannot survive in closed-canopy rain forests because the lack of light restricts the availability of grasses, so grazing species tend to be associated with aquatic habitats or with large forest gaps and the forest edge. Some grazers, such as the African buffalo (*Syncerus caffer*, Bovidae) (Fig. 4.19), are so dependent on the existence of swamp grassland patches that they could reasonably be said

Shelly Masi

Fig. 4.19 African forest buffalo (*Syncerus caffer nanus*) in the Central African Republic. These grazers depend on swampy grassland patches within the forest. (Courtesy of Shelly Masi.)

to occupy the region despite, not because of, the rain forest. Forest buffaloes are very different from their larger, more robust savanna relatives and might be considered a separate species – like the forest elephants – were it not for the existence of intermediate forms. While the savanna buffalo can form herds with hundreds to thousands of individuals, forest buffalo live in groups of 5–25, presumably because of the patchy distribution of food sources (Melletti et al. 2008). Another African bovid, the sitatunga (*Tragelaphus spekei*), a spiral-horned relative of the bushbuck and bongo, is semiaquatic, grazing on reeds, sedges, and grasses in swamps. The pygmy hippopotamus (*Hexaprotodon liberiensis*) of West Africa, though less aquatic than the more familiar hippopotamus (*Hippopotamus amphibius*), relies heavily on water plants and grasses, as well as taking tender shoots, leaves, and fruit, and is found along streams and in wet forests and swamps (Kingdon 1997). Hippos are artiodactyls, but not ruminants.

As in Africa, Asian grazers tend to use the forest for shelter and limited feeding, and move into more open areas to feed intensively. The tamaraw (*Bubulus mindorensis*) of the Philippines and the lowland anoa (*B. depressicornis*) of Sulawesi are mid-sized (up to 300 kg, 660 lb) buffalo-like bovids that eat grasses, ferns, saplings, and fallen fruits. The most impressive grazer is the gaur (*Bos gaurus*), a massive ox that can reach 2.2 m (c. 7 feet) at the shoulder and weighs 1000 kg. Herds of these huge beasts are led by a dominant bull through the forests of Asia, grazing on grasses and browsing shrubs and climbers in clearings, at the forest edge or open riverbanks. Only slightly smaller, the banteng (*B. javanicus*) prefers slightly drier habitats than the guar, but can also be found in the forest, occasionally in association with its larger relative.

As a result of Pleistocene extinctions, such large grazers are now completely absent from South America. The largest surviving grazer is the semiaquatic capybara (*Hydrochaeris hydrochaeris*), the largest rodent in the world, weighing up to 65 kg. This stocky animal has a large head and rectangular muzzle and shows several adaptations to its semiaquatic lifestyle: the ears and eyes are small and set high on the head, and the feet are partly webbed. It feeds on grass and browse, particularly aquatic plants, and is always found near water. Ecologically and morphologically it is most similar to the pygmy hippopotamus of Africa. Fossils of a much larger rodent, the 1000 kg *Josephoartigasia monesi*, have been found in Plio-Pleistocene Uruguay (Rinderknecht & Blanco 2008), showing that the absence of large grazing herbivores in modern Neotropical forests is not simply a reflection of size limits on rodent evolution. The largest herbivore still remaining in the Amazon is not even terrestrial; it is the Amazon manatee (*Trichechus innuguis*), a completely aquatic animal that feeds on water plants and weighs up to 500 kg.

Australia and New Guinea

As we move to Australia and New Guinea, there is another dramatic shift in the dominant mammal group on the rain forest floor. Here, it is the members of the marsupial family Macropodidae, the kangaroos and wallabies, which slip through the understorey in search of edible plant materials (Flannery 1995a). These animals do not fit quite as readily into our classification of frugivores, browsers, and grazers and, indeed, for many species the diet is virtually unknown. The rain forests of New Guinea and Australia also support giant flightless birds,

Fig. 4.20 The musky rat-kangaroo (*Hypsiprymnodon moschatus*), here on the Atherton Tablelands, Queensland, Australia, eats mostly fruit. (Courtesy of Dennis Hansen.)

the cassowaries, which share the terrestrial frugivore niche filled by mammals elsewhere (see Chapter 5).

The best-studied rain forest macropodid is also the most peculiar: the musky rat-kangaroo (*Hypsiprymnodon moschatus*) (Fig. 4.20) is the smallest living macropodid, at around 0.5 kg (1.1 lb), and it apparently diverged from the rest of the family before the evolution of such characteristic kangaroo features as bipedal hopping and the complex stomach needed to digest a fiber-rich diet. Musky rat-kangaroos are confined to the wet rain forests of northeast Queensland, where they occur at high densities and feed largely on easily digested fruit and seeds. This species has been shown to be a key dispersal agent (with cassowaries and white-tailed rats) for large seeds in these forests by burying them for future consumption, just like the agoutis of South American rain forests (Dennis 2003). A more recent radiation of rain forest wallabies in the montane forest of New Guinea has produced several species in the genera *Dorcopsis* and *Dorcupsulus*, ranging in size from around 2 kg – the smallest New Guinea macropodid – to around 5 kg (Flannery 1995a). The diets of these small forest wallabies are assumed to be largely vegetarian, but little else is known. Another group of small wallabies, the pademelons (*Thylogale* spp.), occur in both New Guinea and Australia. Most species seem to prefer more open habitats, where they can graze as well as browse.

As in South America, the fauna of browsing marsupials has been affected by recent megafaunal extinctions associated with the arrival of humans. The Late Pleistocene fauna of the New Guinea highlands included several now-extinct marsupials that were bigger than any of the survivors, including large (> 45 kg) browsing kangaroos (*Protemnodon* spp.) and diprotodontids (*Hulitherium*, *Maokopia*,

Zygomaturus), that overlapped with humans (Long et al. 2002; Fairbairn et al. 2006). *Hulitherium thomasetti* – an animal like a giant wombat or hairy hippopotamus – may have fed on bamboo: a sort of marsupial giant panda. Related species occurred in the Miocene rain forest of Australia (Fig. 1.5), although it is not clear if any of the more recent Australian diprotodons lived in the rain forest.

Madagascar

Modern Madagascan rain forests seem to be an exception to the general rule that the mammals in each rain forest region have evolved species to utilize the sparse but diverse resources available near ground level. Part of this reflects Madagascar's long isolation, which means that many groups of mammals, including the artiodactyls, failed to reach it. Bush pigs now forage in Madagascan rain forests, but these were a recent introduction from Africa. Another factor is the mass extinctions of the last few thousand years, which have removed a large proportion of Madagascar's bigger native mammals as well as the giant, flightless elephant bird (Orlando et al. 2008). Some of the extinct lemurs were probably at least partly terrestrial, and the largest species, the gorilla-sized *Archaeoindris*, must have been largely so, since only the biggest tree branches would support its weight. It is not unreasonable to suggest that some of these lemur species made use of the resources that support other terrestrial mammals in other rain forest regions. Three species of hippopotamus also survived until recently in Madagascar, including one that was very similar to the African pygmy hippopotamus and that may have had similar grazing habits.

Conclusions and future research directions

In contrast to the primate communities discussed in Chapter 3, research and analysis of carnivore and plant-eater biology in tropical rain forests is still in an early stage of development. Certain patterns do emerge, but there is a compelling need for more research to verify and expand what has been learned to date. Our knowledge of primates has come largely from the habituation of unhunted groups to the presence of humans, so they can be observed continuously from close quarters. The nocturnal, secretive habitats and low population densities of most rain forest carnivores make this approach impossible and it is only slightly more practical with most forest floor herbivores. Much of what we know about these two groups of mammals, therefore, is based on fleeting glimpses of fleeing animals or on the interpretation of tracks, signs, and the contents of their droppings. New techniques are changing this, however. Automatic cameras triggered by animal movements can give more reliable estimates of densities and radiotracking or satellite telemetry give information on how the available habitat is used. The application of molecular techniques, such as DNA barcoding, to the analysis of feces can give far more information than the traditional "sieve and count" methods. The tools are now available to check what we think we know already and to answer questions that have previously not been possible to ask.

Carnivores may be the least understood of rain forest mammals, not only because of the difficulties of studying them, but also because their opportunistic and variable behavior makes it hard to interpret the limited data we have. Comparative studies of ecologically similar groups across continents should be undertaken, making use of the few remaining sites where carnivores exist at near natural densities, as well as forests disturbed by logging and other human activities. There is no shortage of questions. For example, why should Asian and American rain forests have more carnivore species than African forests when the density and diversity of potential prey is at least as high in Africa? Why is there no jaguar- or tiger-sized carnivore in African rain forests? To what extent are the procyonids of Neotropical forests the ecological equivalents of the civets and mongooses of Africa and Southeast Asia?

Some questions about carnivores are best addressed in places where particular species have been removed, or exotic species introduced. It would be valuable to know how the removal of carnivore species affects the abundance of prey species and how this impact then propagates through the food web. When carnivores are eliminated by hunting, do their prey species increase, and if so, what are the consequences for the forest community? And when prey species are eliminated by human hunting, what impact does that have on carnivore behavior and numbers? Does that force the carnivores to leave the forest and prey on domestic animals?

Other questions cannot be addressed by considering mammalian carnivores alone. A monkey is just as dead if it is killed by a leopard, an eagle, or a python, and smaller vertebrates are prey to dozens of competing mammals, birds, and reptiles. How can so many species of carnivores coexist in one rain forest area? To what extent can an increased abundance of one species "compensate" for the absence of another as a result of natural factors (e.g. on islands) or human impacts?

The role of browsers and grazers also needs further investigation. In some cases, the exclusion of herbivores by fencing off an area of forest may be the most practical way to identify their impact. Rodents are important understorey herbivores in New World forests, whereas artiodactyls (deer and/or bovids) are more important in Old World forests. These animals have very different teeth and digestive systems, so their supposed ecological equivalence needs more detailed study. It remains to be determined why Old World grazer and browser communities are richer in species than New World communities. Is this due to greater ecological specialization in Old World species? In many cases, African forest species appear to have evolved from savanna relatives. Why has this not happened to the same extent in New World forests, which also border extensive savanna and woodland zones?

There is increasing evidence that terrestrial herbivores play a key role in seed dispersal in all the major rain forest regions (except Madagascar), but also that this role may be very different in different forests. In the Neotropics, caviomorph rodents disperse seeds by caching them for future consumption. This puts the plant and rodent in direct competition for the seed reserves, since only unretrieved caches can give rise to seedlings. In Africa and Asia, in contrast, terrestrial herbivores ranging in size from mouse deer to elephants consume specialized fleshy or fibrous fruits that fall to the ground when ripe, dispersing the seeds when they are later defecated or regurgitated. In one case the seed itself is the reward, while in the other the seed is just ballast to be disposed of as quickly as possible. The ecological and evolutionary consequences of these two mechanisms of seed dispersal are likely to be very different. However, there is also some evidence of scatter-hoarding by Old World rodents, while both deer and tapirs disperse seeds in fallen fruits in the Neotropics. Is it possible that the contrast is not as great as the current literature implies? Pantropical studies with standard methods are needed to resolve this question.

Finally, we urgently need to understand the role of grazing and browsing herbivores in creating the very structure of the vegetation. While the role of African elephants in creating the openings in the forest required by many other species of animals and plants has received considerable attention, the role of the other grazers and browsers needs to be investigated. Asian elephants are a similar size to African forest elephants: do they have a similar impact? It would be valuable to learn how elephant activity affects the full range of species in the forest, such as birds, insects, and small rodents. We also need to know what happens to the vegetation when the browsers and grazers have been removed by hunting. Does the amount of ground-level vegetation increase, and how does the plant species composition change?

The extinction of all the megaherbivores in the Neotropics at the end of the Pleistocene must have had a major impact on both the structure and species composition of the rain forest, if – and this is still uncertain – they occurred in rain forests. Could we introduce forest elephants from Africa or Asia into Neotropical forests as a substitute? Should we do so? What would be the consequences? Would elephants eat the same fruits and browse the same plants as the extinct gomphotheres? The idea of bringing elephants to Neotropical rain forests may seem both impractical and potentially dangerous, since the ecological impact of elephant introduction is unpredictable, but just thinking about it highlights the fact that even the most "pristine" rain forests of the Neotropics have already been transformed by past human impacts. Unfortunately, this is also increasingly true of rain forests elsewhere, not just in the Neotropics. Forest floor vertebrates are particularly vulnerable to hunting and the densities of both large carnivores and large herbivores are almost everywhere lower than they would have been before the arrival of humans. In most rain forests, humans have now replaced large carnivores as the top predators and, in many areas, they have played this role for so long that we can only speculate on the nature of the "natural" predator–prey relationships. Understanding the role of past and present human hunters in rain forest ecology may be our biggest challenge.

Chapter 5

Birds: Linkages in the Rain Forest Community

Birds have many important roles in the ecology of all tropical forest communities: they eat fruits and disperse seeds, pollinate the flowers they visit for nectar, and prey upon insects, spiders, and small vertebrates living on the leaves, twigs, and trunks of trees. Raptors are major predators of many larger forest animals, including frogs, reptiles, rats, squirrels, other birds, and even primates. Birds have received more attention from biologists than any other animal group, except perhaps the primates, so they provide a rich source of material for assessing the relative importance of past biogeographical histories and current ecological conditions in shaping rain forest communities. Birds are particularly instructive in this regard because they can be readily divided into ecological guilds, such as frugivores, nectarivores, insectivores, scavengers, and ground-feeders, making possible comparisons between bird communities that have no species in common. They are also excellent conservation indicators: only 0.6% (63) of all bird species are classified as Data Deficient in the IUCN Red List, compared with 15% of mammals and 25% of amphibians (Butchart & Bird 2010).

The laws of physics place strict constraints on flying animals, so we might expect the birds (and bats) of the different rain forest regions to show more convergence in both form and function than the nonflying mammals considered in the previous two chapters. Rain forest bird communities in different regions do indeed show some striking examples of convergent evolution – one good example is the strong physical resemblance of New World toucans and Old World hornbills, which are completely unrelated families – but there are also many real differences in bird sizes, foraging behavior, and community organization. In some cases, a superficial resemblance between geographically separated families masks significant ecological differences. Seemingly minor differences in the foraging behaviors of flower-visiting birds, for example, have had significant impacts on the way that plants display their flowers and fruits. New World hummingbirds hover while they feed on flowers, whereas Old World sunbirds and Australian honeyeaters usually perch while feeding. As a result, the flowers – and subsequent fruits – of the many New World shrubs, epiphytes, herbs, vines, and small trees that are pollinated by hummingbirds are produced outside of the

Tropical Rain Forests: An Ecological and Biogeographical Comparison, Second edition.
© Richard T. Corlett and Richard B. Primack. Published 2011 by Blackwell Publishing Ltd.

foliage, where the hummingbirds can get at them more easily. In contrast, Old World bird-pollinated plants typically produce flowers (and fruits) inside the foliage, where birds perched on twigs can reach the flowers readily.

Biogeography

Most birds can fly, so we might expect that their distributions would be less affected by the geographical barriers between major areas of rain forest than those of other vertebrates. However, most rain forest birds – particularly those that live in the forest understorey – are either unable or unwilling to cross large open spaces, never mind the sea. Moreover, most modern groups of birds arose in the Cretaceous and radiated in the Tertiary, 65–35 million years ago, at a time when both South America and Australia were more isolated than they are today, and Madagascar was already separated from Africa by hundreds of kilometers of open sea. The tropics of Africa and Asia have been connected by dry land for at least 20 million years, but the connecting habitat has usually been savanna or desert. Thus, forest-dependent species have only rarely been able to disperse between Asia and Africa. A large proportion of rain forest bird diversity therefore is due to independent radiations of bird species in South America, New Guinea and Australia, Madagascar, and, to a lesser extent, Asia and Africa (Fig. 5.1). Although differences among these radiations were reduced after North and South America became connected through Central America, and Southeast Asia and the New Guinea–Australia region were linked by the stepping-stones of the Indonesian

(a)

Fig. 5.1 Despite their ability to fly, many bird groups are restricted to different areas of the world. (a) Violet turaco (*Musophaga violacea*), a member of an endemic African family, in the Gambia. (Courtesy of Martin Goodey.)

(b)

(c)

Fig. 5.1 (*cont'd*) (b) Green honeycreeper (*Chlorophanes spiza*), a frugivorous member of the New World tanager family, in Colombia. (Courtesy of Gustavo Londoño.) (c) Sickle-billed vanga (*Falculea palliata*), a member of the endemic vanga family, from Madagascar. (Courtesy of Harald Schuetz.)

Archipelago, they are still very noticeable today. Only a few bird families – the swifts (Apodidae), swallows (Hirundinidae), pigeons (Columbidae), parrots (Psittacidae), cuckoos (Cuculidae), hawks (Accipitridae), falcons (Falconidae), owls (Strigidae), and nightjars (Caprimulgidae) – are found in all major rain forest areas and, even in these families, the species in one region are almost always more closely related to each other than to species in other regions.

These separate radiations in different rain forest regions have sometimes produced birds that are so similar in appearance that only the DNA studies of the last 20 years have shown that these similarities result from parallel evolution, rather than from a shared ancestry. Past misconceptions are still reflected in the English common names, with terms like warbler, wren, treecreeper, and babbler applied to unrelated birds on different continents. Additional confusion arises from the tendency for the common names of tropical birds, particularly in South America and Australia, to reflect their supposed resemblance to more familiar species, so the ant-pittas, ant-shrikes, ant-thrushes, and ant-wrens are closely related to each other, but not to the birds after which they are named. The ant-thrushes of the Neotropics are thrush-like antbirds of the family Formicariidae, while the ant-thrushes of African rain forests are ant-following true thrushes in the family Turdidae. Ecological convergence between unrelated birds in different regions is most obvious for seed-eaters and for birds that catch insects in mid-air, known generally as flycatchers. Small, foliage-gleaning insectivores from different regions are also very similar. In contrast, fruit-eaters, flower visitors, scavengers, and other bird groups vary considerably among the rain forest regions and some have no obvious counterparts on other continents.

Molecular studies are rapidly sorting out the evolutionary relationships between the major groups of birds, but the poor fossil record, due to their thin bones and lack of teeth, makes it difficult to relate this increased understanding of avian phylogeny to present-day biogeography. The bird fauna of American rain forests is the most distinctive and also the richest in terms of species by a great degree; a thousand bird species have been recorded in the enormous Manú Biosphere Reserve in Peru, which is more than 10% of the global bird fauna. Despite the richness at the species level, fewer families of birds are represented in New World rain forests than in other rain forest areas, because each of a small number of large New World families occupies the ecological roles of several Old World families. Thus the extraordinarily diverse ovenbird family (Funariidae) includes species that look and behave somewhat like Old World larks, jays, tits, nuthatches, wrens, thrushes, dippers, treecreepers, and warblers. The ovenbirds, along with the antbirds (Formicariidae and Thamnophilidae), woodcreepers (Dendrocolaptidae), tyrant flycatchers (Tyrannidae), cotingas (Cotingidae), manakins (Pipridae), and several smaller groups, form part of an exclusively New World radiation of around 1100 suboscine passerines: a group of perching birds distinguished from the oscine songbirds by their more simple vocal apparatus and relatively limited singing ability. Only 52 species of suboscines occur in the Old World, representing a separate radiation that includes the Asian and African pittas and broadbills, and the Madagascan asities, plus a single Neotropical species, the broad-billed sapayoa (*Sapayoa aenigma*), found in the rain forests of Panama and northwest South America (Moyle et al. 2006). The suboscines most likely entered South America from the south, via Antarctica, except for *Sapayoa*, which probably came later via a northern route when the Eocene thermal maximum permitted high-latitude tropical forests in Eurasia. The suboscine flycatchers colonized islands in the Caribbean and invaded North America well before the completion of the Panama land bridge 3 million years ago, but the forest-dependent antbirds and woodcreepers did not move north until around the time the bridge was formed (Weir et al. 2009). Suboscines still dominate the Neotropical rain forest understorey, where they have been joined more recently by a variety of oscine groups, the most important of which are the wrens (Troglodytidae).

Oscines now dominate in the canopy of South American rain forests and in nonforest habitats, although the suboscine tyrant flycatchers are also prominent. Many of the Neotropical rain forest oscines belong to another major New World radiation, the "nine-primaried oscines," which apparently entered from the north more recently. However, many representatives of this group are strong fliers and apparently spread to South America well before the formation of the Panama land bridge 3 million years ago (Weir et al. 2009). These birds are distinguished by having only nine easily visible primary flight feathers (the feathers on the outer part of each wing), compared with 10 in most other passerines. Their relatively recent and rapid diversification has made it difficult to sort out relationships within this radiation, but the rain forest members include the tanagers and their relatives (Thraupini) (Fig. 5.1b), the oropendolas, caciques, and orioles (Icterini), and also the woodwarblers (Parulini), many of which migrate in the spring to temperate North America to breed. Other major bird groups confined to the New World include the toucans (Ramphastidae) and humming-birds (Trochilidae) (Table 5.1).

In contrast to the distinctiveness of Neotropical birds, the rain forest bird faunas of African and Asian rain forests are similar to each other at the family level because of dispersal via intermittent forest connections. Africa and Asia consequently share such important rain forest families as the hornbills (Bucerotidae), bulbuls (Pycnonotidae), and sunbirds (Nectariniidae). But both areas also have small endemic families – the turacos (Musophagidae) in Africa (Fig. 5.1a) and three, more or less closely related, small families in Asia, the leafbirds (Chloropseidae), ioras (Aegithinidae), and fairy bluebirds (Irenidae). Moreover, the relative importance of the families they do share differs considerably. The pittas (Pittidae), for example, are a largely Asian group, with only two species in African rain forests, whereas the weavers (Ploceidae) are a largely African group, with many rain forest species in Africa but only a few species in Asia, none of which inhabits the rain forest.

Many major Old World bird families, including the woodpeckers, barbets, shrikes, tits, babblers, and hornbills, failed to colonize Madagascar, but an endemic group of passerines has radiated to fill an amazing diversity of foraging niches. The vangas (Vangidae) are a family of around 22 species found only on the island of Madagascar. Although the relationship of the vangas to other bird families is unclear, they appear to represent an evolutionary radiation from a single shrike-like ancestor that colonized Madagascar in the past. Despite their common ancestry, the vangas have an exceptional diversity of morphologies, particularly in bill size and shape, but also in wing and tail size and shape, reflecting differences in foraging behavior and prey choice. Indeed, some of the members of this radiation are so distinctive that they were, until recent molecular studies, included in other families (Yamagishi et al. 2001; Johansson et al. 2008a). The helmet vanga (*Euryceros prevostii*) looks like a barbet or toucan, using its massive beak to grasp large insects, lizards, and frogs. The nuthatch vanga (*Hypositta*

Table 5.1
Endemic and dominant families of rain forest birds in each major tropical region.

Region	Endemic families	Dominant families
Tropical America	Tinamidae (tinamous) Cracidae (curassows) Psophiidae (trumpeters) Momotidae (motmots) Ramphastidae (toucans) Galbulidae (jacamars) Bucconidae (puffbirds) Dendrocolaptidae (woodcreepers) Formicariidae (ground antbirds) Thamnophilidae (typical antbirds) Furnariidae (ovenbirds) Cotingidae (cotingas) Pipridae (manakins)	Columbidae (pigeons) Trochilidae (hummingbirds) Ramphastidae (toucans) Dendrocolaptidae (woodcreepers) Formicariidae (ground antbirds) Thamnophilidae (typical antbirds) Cotingidae (cotingas) Pipridae (manakins) Tyrannidae (tyrant flycatchers) Thraupini (tanagers)* Parulini (woodwarblers)*
Africa	Musophagidae (turacos)	Cuculidae (cockoos) Alcedinidae (kingfishers) Bucerotidae (hornbills) Pycnonotidae (bulbuls) Laniidae (shrikes) Sylviidae (Old World warblers) Muscicapidae (Old World flycatchers) Nectariniidae (sunbirds) Ploceidae (weavers)
Southeast Asia	Aegithinidae (ioras) Chloropseidae (leafbirds) Irenidae (fairy bluebirds)	Columbidae (pigeons) Cuculidae (cockoos) Alcedinidae (kingfishers) Picidae (woodpeckers) Pycnonotidae (bulbuls) Timaliidae (babbers) Sylviidae (Old World warblers) Muscicapidae (Old World flycatchers) Nectariniidae (sunbirds)
Australia/ New Guinea	Casuariidae (cassowaries) Megapodidae (megapodes) Menuridae (lyrebirds) Ptilonorhynchidae (bowerbirds) Maluridae (fairy-wrens) Meliphagidae (honeyeaters) Acanthizidae (Australian warblers) Pomatostomidae (Australian babblers) Orthonychidae (logrunners) Paradisaeidae (birds of paradise)	Columbidae (pigeons) Psittacidae (parrots) Cuculidae (cockoos) Muscicapidae (Old World flycatchers) Ptilonorhynchidae (bower birds) Paradisaeidae (birds of paradise) Meliphagidae (honeyeaters) Campephagidae (caterpillar birds) Pachycephalidae (pitohuis)
Madagascar	Mesitornithidae (mesites) Brachypteraciidae (ground-rollers) Philepittidae (asities) Vangidae (vangas)	

* Treated as tribes of the family Fringillidae in this book.
Modified from Karr (1990).

corallirostris) forages like a nuthatch or treecreeper, climbing up tree trunks and catching insects with its short, sharp beak. The sickle-billed vanga (*Falculea palliata*) has a long, downward-curving bill that it uses for probing into tree holes and other cavities, like a woodpecker, in search of insects and other animals (see Fig. 5.1c). The warbler-like newtonias (*Newtonia*) glean small insects from vegetation, Ward's flycatcher-vanga (*Pseudobias wardi*) flycatches in the canopy, and Crossley's babbler-vanga (*Mystacornis crossleyi*) forages for small insects on the ground. Several smaller endemic bird radiations have also been found.

In New Guinea, there are many species from separate New Guinea–Australian radiations of passerine birds, including the bowerbirds (Ptilonorhynchidae), honeyeaters (Meliphagidae), birds of paradise (Paradisaeidae), and several smaller families. The species diversity of pigeons and parrots, two widespread families, is also very high. There are also a few representatives from many Old World bird families, such as the hornbills, pittas, sunbirds, white-eyes, and starlings, which have all presumably entered New Guinea relatively recently from the west. However, many other major bird families of Old World rain forests, including the bulbuls, babblers, barbets, and woodpeckers, are entirely absent. Some Old World families, such as the hornbills and tree-swifts, have reached New Guinea but not Australia.

Little, brown, insect-eating birds

The best-known birds of tropical rain forests are large or brightly colored, or, more rarely, both. In reality, however, the majority of rain forest bird species are small, brown, and relatively inconspicuous. Small brown birds are particularly common in the understorey and lower canopy. Most of these species actively search for insects and other small arthropods on living foliage, twigs, and tree trunks, in vine tangles, and in bunches of dead leaves. An indirect impact on plants has been demonstrated by experimentally excluding birds from tree branches in semideciduous rain forest in Panama (Van Bael et al. 2003). In the canopy, both arthropod densities and leaf damage greatly increased when birds were excluded, but there was no significant impact in the understorey. In Neotropical rain forests, the majority of these birds belong to the endemic radiation of suboscine passerines, while in the Old World these niches are occupied by many different families of oscine passerines.

Mixed-species flocks

Insectivorous suboscines, including ovenbirds, antbirds, woodcreepers, and tyrant flycatchers, form a major part of the mixed-species flocks that are such a conspicuous feature of the New World understorey bird fauna (Munn 1985; Powell 1985). The suboscines are joined in these flocks by both nonpasserines, such as woodpeckers, and a variety of oscine passerines, including vireos (Vireonidae) and tanagers (Thraupini). Tanagers, tyrant flycatchers, trogons, nunbirds (Bucconidae), and cotingas dominate the more variable mixed-species flocks of the upper canopy. For the rain forest birdwatcher, long periods with no birds visible alternate with short bursts of frantic activity as a flock passes through. The understorey flocks consist of a dozen or more core species, each species represented by one breeding pair, which remains with the flock for several years. A variable number of more or less regular or occasional participants join the flock, resulting in up to 50 species foraging together as a group. The communal territory of the flock is defended by all the core species where it borders with the territories of other, similar flocks (Jullien & Thiollay 1998). At least in Amazonia, flock size, composition, and home range are highly stable between seasons and between years (Powell 1985).

Two basic hypotheses have been proposed to explain the advantages accruing to members of these mixed-species flocks. One hypothesis is that the main advantage is increased safety from predators, such as hawks, cats, and snakes, because individual birds cannot keep an adequate lookout by themselves while they are actively searching for insects. This "many eyes" hypothesis has been supported by studies that compare the foraging behavior of flocking and solitary insectivores in the Neotropics (Thiollay 2003). Obligate flock members are more likely than solitary species to use active foraging techniques that make them conspicuous and are incompatible with sustained vigilance. An alternative, or additional, hypothesis is that flock members benefit from the prey found or flushed out by the flock as a whole.

Whatever the explanation, mixed flocks of understorey insectivores and omnivores are found in rain forests all over the world, although the families of birds involved show almost no overlap with those in the Neotropics. In Africa, mixed flocks include bulbuls, babblers, Old World warblers, Old World flycatchers (Muscicapidae), drongos (Corvidae), woodpeckers, white-eyes, sunbirds, and weavers. Woodpeckers are absent from Madagascar, where the endemic vangas join, and sometimes dominate, the mixed flock party. In Southeast Asia, mixed-species flocks are similar to those in Africa, but lack weavers and can contain up to 15 species of babblers – the most diverse group of birds in the region – as well as the endemic leafbirds. In Australia and New Guinea, these flocks attract many species from groups endemic to the region, including Australian babblers (Pomatostomidae), Australian warblers (Acanthizidae), pitohuis (Pachycephalidae), fairy-wrens (Maluridae), and birds of paradise. Although superficially very similar, it is not yet clear if any of these flocks are as highly organized and stable as those in the Neotropics.

Birds and army ants

Mixed-species flocks also form around columns of army ants in American and African rain forests, but not in other regions, where the ant swarms are apparently too small to attract followers. Indeed, only two army ant species in the Neotropics, *Labidus praedator* and *Eciton burchelli*, and a few driver ant species (*Dorylus*, subgenus *Anomma*) in Africa, form swarms that are large enough and dependable enough to regularly attract birds (Gotwald 1995) (see Chapter 7). These army ants raid in huge columns or waves across the forest floor, killing any arthropods and small vertebrates they encounter, and bringing their prey back to their nests to eat. In Neotropical forests, a number of bird species are professional ant followers, rarely feeding away from ant swarms, while others are regular or merely occasional associates. Exclusion experiments show that these birds actually reduce the number of prey caught by the ants because they are snatching up prey that the ant swarms would be catching (Wrege et al. 2005).

Fig. 5.2 Bicolored antbird (*Gymnopithys leucaspis*) in Parque Nacional Darién, Panama. This species is a "professional" follower of army ants and has the strong legs needed to cling to vertical perches in the rain forest understorey. (From Wikimedia Commons: http://commons.wikimedia.org/wiki/File:Gymnopithys-leucaspis-001.jpg.)

As their family name suggests, some species of antbird are particularly prominent among the ant followers, but certain species of woodcreepers, ground cuckoos, and tanagers are also "professionals" – that is, they make most of their living following ants. Professional ant followers can cling to vertical perches, such as the slender saplings that are common in the forest understorey, while watching for likely prey fleeing the army ants (Fig. 5.2). The bird species that are most dependent on army ants track the location of several ant colonies and check the temporary ant nests – known as bivouacs – each morning to assess their activity (Swartz 2001). By keeping track of several ant colonies as they go through their cycles of raiding and resting, the birds can be assured of having an active colony to follow each day.

The army ants of Africa are followed by flocks of birds unrelated to those in the Neotropics but otherwise comparable in size and diversity (Peters et al. 2008). Thrushes (Turdidae) are particularly prominent at African ant swarms and several species of large bulbul cling to vertical perches in a manner similar to Neotropical antbirds. Ant-following birds of African rain forests have much higher incidences of blood parasites than other rain forests birds, suggesting that these species pay a price for aggregating at this rich food source (Peters 2010).

Antbirds

The antbirds of the American tropics are small to medium-sized birds that occupy lowland forests. They belong to two related families of suboscine passerines: a smaller family of ground antbirds (Formicariidae) and a larger family of typical antbirds (Thamnophilidae) (Fig. 5.2), with a total of around 250 species in 53 genera. Antbirds are adapted for catching insects in the forest interior; their long, thick, hooked bills with notches on the edge are effective at holding and slicing up small animals. Antbirds often have moderately rounded wings and long feet, an advantage when foraging in the clutter of the forest interior. Many antbirds are members of mixed-species flocks and some are found in close association with columns of army ants. Molecular dating suggests that this association with ants has persisted in the typical antbirds since the late Miocene (Brumfield et al. 2007). The plumage of antbirds is patterned brown, black, or white, and often has streaked markings. Many species are hard to see in the forest understorey and are more easily recognized by their songs.

Babblers and warblers

Most of the little brown birds in the Old World have traditionally been included in either the babblers (Timaliidae) (Fig. 5.3) or the Old World warblers ("Sylviidae"). Together these two groups form part of a largely Old World bird radiation, the Sylvioidea, which also includes the larks, swallows, and bulbuls (Johansson et al. 2008b). The New World warblers or woodwarblers (Parulini) are not related to the Old World warblers, but are nine-primaried oscine relatives of the tanagers (Thraupini). The babblers show a great diversity in size, bill shape, and other characteristics, but differ from the warblers in their generally larger size, high sociability, weak flight, and the absence of migratory behavior. The noisy "babbling" calls between flock members of some species are the origin of the common name for the family and contrast with the pleasant "warbling" song of many warblers.

Recent advances in DNA-based systematics have played havoc with the traditional classification of both the babblers and the Old World warblers (Gelang et al. 2009). All the "babblers" of Australia, New Guinea, and Madagascar have been shown to belong to other families, as have a number of Asian and African species previously included in the Timaliidae, such as the "wren-babblers" (*Pnoepyga*). Conversely, both the *Sylvia* warblers of the western Palaearctic and white-eyes (Zosteropidae) of the Old World tropics have been shown to belong within the babblers. Molecular studies show that the diversification and spread of white-eyes to occupy the entire Old World tropics took place largely within the last 2 million years: among the highest rates of diversification reported for vertebrates (Moyle et al. 2009). These studies have also confirmed that a single North American species, the wren-tit (*Chamaea fasciata*), is a babbler. The newly

Fig. 5.3 Most babblers are little brown birds. This relatively colorful species, the golden-crowned babbler (*Sterrhoptilus dennistouni*), here in Luzon, is endemic to the Philippines. (Courtesy of Tim Laman.)

delimited babblers are still the most diverse bird family in the rain forests of tropical and subtropical Asia, with a secondary center of diversity in West and Central Africa, but are now also represented in the rain forests of Australia, New Guinea, and Madagascar, as well as many Pacific islands, by *Zosterops* white-eyes. Meanwhile the "Old World warblers" have been split among a number of separate families (Johansson et al. 2008b).

Forest frugivores

In striking contrast to the drabness and superficial uniformity of most insectivores, many frugivores are big and/or brightly colored, and differences among rain forest regions are obvious even to the casual observer. Indeed, most birds seen by the casual observer are likely to be frugivores. Although insectivores dominate in terms of the numbers of individuals and numbers of species in all rain forests, fruit- and seed-eaters contribute far more to the biomass, because individual birds tend to be bigger. Larger size allows birds to handle larger food items – most fruits are bigger than most insects – and also lets them swallow more food during intensive feeding bouts on fruiting plants, so that time spent exposed to predators while feeding is minimized. Large size also directly reduces the risk of predation by decreasing the number of predators that can attack a particular bird. Lower predation risk allows these birds to make more use of color for signaling to potential mates and rivals although, in many cases, it is only the male that does so. Many of the most spectacularly colored rain forest birds are frugivores, including the resplendent quetzal (*Pharomachrus mocinno*) in Central America, the violet turaco (*Musophaga violacea*) in West Africa (Fig. 5.1a), the fairy bluebird (*Irena puella*) in Southeast Asia, and the raggiana bird of paradise (*Paradisaea raggiana*) in New Guinea.

Pigeons and parrots take fruits in all rain forest regions, but there are noticeable differences between regions in the other bird groups involved. A Neotropical fig tree with a large crop of ripe figs will attract avian fig-lovers ranging in size from tiny manakins and tanagers to large toucans, and also including guans, trogons, New World barbets, cotingas, tyrant flycatchers, and orioles (Shanahan et al. 2001). Trumpeters (Psophiidae) feed on any figs that are knocked to the ground. In African rain forests, a similar tree would attract mostly birds from families that are absent from the New World: hornbills, bulbuls, African barbets (Lybiidae), white-eyes, cuckoo-shrikes, and turacos. Asian rain forests lack turacos and African barbets, and add Asian barbets (Megalaimidae) (Fig. 5.15b), frugivorous broadbills, leafbirds, fairy bluebirds, and flowerpeckers (Dicaeidae) (Fig. 5.6), but the list would be otherwise similar. In New Guinea, where there are no barbets or bulbuls, our fig tree would again draw pigeons and parrots, in greater variety than in Asia, plus mynahs, white-eyes, flowerpeckers, and cuckoo-shrikes, and the single species of hornbill, but also bowerbirds (Fig. 5.9c), birds of paradise (Fig. 5.9a,b), berrypeckers (Melanocharitidae), whistlers (Pachycephalidae), and honeyeaters. Cassowaries (Casuariidae) (Fig. 5.12) eat fallen figs from the ground. The assemblage in an Australian rain forest would be similar, but less diverse. Madagascan rain forests have relatively few frugivorous birds and our fig tree would attract mostly pigeons, parrots, and bulbuls, as well as the velvet asity (*Philepitta castanea*), a member of an endemic group of suboscines that is related to the broadbills. Some of the rain forest vangas also include some fruit in their diet.

Turacos

The turacos (Musophagidae) (Figs. 5.1a, 5.4) are a very distinctive and exclusively African group of fruit-eating birds, with 23 species in six genera. The largest species, the great blue turaco (*Corythaeola cristata*), is 75 cm (30 inches) in length but most of the others are somewhat smaller, around the size of a pigeon. Turacos have a long tail, small, rounded wings, and strong long legs which enable them to move readily in the tree canopy. They may have a conspicuous crest on their head and soft, often hair-like feathers. The bright colors of some species rival those of the parrots. One particularly beautiful species is the Livingstone's turaco, *Tauraco livingstonii*, which has delicate apple-green feathers covering its breast and head and a pronounced green crest. A red eye ring and a white eye stripe further enhance its appearance. The red and green pigments of turaco feathers are copper-based metalloporphyrins, apparently unique to the family (McGraw 2006). Green pigments are extremely rare in birds and this color is usually generated, at least partly, by structural mechanisms. We do not currently know how or why turacos accumulate the necessary copper from their diet.

Toucans and hornbills

In tropical rain forests from South and Central America, through Africa and Asia, to New Guinea, there are large fruit-eating birds with huge bills, loud calls, and

Fig. 5.4 Yellow-billed turaco (*Tauraco macrorhynchus*) on Pico Basile, Bioko Island, Equatorial Guinea. The green color of turaco feathers comes from a unique, copper-based pigment. (Courtesy of Tim Laman.)

often striking coloration on the bill, head, and neck. Although superficially very similar, these birds belong to two unrelated families, the toucans (Ramphastidae) in South and Central America and the hornbills (Bucerotidae) in the tropics of the Old World (Fig. 5.5a,b). Of the major rain forest areas, only those of Madagascar and Australia are without one or the other family, although New Guinea has only a single species of hornbill.

The long and deep bills of toucans and hornbills have multiple functions in feeding, display, defense, and nesting, but the primary one is probably to extend their reach when perching (Bühler 1997). This is particularly important when feeding on fruits, which are often borne on twigs that are much too thin to support the weight of these heavy birds. Instead, they can grasp a strong support with their feet and use their long bills and long, flexible necks to reach out for fruits that would otherwise be beyond their reach. The same bill design is also very useful when reaching into deep nests to pluck out young birds to eat. Recent studies have shown that in toucans, at least, the bill is also an extremely efficient shedder of excess heat (Tattersall et al. 2009).

These bills look massive and powerful, but appearances are misleading. The bills are actually thin and light. The shape is dictated by design principles. A long bill for firmly holding food items must also be deep to prevent bending. The muscles that close the bill are attached near the base, so only a relatively small force can be exerted at the tip, which is the only place where the upper and lower parts of the bill meet. The bill is used like forceps to pick fruits one at a time, before tilting the head back to toss them directly into the throat: a unique mechanism called "ballistic feeding" that by-passes the tongue (Baussart et al.

(a)

(b)

Fig. 5.5 Toucans and hornbills are similar-looking birds of the Old and New World. (a) This keel-billed toucan (*Ramphastos sulfuratus*) is from the Children's Eternal Rainforest in Costa Rica. This is one of the largest toucans but is only two-thirds the size of the black hornbill. (Courtesy of Dale Morris.) (b) A black hornbill (*Anthracoceros malayanus*) eating fruit in Gunung Palung National Park, Indonesia. (Courtesy of Tim Laman.)

2007). Larger forces can be applied further back from the tip and this open region, which is serrated in many species, is used to crush large items of animal food, such as lizards, snakes, nestlings, and insects, which make up a significant proportion of the diet in all species and dominate it in some hornbills.

Despite their many similarities, toucans and hornbills also differ in important ways. Most hornbills are a great deal bigger than even the largest toucans and can eat correspondingly larger food items. Indeed, the biggest hornbills are the largest flying frugivores on Earth. Hornbills are strong flyers and some species cover large distances every day, while toucans have been known to fall into the water while attempting to cross large rivers (Cristoffer & Peres 2003). Hornbills also differ from toucans by the presence of a casque, an outgrowth from the top of the already large bill. The shape and prominence of the casque varies among species; it is virtually absent in young birds. The function of the casque is still debated, but it is probably involved in both visual communication, through its size, shape, and color, and acoustic communication, through the ability of the large air cavity to amplify calls of appropriate frequencies. There are also reports of casque-butting between perched or flying hornbills, although the significance of this behavior is unclear.

Both hornbills and toucans nest in natural cavities in trees, but the hornbills are unique among birds in that the breeding female of most species seals the entrance to the cavity from the inside, with mud, sticky food, and her own droppings, leaving only a narrow vertical slit. Through this, the male feeds her and later her chicks, for 6 or more weeks. Droppings are voided out of the slit and food remains are thrown out, thus maintaining nest hygiene and leaving a useful record, below the nest, of the food brought by the male. The female probably builds such a maternity prison to keep out unwanted visitors, such as snakes and other predators.

Mistletoe birds

The fruits of most mistletoes – parasitic plants in the families Loranthaceae and Viscaceae that grow on the branches of trees and shrubs – are small enough to be eaten by any frugivore. In tropical Asia, however, mistletoe fruits are largely consumed by tiny flowerpeckers (Dicaeidae), the smallest frugivores in the region (Fig. 5.6). The 40 or so species of flowerpecker are brightly colored birds with short tails and relatively short beaks. The foraging of mistletoe birds is closely linked to the biology of the mistletoe plants (Reid 1991). The year-round availability of mistletoe fruits allows these birds to specialize on them, while the unique biology of the mistletoe plants means that the fruits eaten by other frugivorous birds are usually wasted. Unlike most plants, where the seeds can be scattered over the ground, successful dispersal of mistletoe seeds requires that they be attached to a branch of a compatible host species, with a suitable diameter for establishment. All other seeds are wasted. Mistletoe seeds are extremely sticky and it is this stickiness that provides both the means of attachment to a branch and the incentive for a bird to remove them from their beak or cloaca. Flowerpeckers typically defecate strings of sticky seeds, which they wipe off on a branch. The preferred perch diameter of these tiny birds apparently matches the size of branch on which the mistletoe seeds can best germinate and establish a new plant.

There are other mistletoe specialists in other rain forest regions, but the mistletoe–flowerpecker association in tropical Asia seems to be uniquely strong. In New Guinea and Australia, some honeyeaters are important consumers of mistletoe berries, but the Australian mistletoe bird (*Dicaeum hirundinaceum*) is

Fig. 5.6 Crimson-breasted flowerpecker (*Prionochilus percussus*) in Gunung Palung National Park, Indonesia. (Courtesy of Tim Laman.)

another flowerpecker, the only species that reached the continent. In Africa, some species of tinkerbird (*Pogoniulus*) are specialists, as are several species of *Euphonia* tanager in the Neotropics, as well as the paltry tyrannulet (*Zimmerius vilissimus*) of Central America and several species of cotinga.

Parrots

Not all fruit-eaters disperse seeds, or even drop them undamaged under the parent plant. Some birds, including most parrots (Fig. 5.7a,b,c), destroy seeds in the beak and then swallow the pieces. Long-tailed parakeets (*Psittacula longicauda*) are major predators of dipterocarp seeds in Southeast Asia, whereas the huge blue hyacinth macaw (*Anodorhynchus hyacinthinus*) (Fig. 5.7a) uses its powerful bill to open palm nuts in the Amazonian region. Even the tiny seeds of figs are not safe from parrots, with species such as the miniscule double-eyed fig-parrot (*Cyclopsitta diopthalma*) in Australia and the orange-chinned parakeet (*Brotogeris jugularis*) in Costa Rica able to break them individually in the bill (Shanahan et al. 2001). Although parrots destroy most seeds they encounter, they are also notoriously messy eaters. A significant proportion of the fruits they try to eat are dropped to the ground unharmed. Here terrestrial rodents, browsing mammals, and more benign bird frugivores may eat the fruits and disperse the seeds that they would otherwise not have had access to in the high canopy. In New Guinea, the vulturine parrot (*Psittrichas fulgidus*) appears to specialize on a few strangling fig species (*Ficus*) that produce exceptionally hard-walled

(a)

(b)

(c)

Fig. 5.7 Parrots are noisy, colorful rain forest birds. (a) Hyacinth macaws (*Anodorhynchus hyacinthinus*) from Brazil. (Courtesy of Tim Laman.) (b) Crimson rosella (*Platycercus elegans*) in subtropical rain forest, Lamington National Park, Australia. (Courtesy of Tim Laman.) (c) Rainbow lorikeet (*Trichoglossus haematodus*) from Australia. (Courtesy of Tim Laman, taken in captivity.)

figs (Mack & Wright 1998). These parrots may disperse the seeds directly, as well as making the pulp and seeds accessible to smaller and weaker-billed seed-dispersing birds.

Most authorities now recognize two families of parrots: the Psittacidae, with more than 330 species spread over much of the world, and the much smaller Cacatuidae, the 22 species of cockatoos, which range from Australia to the Philippines. Parrots vary widely in size and often are brightly colored. They are easily recognized by a number of features such as the short, thick, and hooked upper bill that can move up and down on a specialized joint (Fig. 5.7). In addition, the upper bill has an enlarged fleshy covering where it joins the head. Parrots often have plump bodies and rounded wings, and are known for their loud, screeching calls.

Parrot diets often overlap with those of another group of muscular-jawed seed destroyers, the squirrels, and it is probably no coincidence that rain forest parrot diversity is highest in the Neotropics (with 50 species in the Amazon basin alone) where there are only seven squirrel species, and New Guinea (with 46 species) where there are no squirrels at all (see Table 3.1). In contrast, rain forests in Africa (8 species) and mainland Asia (6 species) have many coexisting species of squirrels and few parrots. The one region in East Asia where parrot diversity is somewhat higher is the Philippines, with 12 species, including one cockatoo; it also a region with few squirrels – only two species. Madagascar spoils the pattern, by lacking squirrels but having only three parrots, but this huge island is peculiar in many ways because of its long period of isolation. Note, however, that the pattern of parrot diversity is also consistent with a historical rather than ecological explanation: an origin on Gondwana in the late Cretaceous, after the separation of Africa and the India/Madagascar block, and a more recent spread into Asia, Africa, and Madagascar (Wright et al. 2008).

Pigeons and doves

The pigeons and doves (Columbidae), like the parrots, are a pantropical group of birds in which most species probably destroy more seeds than they disperse. Unlike the parrots, this destruction takes place in a muscular, grit-filled gizzard, rather than the bill, and they feed mostly on relatively small seeds. However, by no means do all pigeons and doves destroy the seeds they swallow. In the rain forests of Asia, New Guinea, Australia, and the Pacific, there is a distinctive radiation of "fruit pigeons," with thin-walled gizzards and short, wide guts, through which even the largest seeds can pass undamaged (Corlett 1998). This group includes the imperial pigeons (*Ducula*) (Fig. 5.8), fruit doves (*Ptilinopus*), and mountain pigeons (*Gymnophaps*), as well as the Australian topknot pigeon (*Lopholaimus antarcticus*), the New Caledonian cloven-feathered dove (*Drepanoptila holosericea*), and the Madagascan blue pigeon (*Alectroenas madagascariensis*) (Pereira et al. 2007). Imperial pigeons, in particular, are among the largest frugivores in the rain forests of this region and can swallow very large fruits (up to 4 cm diameter), which few other birds can handle. This makes the fruit pigeons extremely important seed dispersal agents for plants that bear large fruits with large seeds, such as nutmegs (Myristicaceae) and palms (Palmae).

Pigeons feed on fruits in the canopy of all rain forests, but the role of species other than fruit pigeons in seed dispersal is unclear. In contrast to the fruit pigeons,

Fig. 5.8 Green imperial pigeon (*Ducula aenea*) on a fig tree in Palawan, the Philippines. (Courtesy of Tim Laman.) Unlike many species in this family, imperial pigeons are excellent seed dispersal agents.

these other pigeon species feed mostly on small-seeded fruits, such as figs and, in the Neotropics, *Cecropia*. Seed fate seems to vary, even between bird species in the same genus, from complete destruction to high-quality dispersal, but too few species have been investigated for any clear patterns to emerge.

Frugivory and extreme courtship behavior

One of the most striking examples of convergence between unrelated bird families in different rain forest regions is the remarkable similarity in both appearance and behavior between the birds of paradise in New Guinea and a number of species of cotingas (Cotingidae) and manakins (Pipridae) in the Neotropics (Johnsgard 1994). The birds of paradise (Paradisaeidae) are probably the best-known bird group in the rain forests of New Guinea (Fig. 5.9a,b). They are related to the crow family (Corvidae), but there is hardly anything crow-like in either their appearance or behavior. The 42 species of birds of paradise vary from starling-like to crow-like in size (Frith & Beehler 1998). What is most astonishing about birds of paradise is the elaboration in males of feathers used in courtship displays. In some species, feathers on the tails, wings, body, and head of males are extraordinarily elongated and often brightly colored or ruffled; presumably, such feathers in some way signal male fitness to the watching female birds and thus desirability of individual males as mates. Male birds often shake or ruffle these feathers while hopping about, with the females watching and evaluating nearby – a display that must be seen to be believed. The King of Saxony bird of paradise, *Pteridophora alberti*, has two exquisite plumes on its

(a)

(b)

Fig. 5.9 Birds of paradise and bowerbirds both exhibit extreme forms of courtship behavior. (a) Red bird of paradise (*Paradisaea rubra*) male on Batanta Island, Indonesia. (Courtesy of Tim Laman.) (b) Blue bird of paradise (*Paradisaea rudolphi*) male foraging in Southern Highlands Province, Papua New Guinea. (Courtesy of Tim Laman.)

(c)

head that are several times longer than its body; males impress females by
raising these feathers while bouncing on branches. Some species even display
while hanging upside down, spreading their long feathers over their head.

These fantastic displays by male birds of paradise are a reflection of an extreme
type of mating system in which male and female birds lead entirely separate lives,
except when they briefly come together to copulate. To put it simply, males
display and females nest (Frith & Beehler 1998). An adult male spends a large
proportion of its time on its display site, which may be solitary or clustered together
with other males in a "lek." A female visits one or more of these sites briefly to
mate and then she alone is responsible for building the nest and rearing the
nestlings. A successful male may mate with many females, but most males will
mate with none. What has made the birds of paradise of even more interest to
biologists is that different members of the family display the full range of mating
systems, from monogamous pairs that forage and raise their young together, to
the extreme divergence between sexual roles described above.

The bowerbirds (Ptilonorhynchidae) are a rather distantly related group of birds
found in the rain forests of New Guinea and Australia, with a mating system
similar to that of the birds of paradise. Bowerbirds are named for the elaborate
courtship sites that the males construct, in which they display to females
(Fig. 5.9c) (Frith et al. 2004). The courtship area, or bower, might consist of
a cleared circular area on the forest floor, in which a domed tunnel of sticks has
been constructed, decorated with colored stones, flowers, snail shells, fruits, and
insect parts. When a female is near, a male sings and dances above and on his
bower, until the female is ready to mate. In some species, the adult male has
spectacular plumage and the bower is simple, while in others the male is dull

but the bower is complex. This shows that the bower represents a transfer of sexual attraction from the male's own appearance to the appearance of his bower.

In the Neotropics, similar extraordinary mating displays are known in the cocks-of-the-rock (*Rupicola* spp.), the red cotingas (*Phoenicircus* spp.), and some manakins (Pipridae) (Kricher 1997). As with the birds of paradise, males of these species have spectacular plumage, which they display to the dull-looking females in elaborate courtship displays, either alone or, more often, in leks. The displays have both visual and sound elements, with one species, the club-winged manakin (*Machaeropterus deliciosus*), producing sustained musical sounds by rubbing modified feathers together – a mechanism otherwise only known in invertebrates (Bostwick et al. 2010). Males of these species spend most of their adult lives at the lek and, as with birds of paradise, the females raise the young alone.

All the bird species with these extreme mating displays are at least partial frugivores and most depend heavily on fruits. This is clearly no coincidence. Fruit is a relatively reliable food source in tropical rain forests so it is possible for a frugivore to meet its daily food requirements in a small proportion of the time available. This gives the males a lot of free time to display and makes it easier for a female to feed both herself and her young. This cannot be the entire story, however, since most frugivores, even in New Guinea and the Neotropics, do not go to these extremes. There is clearly also a very large phylogenetic component, with extreme mating displays confined to a very few evolutionary lineages (Prum 1994). Indeed, once these huge differences between the sexes become established, they may be very hard to reverse, even over evolutionary timescales, because the features that make a male successful in competitive displays are probably incompatible with them contributing to parental care.

Why are these bold courtship displays found in the rain forests of New Guinea and the Neotropics, but not in Southeast Asia, Africa, or Madagascar? It could simply be chance: the combination of factors needed for these displays to evolve may have come together only in bird lineages that did not reach these other rain forests. It could also be a reflection of the reliability of the fruit supply. In particular, the multiyear fruiting cycles in Southeast Asian rain forests may have made it impossible for birds to devote so much of their time to non-feeding activities. Or it could reflect differences in predation risks. Displaying males are extremely conspicuous and multimale leks in particular must be very attractive to predators. Although most recorded attacks on leks involve raptors, most displays are close to the ground and thus vulnerable also to mammalian carnivores. Birds of paradise, bowerbirds, cotingas, and manakins evolved and diversified in rain forests that lacked placental carnivores (see Chapter 4), although the cotingas and manakins have had to contend with cats, dogs, mustelids, and procyonids for 2–3 million years since the appearance of the Panama land bridge.

Birds as seed dispersal agents

In most rain forests, birds are the single most important group of seed dispersal agents. They are particularly important for plants that bear small fruits, while mammals – especially primates – are often more important for large fruits. Birds seem to be most important in New Guinea and Australia, where there are no primates and few mammalian frugivores apart from bats. They may be least important in Madagascan rain forests, where frugivorous birds are few and primates are abundant (see Chapter 3).

Although frugivory might seem to be a simple act, the fate of the seed – and thus the success of seed dispersal – depends critically on how the frugivorous animal selects and processes the fruit, and what it does afterwards. Fruits in each rain forest region are eaten by birds that differ in size and behavior, as well as fruit acquisition and processing techniques, from those in other regions. The major groups of fruit-eating birds in Asia and Africa usually take fruits from a perch and swallow them whole. Fruit consumption and seed dispersal by these birds is therefore limited principally by their maximum gape width (i.e. the maximum size they can swallow), which is generally larger in bigger birds. Seed-dispersing birds in Asian rain forests range in size from 5 g (0.2 oz) flowerpeckers to 3 kg (6.5 lb) hornbills, and the largest fruits that can be swallowed whole range from less than 8 mm to over 40 mm (0.3–1.6 inches) in diameter (Corlett 1998). African rain forests have a similar bird fauna, but a somewhat smaller range of bird sizes.

The New Guinea avifauna includes flowerpeckers and a single hornbill, but also the world's largest frugivorous birds, the flightless cassowaries (*Casuarius*), the biggest of which can swallow whole fruits more than 5 cm in diameter (Bradford et al. 2008). Some birds of paradise, another endemic group, can use their feet and bills together to break open the woody capsules of some Myristicaceae and Meliaceae to get at the arillate seeds inside (Beehler & Dumbacher 1996). This behavior may have evolved in response to the absence of primates and squirrels, which open many such fruits elsewhere, and makes the birds of paradise one of the most important groups of vertebrate seed dispersal agents in New Guinea.

Many seeds in Neotropical rain forests are dispersed by endemic bird families with fruit acquisition and processing behaviors that are rarely seen in the Old World. Most fruit-eating suboscines, including manakins, cotingas, and tyrant flycatchers, have wide gapes, take fruits on the wing (rather than from a perch), and swallow them whole (Levey et al. 1994). Tanagers and related groups of nine-primaried oscines have narrow gapes, take fruits while perching, and crush them in the bill, squeezing out all but the smallest seeds, before swallowing the pulp. These "mashers" drop most large seeds under the parent plant, but are important dispersal agents for species with seeds less than 2 mm in length, which they cannot separate from the pulp.

What are the consequences of these differences between rain forest regions in the birds that eat fruits and disperse seeds? Unfortunately, the necessary inter-regional comparisons with standardized methodology have yet to be made. We would predict that fruits targeted at birds would be more often borne on the outside of tree canopies in the Neotropics, where they would be more accessible to those birds that pluck them in flight, while in the Old World they should always be accessible from a perch. This prediction parallels the observed differences in the positions of bird-pollinated flowers, but birds are far more important as seed dispersal agents than as pollinators, so the consequences of bird pollination and bird dispersal should be easy to separate. We might also predict that the small fruits eaten by fruit-mashing tanagers would have smaller seeds than similar-size fruits eaten by similar-size bulbuls. At least for the common pioneer trees, this does seem to be the case, with those in the Neotropics typically having much smaller seeds than those in Africa and Asia, although the greater importance of fruit bats as dispersers of Neotropical pioneers is an alternative explanation for this pattern.

Fruit size and body size

One of the more striking contrasts among regions is the difference in the sizes of both fruits and fruit-eaters between rain forests of the Old and New Worlds. Evidence that Old World fruits are consistently larger than New World fruits comes from a comparison of 1642 species in 236 genera of eight large, flowering plant families that occur in both the New and Old World tropics (Mack 1993): Anacardiaceae, Burseraceae, Lauraceae, Meliaceae, Moraceae, Myristicaceae, Palmae, and Simaroubaceae. In every family, fruit lengths were longer in species from Africa and Southeast Asia than in New World species, with Old World fruit length being twice as long as New World fruit length in species in the families Anacardiaceae and Moraceae. Within the huge fig genus (*Ficus*), Old World species had significantly longer fruits than New World species (average 2.45 cm vs. 1.86 cm). In another comparison, the fruits of the Southeast Asian nutmegs in the genus *Myristica* were significantly longer than the similar Neotropical nutmegs in the genus *Virola* (average 5.9 cm vs. 3.0 cm).

Along with larger fruits, Old World rain forests also have larger frugivores (Cristoffer & Peres 2003). This is clearest for the primates (see Chapter 3), terrestrial mammals (see Chapter 4), and fruit bats (see Chapter 6), but is also true for birds when similar groups are compared, such as the toucans and hornbills (see above). The largest flying frugivores are Southeast Asian hornbills and the largest of all avian frugivores are the flightless cassowaries of New Guinea and Australia (see below). Larger frugivores presumably have larger mouths that can process larger fruits more efficiently. Birds have no teeth, so the maximum size of fruit that a bird can process is set by its maximum gape width. Fruit bats have teeth, but are also gape-limited when carrying large fruits away from fruiting trees to process elsewhere.

It is clear that the contrasts in fruit and frugivore sizes must have a common explanation, but it is far from obvious what this explanation is. Two broad types of explanation for differences between rain forest regions were mentioned in the first chapter – historical and ecological. The long period of isolation experienced by the South American continent during the Tertiary suggests a possible historical explanation. Although large fruits are found in many families of plants in the Old World, the large frugivores are mostly in just a few vertebrate lineages. It is thus conceivable that the smaller size of Neotropical frugivores has resulted largely from the failure of apes and cercopithecine monkeys (see Chapter 3), pteropodid fruit bats (see Chapter 6), and hornbills and fruit pigeons (this chapter) to get there. Smaller frugivores would auto-matically lead to smaller fruits, since plants that produced large fruits would have less chance of their fruits being eaten and the seeds within them dispersed. There are no comparable sets of fruit data from New Guinea, but the presence there of cassowaries, hornbills, and large pteropodid fruit bats fits with anecdotal evidence that large fruits are common.

Historical explanations are less convincing when it comes to terrestrial frugivores, since South America had its own fauna of megaherbivores until quite recently (see Chapter 4). Indeed, in a classic paper entitled "Neotropical anachronisms: the fruits the gomphotheres ate," ecologists Dan Janzen and Paul Martin argued that a "megafaunal dispersal syndrome" was still recognizable among the fruits produced by Costa Rican trees (Janzen & Martin 1982). They listed a number of tree species in both deciduous forest and rain forest that produce large, indehiscent fruits with well-protected seeds of the type that are dispersed by elephants and other large herbivores in Africa and Asia.

Even for arboreal frugivores and the fruits they consume, a purely historical explanation for the distinctiveness of the Neotropics is not fully convincing. Not only does it require rather a lot of historical accidents to explain the absence of large frugivores, but it also assumes that no indigenous vertebrate groups were capable of evolving large species. What stops toucans or phyllostomid bats from getting as big as hornbills and pteropodid bats? However, ecological explanations for smaller fruits and/or smaller frugivores in the New World are no more convincing. Cristoffer (1987) suggests that small size was "forced" on arboreal frugivores by more fragile vegetation, but offers little evidence for the consistent differences in vegetation structure that this explanation would need. Looking instead for factors favoring large animals or large fruits in the Old World is no easier. The unreliable fruit supply in Southeast Asian dipterocarp forests (see Chapter 2) may favor large frugivores that can cope with a lower-quality diet in the long periods between mast years when fruit is unavailable, but this explanation would not apply to the more seasonal rain forests of Africa. The height and density of Southeast Asian forests may also favor the production of larger seeds that produce large, shade-tolerant seedlings and require larger fruits to contain them (Primack 1987). Again, this explanation does not fit well with the greater openness of forest in Africa. Whatever the explanation, or explanations, it is clear that this striking pattern needs further investigation.

Flower visitors

Any small, brilliantly colored bird that is not a frugivore is likely to be a nectar-feeder. Only in the tropics has the year-round availability of nectar made it possible for birds to specialize on flowers as an energy source, and thus for plants to use them as long-distance dispersers of pollen. Flower-birds have evolved independently in the major rain forest regions, as have bird-flowers, resulting in both some striking parallels and major differences among the regions (Fleming & Muchhala 2008). The most obvious difference is the greater species richness and morphological diversity of both flower-birds and bird-flowers in the Neotropics, where a more predictable supply of nectar appears to have allowed greater specialization in bird–flower relationships.

Nectar is a great source of easily digestible energy, but it usually contains negligible amounts of protein and lipids. Nectar specialists therefore must also eat invertebrates to balance their diets. Long, thin bills, adapted to obtaining nectar from flowers, are probably not ideal tools for catching invertebrates, and nectar specialists seem to be restricted to small insects and spiders, of which they must eat very large numbers. The members of one genus of long-billed Asian sunbirds (*Arachnothera*) are called spiderhunters, because they pick spiders from their webs. Pollen from flowers is also rich in protein, but it is not easy to harvest efficiently, and nectar-feeders seem to vary greatly in the extent to which they can digest it.

Hummingbirds

The best-known of the avian flower specialists are the more than 300 species of hummingbirds (Trochilidae), which are the dominant nectar-feeding birds of the New World (Fig. 5.10). Their name is derived from the humming sound that their

Fig. 5.10 Often seen drinking nectar from flowers, hummingbirds are one of the most distinctive families in the Neotropics. (a) A purple-crowned fairy hummingbird (*Heliothrix barroti*) hovering over tubular white flowers in Peru. (b) A white-necked Jacobin hummingbird (*Florisuga mellivora*) feeding at an orange legume flower in Peru. (Both courtesy of Luis A. Mazariegos.)

(a)

(b)

wings make as they rapidly beat the air, up to 70 times a second in the smallest species, creating a sound like a small high-speed fan. Most hummingbirds are totally aerial in foraging, moving and darting rapidly in flight and hovering upright when removing nectar from flowers. The brilliant, iridescent plumage of many species makes them a conspicuous feature of rain forests from the lowlands up to the alpine zone in the Andes, with as many as 25 species at the richest sites. Their small legs and feet are suited just for perching on twigs, and they never walk on the ground.

Hummingbirds are attracted to flowers with plentiful, relatively dilute nectar. In the most specialized flowers, the nectar is located at the base of a long corolla tube, often red in color, with a narrow opening, and there is no landing platform for visitors that cannot hover. The birds extract nectar by rapid licking with their long, grooved tongues, which are forked near the tip. Hummingbirds may defend clumps of flowering plants against members of their own species, as well as chasing away insects and birds of other, larger species. Like the manakins and birds of paradise described above, male and female hummingbirds lead separate lives, with only the females involved in nest building and rearing the young. However, although males must engage in mating displays in leks or individually, the high energy demand of the hummingbird lifestyle does not permit the extravagant courtship displays seen in some frugivores.

The bills of different hummingbird species vary in length and degree of curvature, with the longest and most curved bills matching the shape of the flowers they visit. Thus, the swordbilled hummingbird (*Ensifera ensifera*), with a 10 cm (4 inch) bill – longer than the rest of its body – visits flowers with a similarly long corolla tube, such as the passionflower, *Passiflora mixta*. Perhaps the most extreme case of correspondence between bill and flower shape is found between the white-tipped sicklebill (*Eutoxeres aquila*), which has an extraordinary scythe-shaped bill matching exactly the shape of the flowers on which it depends. The latter include those of a few species of *Heliconia* and, at higher elevations, lobeliad species in the genus *Centropogon*, which have equally curved corollas (Stein 1992). Sicklebills are heavy for hummingbirds (weight around 12 g, 0.5 oz), so hovering is energetically expensive. Consequently, these birds perch while feeding. An even more curious example comes from the rain forests of St Lucia in the West Indies, where the purple-throated carib (*Eulampis jugularis*) is the sole pollinator of two species of *Heliconia* (Temeles et al. 2009). One *Heliconia* species has short flowers and is visited mostly by the male hummingbirds, which have shorter and less curved bills than the females. The other *Heliconia* species, with longer and more curved flowers, is pollinated by the females, which have correspondingly longer and more curved bills.

Hummingbird species in an area form communities with complex ecological interactions, involving competition for nectar resources, coevolution with plants, and predator avoidance. These communities have been intensively investigated for the insights they can provide on community structure. Phylogenetic analyses have revealed that in one geographical region these communities do not represent a single evolutionary radiation; rather, they have formed as a result of a complex mixture of evolutionary radiation, migration, and colonization of species in combination with local species extinctions. Lowland rain forest communities consist of representatives from six to eight of the nine hummingbird clades, while environmental constraints lead to communities with less phylogenetic diversity at high altitudes and in drier habitats (Graham et al. 2009).

Sunbirds and honeyeaters

Sunbirds (Nectariniidae) (Fig. 5.11) are the most important group of flower-visiting birds in Africa and tropical Asia. The greatest concentration of species is in Africa, where communities of sunbirds can be almost as diverse as those of hummingbirds in the Neotropical lowlands. Many species also occur in India and Southeast Asia. Like hummingbirds, they are small and often brightly colored, and sometimes their feathers have a brilliant metallic sheen. The sexes often differ greatly in appearance. In many species, the bill is curved and elongate in a manner reminiscent of hummingbirds, although the bills of these sunbirds are curved down whereas hummingbird bills are sometimes curved upwards. The tongue is tubular for most of its length, then split in half at the tip, as an adaptation for licking up nectar. In contrast to hummingbirds, sunbirds have strong legs and sharp claws for climbing around inside of foliage. They usually take nectar while perching, so plants adapted to encourage their visits must provide some form of perch. When perches are unavailable, sunbirds can hover for a few seconds, though not with the grace and stability of hummingbirds.

Two species of sunbird reached New Guinea, and one of them also occurs in northeast Australia. However, the major flower visitors in both Australia and New Guinea are the honeyeaters (Meliphagidae), which are found as far west as Bali. Honeyeater species vary greatly in their dependence on nectar, with many species having omnivorous diets, including fruits and insects, in which nectar plays only a minor role. Although the bills of many species are adapted for nectar feeding, being elongated and curved, they are rarely as specialized as those of sunbirds or hummingbirds. Honeyeater tongues are elongate, channeled, can

Fig. 5.11 In the Old World, sunbirds are common flower visitors that perch while feeding. This olive-bellied sunbird (*Cinnyris chloropygia*) was photographed in Sierra Leone. (Courtesy of David Monticelli.)

extend beyond the bill, and have a brush-like tip for taking up nectar. Most honeyeaters are larger than other nectar-feeding birds and most are a dull green, brown, or grey, although some striking exceptions exist. The Sulawesi myzomela (*Myzomela chloroptera*) on the island of Sulawesi, which lies midway between New Guinea and mainland Southeast Asia, is amazingly similar in both appearance and behavior to the unrelated crimson sunbird (*Aethopyga siparaja*), with which it coexists.

Other nectar feeders

In addition to the three main groups of nectar-feeding birds, each rain forest region also supports other groups of more or less specialized flower visitors. In the Neotropics, these include a variety of flowerpiercers, honeycreepers (Fig. 5.1b), tanagers, and orioles, which are all nine-primaried oscines. Some species of flower-piercers (*Diglossa*) are the most nectar-dependent of these birds but, as their name suggests, their bills are specially adapted for piercing the long corolla tubes of hummingbird-pollinated flowers and stealing the nectar without transferring any pollen. However, flowerpiercers visit flowers with short corollas in the normal way and can then be important pollinators.

White-eyes are regular flower visitors from Africa to Australia, and many other Old World birds, including leafbirds, bulbuls, and flowerpeckers, take substantial amounts of nectar (Corlett 2004). In Madagascar, the two species of sunbird asity (*Neodrepanis*) resemble sunbirds in size, diet, and their down-curved bills and long, tubular tongues, but are in fact suboscine relatives of the broadbills. In Australia and New Guinea, a specialized group of parrots, the brush-tongued loris and lorikeets, rival the honeyeaters in their dependence on flowers with easily accessible nectar, although they are very destructive feeders, often damaging the flowers (Brown & Hopkins 1996). These small parrots can apparently harvest and digest pollen more efficiently than the slender-billed nectar specialists (Gartrell & Jones 2001).

The Hawaiian Islands were home to an extraordinary evolutionary radia-tion of more than 50 bird species in the Hawaiian honeycreeper subfamily (Drepaniidae) (see Chapter 8). This radiation included many nectar feeders, such as the orange-red I'iwi (*Vestiaria coccinea*) (Fig. 8.3b), which evolved a long, curved bill to feed from the similarly curved flowers of the endemic lobelias. A smaller radiation of honeyeaters (Mohoidae) in Hawaii produced five species (Fig. 8.4), all of which are now extinct, as are many of the honeycreepers. Although the Hawaiian "honeyeaters" looked and behaved like their Australian namesakes, DNA obtained from museum specimens has shown that this is the result of convergent evolution and they are not at all related (Fleischer et al. 2008). Introduced Japanese white-eyes (*Zosterops japonica*) are the now the commonest visitors to Hawaiian flowers with easily accessible nectar.

Birds as pollinators

Flower-birds and bird-flowers are found in all rain forests, but their independent origin in each region has resulted in major differences in the role of birds as pollinators (Fleming & Muchhala 2008). Hummingbirds have the greatest dependence on floral nectar and have produced the most spectacular examples of coevolution between plants and birds. However, a large majority of the plants pollinated by hummingbirds are herbaceous: either forest floor herbs, such as the many species of *Heliconia*, or canopy epiphytes, particularly in the families Bromeliaceae and Gesneriaceae. Hummingbirds also pollinate shrubs and lianas, but very few species of tree. Hummingbirds are probably poor pollinators for trees because territorial species defend such clumped nectar resources against other birds, which must greatly reduce cross-pollination.

At the other extreme, the honeyeaters are mostly more or less omnivorous, with a much lower dependence on nectar but, unlike hummingbirds, honeyeaters visit and probably pollinate many rain forest trees (Brown & Hopkins 1996). These opportunistic nectar-feeders are often aggressive to other birds while feeding, but do not defend one tree throughout its flowering period. Honeyeaters visit not only trees, however, and are also thought to pollinate many species of orchids in montane forests in New Guinea, where bird pollination may have evolved in response to the relative paucity of insects (Schuiteman & de Vogel 2007). The sunbirds seem to be intermediate in their degree of specialization, with most species eating more insects than hummingbirds do, as well as small fruits. Sunbirds pollinate mostly large herbs, including gingers (Zingiberaceae) and bananas (Musaceae), and mistletoes (Loranthaceae), but also a range of shrubs, climbers, and trees (Corlett 2004). Overall, pollination by birds seems to be most important in the Neotropics and New Guinea, and least important in Southeast Asia.

Despite the large differences in size, morphology, and behavior between the three major groups of flower-visiting birds, bird-pollinated ornamental plants from one region attract the flower-visiting birds of other regions when they are planted there. This suggests a universality to the bird-pollination syndrome, despite the very different birds involved. Large, red, tubular or two-lipped flowers, with copious quantities of dilute nectar and no scent, are recognized as "bird-flowers" by hummingbirds, sunbirds, and honeyeaters alike.

Ground-dwellers

Ground-living birds are often large and usually drab in color, although there are some striking exceptions. Invertebrates, seeds, and fallen fruits are the main foods available at ground level, and many of the large ground-feeding birds probably eat all three, although in widely varying proportions. However, the diets of most of these shy and elusive birds are inadequately known.

Cassowaries, elephant birds, and tinamous

By far the largest ground-living birds in tropical rain forests and, indeed, the largest of all forest birds, are the four species of flightless cassowaries (Casuariidae) of New Guinea and Australia, which feed mostly on fruit (Fig. 5.12). The southern cassowary (*Casuarius casuarius*), one of the few bird species shared between the rain forests of New Guinea and Australia, can attain a height of 1.8 m (6 feet) and a weight of 65 kg (145 lb), making it the second largest bird after the ostrich. Even the smallest cassowary species, the dwarf cassowary (*C. bennetti*), can measure more than 1 m in height. The feathers of these species are reduced

Fig. 5.12 The southern cassowary (*Casuarius casuarius*), found in southern New Guinea and northeastern Australia, is the largest rain forest bird. (Courtesy of Harald Schuetz.)

to a coat of quills, which probably protects the body from scratches in the dense vegetation. Cassowaries are also noted for their yellowish casque, a horny crown on the head, which gives them a royal appearance, a bright, often purplish coloration of the head and neck, and bright red wattles. Cassowaries are well able to defend themselves, and all four species have injured people with a kick from their powerful clawed feet. Recent studies of captive cassowaries have shown that their booming calls include a very low-frequency component that may be important for communication over long distances in dense rain forest (Mack & Jones 2003). These birds can swallow large fruits whole (more than 5 cm diameter in the case of the southern cassowary in north Queensland) and even the biggest seeds pass undamaged through their guts, making them very important seed dispersal agents in these forests, which lack large mammalian frugivores (Bradford et al. 2008). The elephant birds (*Aepyornis* spp.) of Madagascar were even larger than cassowaries – the biggest species may have attained a height of 3 m – and they may have played a similar role in the dispersal of large seeds before their extinction some time within the last 2000 years (Davies 2002).

No other rain forest region has anything remotely resembling a cassowary or elephant bird, but molecular evidence shows that the most characteristic ground-dwelling birds of Neotropical forests, the tinamous (Tinamidae), are members of the same clade (Harshman et al. 2008). Cassowaries, elephant birds, and tinamous, along with the nonforest ostriches, rheas, and emus, and the kiwis and extinct moas of New Zealand, make up one of the three major lineages of birds, the paleognaths or "old jaw" birds. The present distribution of these birds strongly suggests a common origin on the southern supercontinent of Gondwana. Tinamous, however, are much smaller than cassowaries, looking somewhat like a partridge, with their plump bodies and rounded wings. Tinamous can fly, but cannot sustain flight for more than a short distance and prefer to walk or run. They often hide from danger, rather than taking flight. They are shy, unspectacular birds, which like many understorey birds are more likely to be heard than seen. Their loud, whistling calls at dusk are among the strongest and most pleasing in the Neotropical forest. Tinamous eat a lot of fallen fruit, along with seeds and invertebrates, but, unlike cassowaries, apparently grind up most seeds in the gizzard.

Pheasants and their relatives

Most tropical forests also support ground-dwelling species from the order Galliformes, which includes the familiar pheasants, partridges, grouse, and quail of temperate regions. The galliforms, along with the ducks, geese, and swans, make up a second major bird lineage, the Galloanserae, which also appears to have had a Cretaceous, Gondwanic origin (Crowe et al. 2006). Most galliforms are terrestrial, but the major representatives in Neotropical forests, in the family Cracidae, are more arboreal. This is particularly true of the guans and chachalacas, which feed at all levels in the forest, but the larger curassows generally feed on the ground. Curassows, like other cracids, appear to be mostly vegetarian, and eat considerable amounts of fallen fruits. They have a powerful gizzard that can crush even large, hard seeds, so they are probably not effective seed dispersal agents for the seeds of most species they consume (Yumoto 1999).

African rain forest galliforms include the endemic guineafowls (Numididae), the mostly African francolins (*Francolinus*) (Fig. 5.13), and the Congo peafowl (*Afropavo congensis*), but the true pheasants (Phasianinae) are largely confined to Asia. Some rain forest pheasants are exceptions to the rule that ground-feeding birds are drab. Males of several species, such as the aptly named peacock pheasants (*Polyplectron*), have spectacular tail feathers with iridescent markings. The tails are kept folded, however, except when the males show them off deliberately during sexual displays. Fruit, both fallen and plucked directly from low-growing plants, seems to be an important part of the diet of many rain forest pheasants, and the larger species, such as the great argus (*Argusianus argus*), can swallow large fruits whole. Very few frugivores occupy the understorey of Asian rain forests, so pheasants are potentially important seed dispersal agents (Corlett 1998).

Forest pheasants are confined to mainland Southeast Asia and the main islands on the Sunda Shelf. Another galliform family, the megapodes (Megapodiidae), has an almost exactly complementary distribution, being found on the eastern Indonesian Islands, New Guinea, and Australia. This distribution may not necessarily have resulted from competitive exclusion between the two groups, however, since the limits of the megapode range also coincide with the absence of some of the most important mammalian predators, such as civets and cats. Megapodes may be particularly vulnerable to ground predators because of the extraordinary ways in which they incubate their eggs: in giant compost heaps warmed by the heat of decomposition or in burrows excavated in soil that is

Fig. 5.13 The handsome francolin (*Francolinus nobilis*) is a large terrestrial bird of African rain forests, here photographed in Rwanda. (Courtesy of Richard Fleming.)

heated by the sun or volcanic activity (Dekker 2007). Megapodes provide no parental care at all for their chicks, which have to dig themselves out from up to 150 cm (60 inches) underground and then learn to find food and avoid predators without any parental guidance. All six New Guinea species are mound-builders, piling damp leaf litter and other materials to a height of 1 m or more and covering several square meters. The mounds of another megapode, the Australian brush-turkey (*Alectura lathami*), can weigh from 2 to 4 tons. In contrast, the maleo (*Macrocephalon maleo*) of Sulawesi does not construct mounds, but lays its eggs either at inland geothermal sites or on sun-warmed beaches. Bizarre breeding habits aside, megapodes resemble the other rain forest Galliformes in most other respects: they are dull-colored, medium to large birds, with a plump body and relatively small head.

More pigeons and doves

Although there are many species in the canopy (see above), the pigeons and doves (Columbidae) are also widely represented on the forest floor. Ground pigeons are particularly diverse in New Guinea, which is home to the biggest of all living pigeon species, the almost turkey-sized crowned pigeons in the genus *Goura*. The elaborate fan of feathers that forms a crest on the head is displayed during courtship bows (Fig. 5.14). New Guinea rain forests also support the magnificent pheasant pigeon (*Otidiphaps nobilis*), which not only looks like a pheasant but moves like one. Most ground-feeding pigeons are probably seed predators rather than dispersers, but the types and proportions of seeds damaged

Fig. 5.14 Victoria crowned pigeon (*Goura victoria*) on Yapen Island, New Guinea. These are the largest living members of the pigeon family. (Courtesy of Roger Le Guen.)

vary greatly between species (Corlett 1998). The emerald dove (*Chalcophaps indica*), named for its emerald-green wings, is found from tropical Asia to northern Australia; it feeds on fallen fruits and seeds with some small seeds surviving passage through its gut. In contrast, the Nicobar pigeon (*Caloenas nicobarica*), which is confined to small tropical islands in the Oriental and Australian regions, has a thick-walled stomach lined with horny plates that are used, together with swallowed stones, to grind up even large and very hard seeds.

DNA extracted from museum specimens of the extinct dodo (*Raphus cucullatus*) of Mauritius (Fig. 8.7) and the related (and equally extinct) solitaire (*Pezophaps solitaria*) of Rodrigues has been used to show that these giant, flightless ground pigeons were most closely related to the Nicobar pigeon, with this group in turn close to the crowned pigeons of New Guinea (Shapiro et al. 2002). We can only speculate on the past role of these birds in the rain forests of Mauritius and Rodrigues.

Other ground-dwelling families

In addition to these widespread families, other ground-living families are characteristic of particular regions. Pittas (Pittidae) are medium-sized, largely insectivorous suboscines of the rain forest floor from Africa to Australia, but with a center of diversity in Southeast Asia. Their often bright colors are confined, in most species, to the underparts, with the upper parts generally having cryptic patterns, so pittas are difficult birds to spot in deep shade. Rain forests in New Guinea and Australia also support the endemic insectivorous logrunners and chowchilla (Orthonychidae). Madagascan forests have the endemic thrush-sized, rail-like mesites (Mesitornithidae) and ground-rollers (Brachypteraciidae). New World rain forests have trumpeters (Psophiidae): long-legged, chicken-sized birds that are named for their loud calls. Trumpeters defend large group territories, in which they feed on fallen fruits and invertebrates.

Woodpeckers

Woodpeckers (Picidae) are the most specialized of the many forest birds that forage on the trunks and larger branches of rain forest trees (Fig. 5.15a). Woodpeckers pick insects and other invertebrates off the trunk surface and probe into holes with the aid of an extensible and often sticky tongue. Woodpeckers can also dig into the bark and wood with their powerful bills, which gives them access to a source of food – wood-boring insects – that is not available to any other bird. In addition, many woodpeckers consume considerable amounts of fruit when it is available. They also use their bills to excavate holes for roosting and breeding. Woodpeckers range in size from tiny piculets (species of *Picumnus* and *Sasia*), 10 cm (4 inches) or less in length, to giants like the crimson-crested woodpecker (*Campephilus melanoleucos*) in South America and the great slaty woodpecker (*Mulleripicus pulverulentus*) in tropical Asia, which is 50 cm long.

The woodpeckers are most diverse and conspicuous in Southeast Asia. Here, up to 16 species can coexist and several species may participate in the same mixed-species flock, which is not usually observed elsewhere (Styring & Ickes 2003). In South American rain forests, up to a dozen species of woodpecker can coexist in an area, along with a similar number of bark-feeding woodcreepers (Dendrocolaptidae). In contrast, only a few woodpecker species, none very large, are found in African rain forests and the family has not reached Madagascar, New Guinea, or Australia. In those areas, other birds pick insects off tree trunks and branches: the sickle-billed vanga (*Falculea palliata*) in Madagascar (Fig. 5.1c) and three species of riflebird (*Ptiloris*) in New Guinea and Australia. None of these birds, however, is able to excavate for wood-boring insects or to make its own nest holes. Rain forests in Madagascar and New Guinea also have specialized mammals – the prosimian aye-aye (*Daubentonia madagascariensis*) in Madagascar and the marsupial striped possums (*Dactylopsila*) in New Guinea – with morphological adaptations, including chisel-like incisors and an elongated middle finger, for extracting wood-boring insect larvae (see Chapter 3).

Holes excavated by woodpeckers are often taken over by other species of birds, such as toucans and starlings, as well as by mammals. Natural tree holes are relatively rare, particularly in young forests, and only a few other bird species can excavate their own, most notably the barbets (Fig. 5.15b) and the trogons. Thus the diversity and abundance of woodpeckers might be expected to have a strong influence on the availability of nesting and roosting sites for cavity-using non-excavators. This possibility does not seem to have been investigated in the tropics, but the diversity of hole-nesting birds and mammals in New Guinea, where both woodpeckers and barbets are absent, suggests that the relationship, if any, is not simple.

(a)

(b)

Fig. 5.15 Birds that can excavate their own nest holes in trees. (a) Pale-billed woodpecker (*Campephilus guatemalensis*) in Guatemala. (Courtesy of Harald Schuetz.) (b) Red-crowned barbet (*Megalaima rafflesii*) in Borneo. (Courtesy of Tim Laman.)

Birds of prey

Birds of prey, also known as raptors, are diverse and abundant in tropical rain forests, with a huge range of sizes and diets. The threat these birds pose to other animals in the forest is reflected in a variety of antipredator adaptations,

Fig. 5.16 Ornate hawk eagle (*Spizaetus ornatus*) in Belize. This small but powerful eagle hunts birds and mammals from a perch inside the forest. (Courtesy of Tim Laman, taken in captivity.)

including cryptic colors, vigilant behavior, and mixed-species associations, which occur in both insectivorous birds (see above) and even some forest primates. For many forest animals, avoiding being eaten by a bird of prey is at least as important an occupation as finding food to eat. Antipredator adaptations make prey very difficult to detect, so the great majority of forest raptors spend most of the day sitting motionless on a perch, watching for any movement (Fig. 5.16). For the human visitor, the frustrating result of that behavior is that neither predator nor prey is easily seen, and the forest often appears completely devoid of vertebrate life. Hiding (or membership of a mixed-species flock; see above) is not the only possible defense, however. Some species of pitohui (*Pitohui*) in New Guinea have evolved defensive toxins that are present in their skin and feathers and have a striking orange and black color scheme that may serve as a warning to birds of prey (Dumbacher et al. 2008).

Rain forest raptors belong to two families: the Accipitridae, which includes the hawks and eagles (Fig. 5.16), and the Falconidae, which includes the falcons and Neotropical caracaras. These two families belong to different bird clades so their many similarities must be the result of convergence on the same raptorial lifestyle (Hackett et al. 2008). The New World vultures (Cathartidae) are now considered closest to the Accipitridae. Eagles large enough to threaten the biggest arboreal animals in the forest are found in most rain forest areas. In the Neotropics, the formidable harpy eagle (*Harpia harpyja*) pursues monkeys through the canopy with surprising speed and agility, and will also capture sloths, opossums, and other mammals (Brown & Amadon 1968). The crowned eagle (*Harpyhaliaetus coronatus*) plays a similar role in African rain forests, where it is a major predator on adult monkeys, including the largest species, as well as taking a range of other animals, such as civets, duikers, and hornbills (Struhsaker & Leakey 1990). There have even been reports of human remains in their nests.

In Asia, the changeable hawk eagle (*Spizaetus cirrhatus*) and the Asian black eagle (*Ictinaetus malayensis*) are big enough to threaten large arboreal mammals, but the former takes mostly terrestrial prey and the latter seems to be a specialist feeder on the contents of birds' nests (Ferguson-Lees & Christie 2001). The critically endangered great Philippine (or monkey-eating) eagle (*Pithecophaga jefferyi*) may have been a primate specialist, but in its few remaining strongholds, where monkeys are scarce, it also eats colugos, civets, and squirrels, as well as large birds and reptiles. Primates do not occur in New Guinea and Australia, but the largest eagles present in these rain forest areas, the New Guinea eagle (*Harpyopsis novaeguineae*) and wedge-tailed eagle (*Aquila audax*), respectively, are capable of killing the biggest arboreal marsupials, including possums and tree kangaroos (Brown & Amadon 1968). The New Guinea eagle also preys on terrestrial mammals, such as forest wallabies, and there is one report of a small child being taken (Ferguson-Lees & Christie 2001).

In contrast, the island of Madagascar does not have birds of prey that are big enough to threaten adults of the largest primate species. The largest forest raptor in Madagascar, Henst's goshawk (*Accipiter henstii*), takes sleeping individuals of only the smaller nocturnal lemurs, but even the larger, diurnal lemurs give alarm calls in the presence of birds of prey (Wright 1998). It has been suggested that these calls have been retained from a time when a larger eagle, as big as the crowned eagle of Africa, was present, but they may also reflect a continuing threat to infants and juveniles from smaller raptors.

Most major rain forest areas also have a specialized raptor preying on snakes and other reptiles that live in the forest canopy: serpent-eagles in Asia, Africa, and Madagascar, and several species of hawks and the laughing falcon (*Herpetotheres cachinnans*) in the Neotropics. Other species in all rain forests specialize on small birds, arboreal mammals, or tree frogs and lizards, but all raptors are opportunists, prepared to take any suitable sized prey if given a chance.

At the other end of the size scale, the smallest birds of prey, such as the tiny falconets, feed mostly on insects. In the Neotropics, however, the thrush-sized tiny hawk (*Accipiter superciliosus*) appears to specialize on hummingbirds, the similar-sized bat falcon (*Falco rufigularis*) on bats captured at dawn and dusk, and the tiny pearl kite (*Gampsonyx swainsonii*) on lizards. Honey buzzards (*Pernis, Henicopernis*) are specialist predators on wasp nests in Asia and Africa, while the red-throated caracara (*Daptrius americanus*) fills the same niche in the New World. Even the most aggressive wasps reportedly stay away from this latter species, suggesting that it has some form of chemical repellent (Thiollay 1991). In the Neotropics, but not in other regions, gregarious aerial insectivores, such as the swallow-tailed kite (*Elanoides forficatus*) and plumbeous kite (*Ictinia plumbea*), hunt in flight above the forest canopy. Another feeding habit confined to this region is shown by the hook-billed kite (*Chondrohierax uncinatus*), which specializes on arboreal snails.

Scavengers

In most tropical habitats, vultures are among the most important scavengers. The birds that have evolved to fill this niche in the Old and New Worlds are not closely related, yet, seen together in a zoo aviary, are almost indistinguishable. They are all large birds, with powerful hooked bills, bare heads and necks, and massive wings used for their energy-saving, soaring flight. When it comes to tropical forests, however, the differences between the two lineages are immediately apparent: the New World vultures (Cathartidae) are the dominant scavengers in Neotropical forests, while Old World vultures are entirely absent from forested regions. The reason for this is straightforward: Old World vultures rely on their acute eyesight to detect carcasses from the air, which is useless when a dense tree canopy intervenes, while at least one genus of their New World counterparts (*Cathartes*) makes use of its well-developed sense of smell. How and why this fundamental difference in scavenging communities arose, however, is not clear.

Dying animals provide a large potential food supply for scavenging birds, but it is necessary to find a corpse before it is consumed by mammals and various groups of insects, or rendered inedible by bacteria. Neotropical turkey vultures (*Cathartes aura*) cannot easily find newly dead animals, but meat decays quickly in the tropics and they are very efficient at locating day-old carcasses. They fly low over the canopy, gaining lift from the updrafts on the windward side of emergent trees, and descending to the ground only when the smell of food is detected. The yellow-headed vultures (*Cathartes* spp.) feed in the same way as the turkey vulture. The king vulture (*Sarcoramphus papa*) appears to lack this acute sense of smell, so it flies high above the canopy, depending on the *Cathartes* vultures to locate carcasses and then following them to join the feast (Houston 1994).

Night birds

As a general rule, the day belongs to the birds in the rain forest while the night belongs to the mammals. But just as two major groups of mammals, the primates and squirrels, have become fully adapted to a diurnal lifestyle, the owls and nightjars have successfully invaded the night, with the owls becoming adapted to larger, vertebrate prey, while the nightjars and several similar families specialize on insects.

Owls (Strigiformes) (Fig. 5.17) are the nocturnal counterparts of the diurnal birds of prey, although they are not very closely related. There are two families of owls: the barn-owls (Tytonidae, 18 species) and typical owls (Strigidae, *c.* 190 species). Barn-owls (in two genera, *Tyto* and *Phodilus*) are found in rain forest only in Madagascar, Southeast Asia, New Guinea, and Australia, and on some small tropical islands. One species, the greater sooty owl (*Tyto tenebricosa*), is a rain forest specialist in New Guinea and Australia, taking arboreal marsupials up to the size of a 900 g (32 oz) ringtail possum, as well as terrestrial mammals and birds.

Little is known about the ecology of most of the typical owls of tropical rain forests, but the great range in size suggests that they are ecologically diverse, despite their similar overall appearances. The smallest are the pygmy owls or owlets, with some species a mere 15 cm (6 inches) in length. Old and New World species

Fig. 5.17 The Malagasy scops-owl (*Otus rutilus*) is widespread in the rain forests of Madagascar but its natural history is largely unknown. (Courtesy of Harald Schuetz.)

are currently placed in the same genus (*Glaucidium*), but both DNA and morphology suggest that this should be split (Wink et al. 2009). These small owls feed on insects and small vertebrates, and some species, such as the collared owlet (*Glaucidium brodiei*) of tropical Asia, kill birds as big as themselves. Some small owls, including the Amazonian pygmy owl (*G. hardyi*), are partly diurnal, and when seen during the daytime are mobbed by small birds. The larger owls are more diverse and can kill larger prey. In Madagascar, owls are important predators of small lemurs. The biggest rain forest owl in Africa is Shelley's eagle owl (*Bubo shelleyi*), which weighs more than 1 kg and can capture large, nocturnal, flying squirrels (anomalures). The closely related forest eagle owl (*B. nipalensis*) in mainland Asia can kill birds as large as a peafowl while, in the Neotropics, the similar-sized spectacled owl (*Pulsatrix perspicillata*) takes birds and mammals up to the size of an opossum, agouti, skunk or even, apparently, a sloth (Voirin et al. 2009). Unrelated owl species that specialize on fish are found in African and Asian rain forests.

The nightjars (Caprimulgidae) are a cosmopolitan and relatively uniform group, while four other nocturnal families have more restricted distributions: the oilbird (Steatornithidae) and potoos (Nyctibiidae) in the Neotropics, the frogmouths (Podargidae) from tropical Asia through to Australia, and the owlet-nightjars (Aegothelidae) in New Guinea and Australia. Although traditionally combined into the same order, Caprimulgiformes, there is no strong evidence to support a close relationship between them and the similarities may be the result of convergence. Some species spend the day hidden from predators in tree holes or dense vegetation, but most roost on branches (potoos and frogmouths) or on the ground (most nightjars), and rely on the highly cryptic (camouflage) colors and variegated patterns of their plumage to avoid detection.

The nightjars are adapted for capturing insects in flight. They have small bills but a very large gape and are able to open their mouths both vertically and horizontally. Nightjars hunt visually, and their large eyes have a reflecting tapetum behind the retina, which increases the efficiency of light gathering in low light intensities, as well as making their eyes shine like a cat's in the beam of a flashlight. The potoos have larger, hooked bills and also hunt insects in flight or from a perch. The aptly named frogmouths have large heads and massive bills, particularly in the genus *Podargus* of Australia and New Guinea. The Papuan frogmouth (*P. papuensis*) is as big as a large owl and hunts from low perches by scanning the ground for insects and small vertebrates. The owlet-nightjars have forward-facing eyes and broad, flat bills, which give them an owl-like appearance. They feed by catching insects in flight or on the ground.

The final member of this group of night birds is the bizarre oilbird (*Steatornis caripensis*), a single Neotropical species in its own family, which is the only known nocturnal frugivorous bird. Oilbirds roost and breed gregariously in caves, where they use low-resolution echolocation from audible clicks to navigate in the dark. When foraging, however, they must depend on their highly light-sensitive (but, again, low-resolution) vision and, perhaps also, their sense of smell and tactile clues from their prominent facial bristles. Oilbirds are totally frugivorous and even rear their young on an exclusive diet of low-protein fruit pulp, which leads to a very slow growth rate. They feed largely on the lipid-rich fruits of species in the laurel family (Lauraceae) and palms, which they typically pluck off in flight and swallow whole. A recent study using GPS tracking technology showed that birds make extended foraging trips away from their caves, roosting in trees during the day (Holland et al. 2009) (Fig. 5.18). The average distance between the last foraging tree and the daytime roost was 10 km, suggesting that they could potentially be the best long-distance seed dispersal agents in the Neotropics!

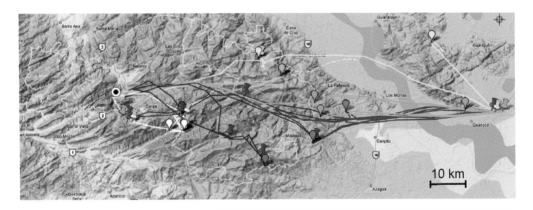

Fig. 5.18 Movements of GPS-tagged oilbirds (*Steatornis caripensis*) foraging from a cave in northeastern Venezuela. The drawing pin markers indicate foraging sites and the balloon markers roosting sites. The average distance from the last foraging tree to the roost tree was 10 km, suggesting that these birds may be the most important long-distance seed dispersal agents in Neotropical forests (Holland et al. 2009). (Courtesy of the authors.)

Migration

In the relative constancy of the tropical rain forest environment, we would expect migration to be unnecessary – and indeed, most rain forest bird species are more or less resident year round. Staying put in a permanent territory with predictable food resources, known predators, and known hiding places has obvious advantages. However, a minority of species – but billions of individual birds – migrate each year between the tropics and the temperate region (Newton & Brockie 2008). Most migrant species are apparently of tropical origin; their members fly north every summer to breed, taking advantage of the annual superabundance of insects and, possibly, lower predation risks. Fewer tropical species fly south to temperate latitudes, which cover a much smaller area of land in the southern hemisphere. Long-distance migrants are found in all rain forest regions but they are least important in Africa, where most migrants winter in the extensive grassland and savanna environments, rather than in the forest, and are most important in Southeast Asia (Fig. 5.19). Indeed, the migrant Siberian blue robin (*Luscinia cyane*) is one of the most common understorey birds in winter in the Malay Peninsula. In the Neotropics, the abundance and diversity of wintering migrants (mostly woodwarblers, tyrant flycatchers, vireos, and thrushes) in rain forests increases northwards from Amazonia, where numbers are few. In Mexico, Central America, and the Caribbean islands migrants can be a seasonally important component of the forest avifauna. In the summer breeding season, these Neotropical migrants dominate some forest bird communities in the eastern and northern United States.

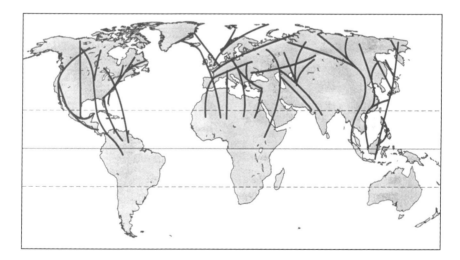

Fig. 5.19 Major bird migration flyways. Note that African flyways are north of the rain forests and that Neotropical flyways begin mainly north of the Amazon basin. New Guinea, Australia, and Madagascar are not part of the flyways. Of the five major rain forest regions, only Southeast Asian rain forests represent an endpoint for a major flyway. (From Brown & Lomolino 1998; after McClure 1974 and Baker 1978.)

Migrations also occur within the tropics on various scales. It is hypothesized that the long-distance migrations described above evolved from these more local movements. The best-known cases occur in Central America, where a number of species have been shown to make more or less regular altitudinal migrations. Most are fruit- or nectar-feeders and may be forced to move by large seasonal fluctuations in the availability of these foods. In the highlands of Costa Rica, Mexico, and Guatemala, for example, the frugivorous resplendent quetzals make seasonal elevational movements that apparently track the fruiting periods of the Lauraceae trees on which they feed. Such movements have obvious consequences for the design of protected area systems, which must encompass the full range of habitats used by such birds (Powell & Bjork 2004).

Comparison of bird communities across continents

Ornithologists have often speculated about the structure of bird communities in different parts of the world. They have been particularly interested in learning if the species in a community are organized in the same way. For example, do birds in different communities have the same range of feeding behaviors? Are the proportions of feeding specialists and generalists similar? The problem with making such comparisons is that field biologists in different parts of the world have typically gathered and analyzed their data in different ways.

A pioneering study

In order to compare bird communities in different rain forests accurately, it is necessary to use the same methods in each place. So far, the most comprehensive attempt to compare rain forest bird communities across continents was undertaken by ornithologist David L. Pearson during the 1970s (Pearson 1977). Using the same methods, he studied bird communities at six sites: three in western Amazonia, in Ecuador, Bolivia, and Peru, and one each at Kutai in Indonesian Borneo, Maprik in Papua New Guinea, and at Makokou in Gabon in Central Africa. At each site 200–700 hours were spent observing birds and recording foraging heights and foraging techniques used, such as gleaning an insect from a plant surface, sallying to catch an insect on the wing, snatching prey from a surface, pecking and probing at bark to get at hidden prey, and eating fruit. The study was conducted during daylight hours, and did not include raptorial birds, such as hawks, or nighttime species, such as nightjars. The observations were made along a circular path 2.5–3 km (1.5–2 miles) long. By observing birds within 25 m (80 feet) on each side of the path, a total of around 15 ha (35 acres) was covered for each forest.

While this study is unique in comparing data on bird communities in six countries using the same methods, it also has several major problems. Only one 15 ha plot was examined at each site, and just one plot was used to represent the entire African and Asian continents and the island of New Guinea. No replicates were used to determine if these plots were typical or abnormal. The observations at each site were made over a period of several months in only one year. It is unknown if the period of observation was typical for the year, or if the year itself was typical.

Key results

Despite the limitations, Pearson's study provided many valuable insights, some of which highlight the regional differences discussed earlier in the chapter. Each of the plots contained a surprisingly high percentage, 53–70%, of the forest birds known to occur in the area. The 15 ha plot in New Guinea, for example, contained 70% of the forest bird species known from the area. The overall richness of birds varied widely among plots, with more than 125 species for each of the three Amazonian plots, and fewer than 90 species for the New Guinea plot (Fig. 5.20).

In analyzing the patterns of foraging in the six forests, the most striking observation is how similar foraging techniques were across forests. The most common foraging technique was gleaning, followed by fruit eating, and then sallying (i.e. taking short flights from a perch to capture insects). Snatching prey and probing/pecking were methods of intermediate frequency, with flower visiting and army ant following less common. Some differences among regions were explained by the presence or absence of particular bird families. The foraging technique that varied most among forests was probing/pecking. At Amazonian sites, where probing/pecking is common, the technique is employed by woodpeckers (Picidae) and woodcreepers (Dendrocolaptidae). Woodcreepers are absent from the Borneo and Gabon plots where probing/pecking was less common and was exhibited primarily by woodpeckers. Both woodcreepers and woodpeckers are absent from New Guinea; at this site, probing/pecking was used at a very low rate only by the world's smallest parrot, the 10.5 g buff-faced pygmy parrot (*Micropsitta pusio*). The technique of following army ant swarms occurred in the South American and West African plots, which had specialized antbirds and huge army ant swarms. This technique was not observed in plots in Borneo and New Guinea, which lack large swarms of foraging army ants.

Certain patterns of bird foraging behavior can be related to the co-occurrence of other groups of animals, most notably mammals. For example, fruit-eating behavior is commonest in New Guinea, where there are no primates or squirrels to compete for fruit. Around 50% of the bird species in New Guinea eat

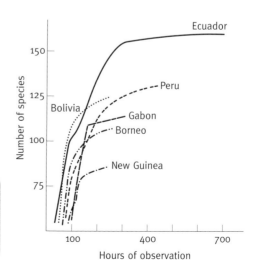

Fig. 5.20 Cumulative number of bird species found at six field sites after a specified number of hours of observation. (From Pearson 1977.)

some fruit, with 19% of the species observed feeding on fruit more than 90% of the time. In contrast, only around 25% of bird species ate fruits at the Bolivia, Borneo, and Gabon sites. The presence of fruit-eating birds was lowest in Gabon, where there are also eight species of fruit-eating squirrels, six species of fruit-eating ruminants, and 13 species of primates.

Conclusions and future research directions

The study by Pearson (1977) illuminated many significant differences in bird communities between the six forests. The differences may have been due to the fact that the forests have different families of birds. An alternative possibility is that the species in the bird communities were responding to competition with members of the mammal communities, principally the abundant squirrels of the Old World and the varying abundance of primates. Also worth considering is the role of the dipterocarp family and multiyear fruiting cycles in reducing the abundance of animal prey, such as insects, in average years in Southeast Asian forests. Furthermore, the forest height and climates of each of the forests differ, with a pronounced dry season and lower forest in Gabon and everwet conditions and taller forest in Borneo. Lastly, hunting by people had depressed the populations of large animals in Ecuador and New Guinea to levels below the other sites, affecting the community structure of the entire forest.

What can be said is that Pearson's groundbreaking study described major differences among six widely separated rain forests. It is up to a new generation of tropical ecologists to expand his approach to include more sites over a period of several years, and to include in the study key groups of mammals such as primates and squirrels. The relative abundance of basic building blocks of the food chain, such as new leaves, flowers, fruit, and insects, also needs to be tracked. Such a cross-continental approach could help to reveal the relative importance of evolution, biogeography, and current environmental conditions in explaining differences in bird communities. These studies could also provide insight into, for example, the structure of mixed-species flocks in different rain forests rather than, as tends to happen at present, simply extrapolating Neotropical studies to the rest of the world.

Such cross-continental studies also need to be combined with investigations of the human impact on bird communities. In many areas of the world, large birds have been intensively hunted by traditional people for food, and some brightly colored birds have been hunted for their feathers. With the arrival of guns, the intensity of this collection has increased. Large frugivorous birds, such as hornbills or cassowaries, are the only dispersal agents for some of the largest fruits, so their loss may have a long-term impact on the plant community. Bird populations have also been decimated by the introduction of diseases and predators, particularly on tropical islands such as Hawaii and Guam (see Chapter 8). So although we need investigations of bird communities, we also need to be aware that the bird communities we see today may not represent the original structure of the community, but may rather reflect species that are most resistant to the ever-increasing human impact.

Moving beyond observations to experiments on their ecological roles may be particularly difficult with forest birds. Experimental manipulations of rain forest bird communities are likely to be both difficult and controversial, while comparisons between intact bird communities (if they exist) and those impacted by hunters are usually confounded by simultaneous impacts on other components of the rain forest community, such as mammals or plants. The exclusion of all birds from branches or small trees with netting is possible (Van Bael et al. 2003), but selective exclusion is probably not practical. The best approach may be to study the impact of reintroducing key bird species to forest areas from which they have been eliminated by hunting. Such studies would obviously require the cooperation (or exclusion) of hunters, but could have major educational and conservation benefits, as well as helping to understand the role of particular bird species in the rain forest community.

Chapter 6

Fruit Bats and Gliding Animals in the Forest Canopy

Birds are the most conspicuous group of animals in the forest canopy, but there are also other animals flying or gliding through the forest. These animals are remarkably different in each forest region, with less evidence for the convergence in form and function that we have seen in rain forest bird communities. Some of these differences appear to be related to patterns of food availability – a key concern for highly mobile animals – while others may reflect differences in forest structure. Some may be simply the results of biogeographical accidents. Bats are one such group of highly mobile animals, replacing the birds as the dominant nighttime vertebrates in the air above and inside the forest canopy. All rain forests have bats and on all but the most isolated of oceanic islands there are species that eat fruits and disperse their seeds. Another group is the gliding vertebrates, which are principally found in Asian forests. These animals use their gliding ability to move widely through the forest without having to descend to the ground. Gliding animals have different diets, but they all solve the challenge of getting enough to eat by moving among many trees rather than staying mostly in one place.

Fruit- and nectar-feeding bats

Bats are the most species-rich group of mammals in tropical forests and they are surprisingly well studied. Many biologists find them fascinating because of their unique adaptations for flight and foraging at night. Although the ancestors of all bats were probably insectivores, modern bats have a huge range of diets: from fish and small terrestrial vertebrates, to blood, insects, nectar and pollen, fruit, and leaves. Here, we consider only the fruit- and nectar-eating bats. Confined to the tropics and subtropics, these plant-visiting bats make up about a third of the world's 1100 or so bat species. There are two main reasons for this focus. First, these bats play an important role in all rain forests as pollinators and seed dispersal agents. In island rain forests, where bats are often the only native mammals, they are keystone species on which the long-term survival of many plant species depends. Second, these bats provide one of the clearest examples of convergent evolution, where unrelated groups of organisms have evolved to play similar ecological roles in the rain forests of the Old and New Worlds.

Fruit bats are abundant in all rain forests, except on the most remote Pacific islands, but their conspicuousness varies between regions. In the Old World tropics, before the sun has set, the first detachment of fruit bats appears on the horizon. High overhead, hundreds fly fast and purposefully over the village and head toward the forest. Their huge size recalls a more prehistoric scene; with wings that stretch up to 1.8 m (6 feet) across, silhouetted against the darkening sky, they look more like pterodactyls than mammals. This nightly ritual as the fruit bats leave their daytime arboreal roosts (Fig. 6.1) prompts not so much as an upward glance by the villagers. The large trees that were black with roosting bats during the day are soon vacated. Although the combined effects of hunting and habitat loss are making such stirring sights far less common, this evening scene is still the most popular image of fruit bats in the Old World tropics and can still be seen from Africa and Madagascar, to India, across Southeast Asia to Indonesia, New Guinea, Australia, and out into the Pacific. Yet such imagery is as alien to the New World tropics as it would be to the temperate zones. New World fruit bats are all considerably smaller in size, do not roost in huge tree colonies, and belong to an entirely different group of bats. Although fruit bats in the Old and New World are united by their consumption of fruits and nectar, they are evolutionarily and ecologically worlds apart.

Fig. 6.1 Daytime roost of straw-colored fruits bats (*Eidolon helvum*) near Accra, Ghana. (Courtesy of Emily Babin.)

Fig. 6.2 Spectacled flying fox (*Pteropus conspicillatus*) in Madang, Papua New Guinea. (Courtesy of Milan Janda.)

In the Old World, all the plant-visiting bats belong to the family Pteropodidae (Figs. 6.1, 6.2), now grouped with the rhinolophoid insectivorous bats in the exclusively Old World suborder Pteropodiformes (or Yinpterochiroptera). All pteropodids – commonly known as the "megabats" – are herbivorous, and most depend on fruit and a variable amount of nectar and young leaves for their diet. All the nectar specialists have traditionally been placed in a separate subfamily, Macroglossinae, but molecular studies have shown that their long narrow muzzles, reduced teeth, and protrusible, brush-tipped tongues are the result of convergent adaptations to nectar feeding that have arisen more than once among the megabats. Pteropodids range in size from around 13 g to 1.5 kg (0.5 oz to 3.3 lb).

The fruit bats of the Neotropics, in contrast, are all members of the family Phyllostomidae (Figs. 6.3, 6.4) in the pantropical and predominately insectivorous suborder Vespertilioniformes (or Yangochiroptera). In addition to frugivores, the Phyllostomidae also includes species feeding on insects, blood, small vertebrates, nectar, or omnivorous mixtures of these items. The diets of most phyllostomids are poorly known, but feeding strategies seem to be flexible in many species, with the proportion of fruit and nectar varying seasonally. Fruit-eating phyllostomids range in size from 5 to 100 g. The most recent common ancestor of the two bat suborders, and thus of the two families containing fruit- and nectar-feeding bats, lived around 65 million years ago (Teeling et al. 2005) and was probably insectivorous. Thus the fruit- and nectar-eating habit has evolved entirely independently in unrelated bat families in the Old and New World.

Fig. 6.3 A typical New World fruit bat, *Artibeus* sp., roosting under a leaf in Tortuguero National Park, Costa Rica. (Photograph by Leyo, Wikimedia Commons, http://upload.wikimedia.org/wikipedia/commons/1/12/Artibeus_sp._Tortuguero_National_Park_crop.jpg.)

Relative sizes

Comparisons between New and Old World fruit bats usually focus on the much greater sizes of the latter. This is not surprising: for a tropical ecologist from the New World, a flock of giant fruit bats is one of the "must-see" sights, along with elephants, apes, and hornbills – other taxa where the Old World representatives are relative giants. This focus on relative size, however, has tended to overshadow both the similarities between the two groups and some other, equally significant, differences. The frequency distribution of body sizes is in fact bimodal in pteropodids, with one peak between 100 and 200 g – heavier than any frugivorous phyllostomid – but another below 20 g, coinciding with the unimodal peak in the phyllostomids (Muscarella & Fleming 2007). These smaller Old World fruit bats have received much less research attention than their larger and more conspicuous relatives, and thus tend to get relatively neglected in comparisons. It is important therefore that we distinguish clearly between differences that are simply a consequence of size and those that have other, more complex, explanations.

Patterns in species diversity

The overall species diversity of the pteropodids and phyllostomids is very similar. The pteropodids include around 190 species in 42 genera, of which 15% are African and the rest are distributed from Asia to New Guinea, Australia, and the western Pacific. All but the dozen or so nectar specialists are predominantly frugivorous. There are around 170 species of phyllostomids distributed within 56 genera, but while most phyllostomid species eat at least some fruit, only around half are predominantly frugivorous.

Although the global diversity of pteropodid frugivores is higher than that of phyllostomid frugivores, the story is reversed at the local level. The number of coexisting frugivorous bat species is at least 2–3 times greater in Neotropical rain forests than it is in the Paleotropics. In lowland rain forest in the Yasuní area of eastern Ecuador, 46 phyllostomid species were captured within a radius of 1.5 km, including 27 frugivorous and 7 nectarivorous species (Rex et al. 2008). Some rare species were certainly overlooked in this short survey and total phyllostomid diversity for the site was estimated at an amazing 72 species, of which more than half are expected to be frugivores. In contrast, the most diverse sites in Africa, Asia, and New Guinea rarely support as many as a dozen species of pteropodids and the large island of Madagascar has only three. How can we reconcile the greater local species diversity of frugivorous phyllostomids with the greater global diversity of frugivorous pteropodids?

The greater global diversity of pteropodid fruit bats partly reflects their wider global distribution, throughout tropical and subtropical Africa, Asia, New Guinea, Australia, and the western Pacific, while phyllostomids are confined to the Americas. More important, however, is the fact that the distribution of Old World fruit bats is not continuous. Not only was there a largely separate radiation of fruit bats in Africa, but also many pteropodid species are confined to islands (Muscarella & Fleming 2007). Phyllostomid species, in contrast, are often widespread throughout the mainland Neotropics and fewer species are associated with islands. The diverse local communities of phyllostomids thus include many widespread species, while the relatively species-poor pteropodid communities contain few.

Plant resources for bats

The much greater local diversity of phyllostomids has usually been attributed to a greater diversity and reliability of food resources for plant-feeding bats in the New World. This in turn is thought to have allowed much greater specialization in the bat fauna. In an 8-month study in secondary vegetation at La Selva Biological Station in Costa Rica, all but one of the 15 species of frugivorous bat captured had a diet dominated by one or two plant genera (Lopez & Vaughan 2007). This appears to reflect a general pattern, with most phyllostomid species specializing, at least locally, on a plant species or genus that produces a year-round food source (often *Cecropia, Ficus, Piper, Solanum,* or *Vismia*). Such dietary specialization reduces food overlap and is expected allow more species to coexist in any one locality. The variable nonfruit – usually insect – component in the diet of many frugivorous phyllostomids may have also contributed to greater specialization in the fruit component of the diet, since the bats can fill gaps in the supply of their preferred fruits – or in their nutritional value – by consuming more invertebrates. Pteropodids, in contrast, do not usually eat insects and tend to be "serial specialists," concentrating on one plant species at any one time, but eating a wide range of plant species, genera, and fruit types over the course of a year.

One major difference between rain forest regions is the greater diversity of fleshy-fruited shrub species in Neotropical rain forests, in comparison with rain forests elsewhere, particularly those in Southeast Asia (see Chapter 2). These understorey shrubs (such as the many species of *Piper*) provide food for an entire guild of frugivorous bats in the Neotropics in a way unmatched in Asia and Africa. At Yasuní, although canopy frugivores were most abundant, understorey frugivores were most diverse, while rain forests in Peninsula Malaysia support only a single understorey specialist (Hodgkison & Kunz 2006). Moreover, even in the forest canopy, where most Malaysian fruit bats feed, fruit production is dominated by the wind-dispersed fruit of dipterocarp species and resource scarcity is compounded by the multiyear seasonality of so many non-dipterocarp species, as described in Chapter 2. Many Neotropical epiphytes (particularly species in the families Araceae and Cyclanthaceae) are also dispersed by bats (Lobova et al. 2009), while this is rare in the Old World. Bat-dispersed shrub and tree species also contribute a major component of secondary forests in the Neotropics, both human-made and those resulting from natural disturbances, and many such species fruit continuously (Muscarella & Fleming 2007). Secondary forests in the Old World are usually dominated by bird-dispersed plants. Thus annual consistency in the availability of fruits is thought to have promoted dietary specialization in the Neotropical bat fauna, while the more generalized feeding habits of Old World fruit bats may be a response to the patchiness of the food supply in space and time in Old World forests.

Over evolutionary time, specialization on food sources has become reflected in the varied morphologies of the skull, jaws, and teeth in the New World phyllostomids, in comparison with the relative conservatism of most pteropodids (Dumont 2003). In general, New World fruit bats have short, wide faces and a nonprotrusible tongue; sharp serrated teeth are used to cut fruits into small bites. Within this general pattern, New World phyllostomids exhibit a broad range of skull and jaw shapes, suggesting that they have become specialists on particular types of fruit or other food items (Dumont et al. 2009) (Fig. 6.4). In contrast, the jaws of most pteropodids are longer and allow bigger bites with a wider gape than phyllostomids; the cheek teeth are relatively blunt. Differences in morphology have also resulted in big differences in the way fruit bats process the fruits on which they feed. Nearly all pteropodids, as well as those phyllostomids that feed on figs or relatively fibrous canopy fruits, chew fruits slowly and thoroughly, swallowing the juices while pressing the fibrous remains of the pulp and most seeds into a pellet ("spat") which they then spit out (Dumont 2003; Muscarella & Fleming 2007). In contrast, most understorey phyllostomids and, apparently, the pteropodid *Syconycteris*, consume relatively non-fibrous fruits, swallow all or most seeds, and do not produce spats.

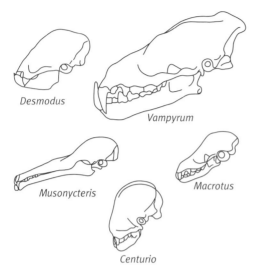

Fig. 6.4 A diversity of skulls from New World phyllostomid bats, illustrating adaptations for feeding on different food types. *Musonycteris* is the most extreme nectarivore, probing into flowers with its long snout. *Centurio* is the most extreme frugivore and probably eats overripe fruit. *Vampyrum* is a large carnivore, eating birds and bats rather than insects. *Desmodus* is a blood-feeder (sanguinivore), while *Macrotus* is a more generalized feeder, eating mostly insects and fruits, and sometimes a bit of nectar. (Courtesy of Patricia W. Freeman.)

Flying behavior

Flight efficiency is related to the shape of the wing. Long, narrow wings give a high aspect ratio and insure energetic efficiency over long distances. Flight speed depends on the relationship between body mass and wing area. A relatively heavy body with a small wing area results in high wing loadings and fast flight. In general, the rain forest pteropodids tend to have high aspect ratios and wing loadings, and to be fast, efficient fliers. Many species travel considerable distances to and from their communal roosts every night. In the extreme case, the large flying fox (*Pteropus vampyrus*) from Southeast Asia has a wingspan of 1.7 m (5.5 feet) and can make nightly foraging trips of more than 50 km (30 miles) at speeds of 65 km per hour (Epstein et al. 2009). This species moves frequently between widely separated roosts and all four male bats satellite-collared at a roost in southern Peninsular Malaysia had ranges that included parts of Sumatra, with one bat moving 363 km in 4 days (Fig. 6.5). Moreover, several species in diverse genera of Asian and African pteropodids have populations that migrate seasonally in those parts of their range where there are distinct wet and dry seasons. One satellite-tracked 300 g individual of the straw-colored fruit bat *Eidolon helvum* (Fig. 6.1) in Central Africa moved 370 km (230 miles) in one night (Richter & Cumming 2008)! These examples are extreme, but many other pteropodids species make are known to fly long distances. However, some small, forest-dwelling pteropodid species have small home ranges. The spotted-winged fruit bat (*Balionycteris maculata*) is one of the smallest pteropodids (13–15 g) and the only one exclusively associated with the forest understorey in Peninsula Malaysia (Hodgkison & Kunz 2006). In contrast to the species mentioned above, it has short, broad wings and highly maneuverable flight. It roosts in small harem groups in hollowed-out ant nests or epiphytic ferns, and forages within a 1 km radius of the roost.

In the Neotropics, the greater diversity of foods available to the phyllostomids, in combination with the greater availability of fruits in the forest interior,

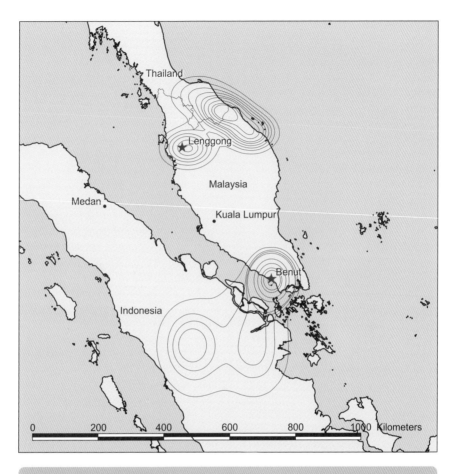

Fig. 6.5 Home ranges of seven male *Pteropus vampyrus* based on satellite telemetry data. Bats are expected to spend 90% of their time within the outermost ring. (From Epstein et al. 2009.)

has had two consequences. First, there is less need for long-distance flight capabilities and no rain forest phyllostomid is known to fly anything like the distances covered by pteropodids. The fact that a nectar-feeding phyllostomid, *Leptonycteris curasoae*, commutes 25–35 km a night to visit the flowers of cacti in the Sonoran Desert, as well as performing seasonal migrations, shows that the lack of such movements in rain forest species reflects resource availability rather than an inability of New World bats to evolve such behavior (Horner et al. 1998). At the same time, rain forest phyllostomids need to be maneuverable in order to forage within the cluttered environment of the forest interior for fruits and floral nectar (Dumont 2003). So the primary selection pressure on the wing is for maneuverability, which relates inversely to both aspect ratio and wing loading. Thus what we expect to see are shorter, broader wings (lower aspect ratio) and lower wing loading (smaller body size). And this is indeed what is found in these bats.

Foraging behavior

A dispersed food base and highly efficient flight has promoted colonial roosting and flock foraging in many Old World fruit bats. Nearly two-thirds of pteropodid species are communal roosters, and colonies in some species can number several thousands of bats. These colonies are predominantly tree roosting and sometimes are found in urban areas if they are protected from hunting. One conspicuous colony is found on the campus of the Indian Institute of Science in the heart of Bangalore in southwest India. In addition, at least 27 species use caves (Marshall 1983).

Some of these cave-roosting bats, members of the genus *Rousettus*, are the only Old World fruit bats known to echolocate. They do so by tongue-clicking, a broadband, low-frequency method that is useful for orientation within dark caves, but it is not clear whether these bats use echolocation in foraging (Raghuram et al. 2009). However, the great majority of pteropodid bats locate fruits and flowers by vision and smell. Old World fruit bats have large eyes and a correspondingly large visual cortex (Altringham & Fenton 2003). In many Old World plant species that target bats, the flowers or fruits are produced in a way so they can be readily detected and used by bats. The flowers and fruits often extend beyond the leaves on branches or are produced on the trunk, and they are often light in color to contrast with the foliage (Dumont 2003). Many of the larger Old World species begin to forage before sunset, both to take advantage of the light during foraging and because of the long commuting time to food sources. They often land, at least briefly, on the plants where they are feeding. For large species, such as *Pteropus*, this may compensate for lack of maneuverability in flight, since a bat can land on the outer branches of a tree and then climb around inside the canopy to find the fruits. Communal roosting may allow bats to share information about the locations of short-lived but superabundant feeding areas. This behavior also makes it possible for individuals to forage together as a flock, further increasing the awareness of potential foods as well as reducing the potential dangers of predators such as hawks. Single-species flocks are most common, but flocks with two or more species of bats are known to visit bat-flowers in West Africa.

In Neotropical bats, foraging within the forest is further facilitated by the use of echolocation for orientation. Vision and olfaction are also used during foraging, as in Old World fruit bats, but at least some Neotropical bats can, if necessary, forage in complete darkness using echolocation alone. Many Neotropical species begin foraging only 1 or 2 h after sunset – considerably later than most pteropodids. Most species take fruits or nectar in flight – often hovering briefly – rather than landing on the plant. These bats can visit plants along a regular route or trapline, in contrast to most of the larger Old World bats that tend to concentrate at large food sources.

Bats as pollinators and seed dispersal agents

Bats are effective but expensive pollinators. They can carry large pollen loads for long distances between widely separated plants, but these flights must be fueled by large volumes of nectar. Not only is the reward expensive, but the flowers themselves must be larger and more robust than flowers targeted at pollinating insects, such as bees. The greater importance of bat pollination in the Neotropics than elsewhere must surely, at least in part, reflect the lower costs of attracting hovering visits from tiny (6–25 g, 0.2–0.9 oz) phyllostomid nectar specialists than accommodating the generally larger and less maneuverable pteropodids. Many Neotropical epiphytes, herbs, and climbers produce bat-pollinated flowers that are smaller and more delicate than the flowers of the trees and shrubs pollinated by Old World bats (Tschapka & Dressler 2002). Interestingly, many of the plants producing flowers targeted at Neotropical nectar specialists belong to endemic or near-endemic families, such as the Bromeliaceae, Cactaceae, and Marcgraviaceae. Many appear to have evolved from hummingbird pollinated ancestors, suggesting that bats may be better pollinators, perhaps because fur can hold more pollen than feathers (Muchhala & Thomson 2010).

Even in the Neotropics, however, there are larger bat-flowers, often produced by trees, which are targeted at the many larger phyllostomids that include some nectar in a more omnivorous diet (Tschapka & Dressler 2002). These flowers look (at least, to human eyes) very like those visited by Old World pteropodids and belong to many of the same families, including the Bignoniaceae, Bombacaceae, Leguminosae, and Lythraceae. The readiness with which pteropodids visit the flowers of Neotropical bat-plants grown as ornamentals in the Old World, and with which phyllostomids visit Old World bat-plants planted in the Neotropics, suggests a universality to the bat-pollination syndrome that is surprising in view of the very different sensory systems of the two bat families.

Bats are often very wasteful seed dispersal agents, depositing most seeds from the fruits they eat under their day roosts or in temporary nighttime "feeding roosts" near the fruiting plants. Seeds that are small enough to be swallowed, however, are scattered widely as bats defecate in flight, in contrast to birds, which usually defecate from perches. The size threshold for swallowing seems to be around 2–3 mm diameter for most pteropodids (Corlett 1998), but is more variable among phyllostomids, some of which swallow and defecate much larger seeds. Wide dispersal across open areas is particularly important for the pioneer shrubs and trees that invade natural or human-made clearings, so it is not surprising that many of the most important Neotropical pioneers are dispersed primarily by bats, including the many species in the genera *Cecropia, Muntingia, Piper, Solanum,* and *Vismia*. What is surprising is that bats are generally much less important than birds as dispersers of woody pioneers in the Old World, although there are a number of important exceptions, such as *Adinandra dumosa* in Singapore. Where Neotropical pioneers, such as *Cecropia* spp., *Piper aduncum*, and *Muntingia calabura*, have been introduced to the Old World, they are rapidly adopted by the local bats, making the paucity of native bat-dispersed pioneers even harder to explain (Fig. 6.6).

Bats have traditionally been viewed as benign seed dispersers, which drop or defecate seeds without damaging them. It is becoming apparent, however, that this is not necessarily true for all species of bats and all seeds. *Pteralopex*, a pteropodid from the Solomon Islands with particularly large and complex teeth, cracks the nuts of *Canarium* species (Flannery 1995b), while two species of *Chiroderma* in Brazil have been shown to destroy the tiny seeds of figs by separating them in the mouth and then audibly crunching them up (Nogueira et al. 2005). As these examples make clear, there is a long way still to go before we have a complete understanding of the complex relationships between bats and plants.

Fig. 6.6 *Cecropia pachystachya* dominating secondary regrowth on wasteland in Singapore. The fruits of this South American pioneer genus are dispersed by New World fruit bats within their native range and by unrelated Old World fruits bats, as well as birds, in Southeast Asia. (Courtesy of Fam Shun Deng.)

Gliding vertebrates

Despite the striking differences between the Old and New World communities, fruit bats are found in all the major rain forest areas. In contrast, if you see animals gliding through the rain forest during the day that are not birds, you are probably in Southeast Asia. This region has a unique abundance and diversity of gliding animals not seen elsewhere. The curious puzzle of gliding animals in Southeast Asia is potentially very revealing of continental differences in rain forest ecology. Southeast Asian forests are inhabited by over 60 species of gliding animals, included in 16 genera, representing at least six independent evolutionary origins of this remarkable trait and suggesting a common ecological pressure. These gliding animals include flying squirrels, flying lemurs (colugos), flying *Draco* lizards, flying geckos, flying frogs, and most incredibly, flying snakes (Dudley et al. 2007).

The 45 of so species of flying lizards (*Draco* spp.) are 12–25 cm (5–10 inches) long and can soar from tree to tree by expanding their ribs to make a brightly colored fan (Fig. 6.7a,b). This fan, along with an expandable pouch on the chin of the males, known as a dewlap, is part of the territorial display of these species. These lizards are highly maneuverable and social, with males gliding around a single tree to find a better spot to make a territorial display. Even though these species are almost exclusively arboreal, feeding on ants, termites, and other insects found on the tree, they descend to the ground to lay their eggs. These lizards are by no means rare, and as many as seven species can coexist in one area of forest.

Fig. 6.7 Gliding lizards (*Draco cornutus*) from Danum Valley, Sabah, Malaysia. (a) Lizard gliding between trees. (Courtesy of Tim Laman.) (b) Close-up of a lizard with its ribs spread out to form a gliding surface. (Courtesy of Tim Laman.)

Gliding is their normal means of getting round a home range that includes several trees. The larger species are poorer gliders than the small species, because they retain the same general shape, and body mass increases with the cube of linear dimensions while the wing area increases only with the square (McGuire & Dudley 2005). The large species only occur in areas with smaller species, suggesting that size gives a competitive advantage that makes up for reduced gliding ability.

The flying geckos (*Ptychozoon* spp.) have flaps of skin along their head, neck, body, and tail, and webbed feet. These geckos are completely arboreal, feeding on insects and laying eggs on the tree bark. When pursued, they jump off and glide to the ground or a lower perch. Another genus of Southeast Asian gecko, *Cosymbotus*, has two species that glide in a similar fashion, as apparently do at least some species of *Luperosaurus* (Dudley et al. 2007).

The five species of flying snakes in the genus *Chrysopelea* flatten their bodies to catch the air after leaping from a branch and then undulate as if they were swimming (Socha et al. 2005). The body width approximately doubles while gliding and the ventral surface becomes slightly concave. The undulations are thought to contribute to lift, but the precise mechanism is not clear. These snakes can maneuver in flight to avoid obstacles and attain glide angles on a par with those of flying squirrels. They are diurnal, exclusively arboreal, and feed on small vertebrates, particularly lizards. They are known to prey on *Draco*, raising the possibility that their respective gliding abilities have been subject to an evolutionary arms race!

The most mobile gliding vertebrates are the giant flying squirrels of the genus *Petaurista*, which are said to glide up to 450 m (1500 feet) between trees. They are nocturnal and arboreal, feeding on young leaves, fruit, nuts, twigs, and insects. When fully extended for gliding, these animals are 1 m long. Cartilage extends from their wrists to expand the skin flaps even further. Giant flying squirrels can forage over several kilometers during a night to visit widely scattered food sources. Several smaller species of flying squirrel coexist with these giants, with a total of 15 species in Borneo alone.

There are also at least two strange species in their own order, the Dermoptera (meaning "wings of skin") (Byrnes et al. 2008). Known misleadingly as flying lemurs – which they are not – or, better, as colugos, the species are *Galeopterus variegatus*, a widespread animal of Southeast Asia that should probably be divided into several species, and the more restricted *Cynocephalus volans*, found in the southern Philippines. Colugos are found from the lowlands to the mountains, in primary and secondary forest, and even in rubber gardens and coconut plantations. Molecular studies suggest that the colugos are the closest relatives of the primates. Weighing between 0.75 and 1.75 kg (2.0–3.8 lb), colugos have a large extensible skin membrane that stretches along the neck and sides of the body to the fingers and toes as well as going to the tail. This allows the animals to glide up to 150 m between trees in search of young leaves to eat. Over the course of an evening, an animal might cover a distance of 1 or 2 km. In flight, the nocturnal animals stretch out their membranes to form a dark square kite, with the head in the middle of the leading edge. Their superiority as gliders is demonstrated by the ability of colugo mothers to carry their infants while foraging. When resting from flight, colugos hang upside down on branches wrapped in their membranes, resembling shy lemurs enfolded in large fur blankets. Their relative clumsiness on the ground, as well as their immediate tendency to climb the nearest tree, bears testament to their tree-dwelling nature.

The animals described here are considered gliders because they all have structures that increase their surface area, allowing them to be better airfoils, but they lack the ability to self-propel. The squirrels, flying lemurs, lizards, and geckos are true gliders that can alter direction in flight, brake to slow down, and make a heads-up landing on a tree trunk. The snakes can glide and maneuver as well as the other gliders, but they land on foliage and branches rather than tree trunks. The flying frogs of the genus *Rhacophorus* are the weakest gliders, using additional air surfaces provided by their long webbed fingers and toes to slow descent and prevent tumbling (Fig. 6.8a,b). Even so, flying frogs can alter the angle of their feet and hands, changing direction in mid-flight, and even glide from tree to tree using adhesive pads to stick on the landing site. These frogs spend their lives in treetops feeding on insects, except when they descend to the ground to mate and lay eggs in pools of water, often the wallows of wild pigs.

This abundance of gliding animals in Southeast Asia contrasts with other rain forests. The outlying Asian rain forests of the Western Ghats in India have two species of flying squirrel, a *Chrysopelea* flying snake, a *Draco* flying lizard, and a *Rhacophorus* gliding frog. The rain forests of New Guinea have only one widespread glider, the tiny marsupial sugar glider (*Petaurus breviceps*) (Fig. 6.9). Africa has six species of gliding rodents, the squirrel-like anomalures, two species of flying lizard (*Holaspis*: Vanhooydonck et al. 2009), and a gliding frog (*Hyperolius castaneus*). Madagascan rain forests lack gliders altogether. In the New World, numerous species of hylid frogs show some ability to control their aerial trajectories, but only a few with well-developed webbed feet and folds along their limbs appear to glide well. This paucity of gliding ability in South American animals is particularly surprising, considering that this continent has some squirrels and a great abundance of lizards, snakes, frogs, and rodents.

Why are gliding vertebrates abundant only in Southeast Asian forests?

What could be the explanation for the great diversity of gliding animals in Southeast Asia and their almost complete absence from South America? The repeated evolution of gliding species in Southeast Asia suggests that there has been some constant ecological pressure driving this process over the millions of years needed to produce such adaptations. Gliding has generally been viewed as either a means of escaping predators, by allowing animals to move between trees without descending to the ground, or as an energetically efficient way of traveling long distances between scattered food resources. But what is special about Southeast Asian rain forests?

Scientists have proposed various theories to explain the diversity of gliding animals in Southeast Asia (Laman 2000; Dudley et al. 2007). The first theory might be called the tall trees hypothesis. The forests of Southeast Asia are taller than forests elsewhere (see Chapter 2), due to domination by tall trees in the dipterocarp family and also, possibly, lower wind speeds, so there is a more advantageous situation for gliding between trees. Taller trees perhaps allow for longer glides and the opportunity to build up speed in a dive before gliding. This argument has several flaws, however. First, gliding animals are found throughout the Southeast Asian region, even in relatively short-stature forests found in the northern range of the rain forest in China, Vietnam, and Thailand. Some

(a)

(b)

Fig. 6.8 Wallace's flying frog (*Rhacophorus nigropalmatus*) has large webbed feet.
(a) Illustration based on a sketch by Wallace in his 1869 book, *The Malay Archipelago*.
(b) Photographed in Danum Valley, Sabah, Malaysia. (Courtesy of Tim Laman.)

Fig. 6.9 Sugar glider (*Petaurus breviceps*), a gliding marsupial, in Australia. (Courtesy of Harald Schuetz.)

species occur in the understorey of tall forests. Some gliders also thrive in low secondary forests, plantations, and even city parks. Clearly, gliding animals do not all require tall trees for their activities.

A second theory, which we might call the broken forest hypothesis, speculates that the tree canopy in Southeast Asian forests has fewer lianas (woody vines) connecting tree crowns than Neotropical and African forests. As a result, animals must risk descending to the ground or glide to move between trees. In addition, the tree canopy is presumed to be more uneven in height in Asian forests due to the presence of scattered, emergent dipterocarp trees with lower trees between them, again favoring gliding animals. Yet ecologists who work in different regions of the world observe tremendous local variation in tree height, canopy structure, and abundance of lianas, depending on the site conditions of soil, climate, slope elevation, and local disturbance. One can find many locations in Southeast Asia where there are abundant lianas and numerous connections between trees, and similarly many Amazonian forests with few lianas. The fact that many primates, civets, and nonflying squirrels move constantly between tree canopies in Asian forests without gliding certainly suggests that forest structure is not the main driving force behind the abundance of gliding animals.

A final theory, that we could call the food desert theory, suggests that it is the presence of the dipterocarp family itself that is driving the evolution of gliding species. The gliding animals consist of two main feeding groups: leaf-eaters and carnivores that eat small prey such as insects and small vertebrates. How would such animals be affected by living in a dipterocarp forest? For leaf-eating animals, the dipterocarp forest is a leaf desert. Dipterocarp trees often comprise 50% or more of the total number of canopy trees in a forest and over 95% of the large trees. Despite their abundance, dipterocarp leaves are unavailable to most vertebrate plant-eaters because of the high concentrations of toxic chemicals in their leaves. Flying squirrels, flying lemurs, and other flying animals all avoid eating dipterocarp leaves; they are similar to the shipwrecked sailors marooned on a raft, floating on an ocean and yet dying of thirst. These vertebrate leaf-eaters must travel widely through the forest, bypassing the dipterocarp trees, to find the leaves they need to eat. And gliding is a more efficient manner of traveling between trees than descending to the ground and walking, or jumping between trees.

In a similar manner, carnivorous animals, such as lizards and geckos, may need to search more widely for food due to the lower abundance of insects and other prey. The low insect abundance in dipterocarp trees is caused by the irregular flowering and fruiting cycles at 2–7-year intervals, causing a scarcity of the flowers, fruits, seeds, and seedlings that are the starting point of so many food chains (see Chapter 2). With fewer insects, there are in turn fewer spiders and centipedes. There also appears to be a lower density of small vertebrates in the tree canopy and on the forest floor in Southeast Asian forests than in rain forests elsewhere (Inger 1980; Allmon 1991). The lower abundance of invertebrate and small vertebrate prey in dipterocarp forests forces animals such as lizards and geckos to move between tree crowns in search of food, with gliding being the most efficient means. Gliding would also facilitate finding breeding partners, a critical consideration when a population is at a relatively low density.

Conclusions and future research directions

In earlier chapters, examples of differences among rain forests were developed using plants, mammals, and birds. In this chapter, we have focused on animals that live in and above the forest canopy. After primates, some of the most striking differences among rain forest areas are found in the frugivorous bats. Old World fruit bats are larger on average than New World fruit bats, and tend to land while feeding on fruits, in contrast to smaller New World bats that usually take fruits on the wing. Old World bats detect fruits by their smell, shape, and color, while New World bats supplement these senses by echolocation. Old World bats tend to be sequential specialists, eating a wide range of fruit species over the year and supplementing their diet with young leaves, while New World bats often specialize on a core group of fruit species throughout the year, but also consume a varying amount of invertebrates. There are similar differences between the Old and New World nectar-feeders.

Despite the many differences on the bat side of the relationship, there is plenty of evidence that Old World bats can locate, eat, and disperse the seeds of New World bat-fruits grown in Old World gardens and locate and pollinate New World bat-flowers. Similarly, New World bats recognize Old World bat-fruits and flowers planted in the New World. A striking consequence of the convergence in fruit characteristics has been the "escape" and spread of several Neotropical pioneer trees and shrubs into disturbed habitats in the Old World. However, despite the large amount of anecdotal evidence, there has been no systematic study of bat–plant relationships across the evolutionary divide. Are fruit and nectar sources really interchangeable? Are all fruits and flowers targeted at one group of bats recognized by the other group, or is it just a few conspicuous examples that have given this impression? Do Old and New World bats show similar behavior when visiting the same fruit or flower species? Are they equally effective as seed dispersal agents or pollinators? We predict that seed dispersal, which does not require a precise match between fruit and frugivore, will be relatively insensitive to the origin of the bat involved. In contrast, pollination, which often does require an accurate fit between bat and flower, is less likely to be successful with the "wrong" bat. We must also emphasize that studies involving exotic plants – particularly potentially invasive pioneers – should be based on the many existing plantings and should never involve the introduction of new plant species to an area.

Another important topic in bat biology is the impact of their extinction on plant communities. In all rain forests, bats are the principal pollinators or seed dispersers of many tree, shrub, and vine species, and in the Neotropics they also pollinate many herbs and epiphytes. Predicting the impact of bat extinctions is difficult, since the degree of dependence by a particular plant on a particular bat is rarely known. Bats are more important as pollinators and seed dispersers in the Neotropics than elsewhere, but the greater diversity of the bat communities there may limit the impact of a single extinction. Plant-feeding bats are particularly important on islands because they have reached many that are too remote or too small to support any other mammals (McConkey & Drake 2006). When a bat species goes extinct on an island, there may be no other species to take its place in pollinating flowers and dispersing seeds. Most island bats are pteropodids, but phyllostomids play an important role on some Caribbean islands. Although many observers have warned of the potential impact of island bat extinctions, there has been little systematic study of their importance and it is vital that such studies start before it is too late to find intact bat communities for comparison.

In the Neotropics, bats are particularly significant in the recovery of forest in cleared and cultivated areas because of their tendency to defecate seeds of pioneer trees while flying, whereas birds usually defecate from perches. Bats are considerably less important in early forest succession in the Old World, but they may be more important in the dispersal of large, late-successional canopy trees (Muscarella & Fleming 2007), although the importance of phyllostomids in dispersing large-seeded, late-successional plants has probably been underestimated (Melo et al. 2009). Differences in the relative importance of birds and bats in rain forest recovery between the Old World and New could have significant implications for the way in which such degraded landscapes are managed. The relative importance of nocturnal bats and diurnal birds in dispersing seeds into degraded and open habitats can be assessed very easily by comparing the "seed rain" into seed traps during the day and during the night.

There are also striking differences in the nature of the threats bats face in different rain forest regions. Hunting is the major current threat to the larger Old World fruit bats (particularly *Pteropus* species) throughout Asia and the Pacific Islands, and also in Madagascar (Epstein et al. 2009; Mickleburgh et al. 2009). The largest fruit bat species, *Eidolon helvum* (Fig. 6.1), is similarly targeted in West and Central Africa. Colonial roosting makes these species particularly vulnerable and they are hunted for both local consumption and trade, as well as for sport and to reduce damage to fruit crops. These large bats are long-lived and have low reproductive rates, so hunting impacts can lead to irreversible population declines and eventual extinction. They are also the species most likely to have an irreplaceable role in seed dispersal, because they consume larger fruits and fly longer distances (Corlett 2009b). In contrast, phyllostomids are not big enough to attract hunters and the major threat to forest-dependent species is from habitat loss (Meyer & Kalko 2008). The same is presumably true of the smaller pteropodids although, as noted above, a smaller proportion of these are adapted to the forest interior habitats.

The abundance of gliding animals in Asian rain forests and their rarity in rain forests elsewhere provides a strong argument that dipterocarp forests have unique properties that have influenced the evolution of its animal community over millions of years. The most likely explanation is that the multiyear periods between episodes of mass flowering and fruiting in dipterocarp forests, and the low diversity of edible leaf material, forces animals to forage more widely via gliding. The shortage of plant food may, in turn, lead to a low density of insects, other canopy invertebrates, and small vertebrates for canopy-dwelling reptiles, frogs, and snakes to eat. The same factors must also influence the ecology of other, nongliding, animal species in these forests, as has already been suggested for the primates (see Chapter 3).

The weakness with this line of reasoning is that comparative studies of the abundance of insects, lizards, frogs, snakes, and other animals in the rain forest canopy have not been undertaken at enough sites to confirm this impression. Do Southeast Asian forests really have a lower density of insects and insect-eating vertebrates than do comparable forests elsewhere? Such studies could yield a wealth of new insights. Particularly valuable would be studies comparing gliding animals in Southeast Asia with the ecologically most similar species in African and Neotropical rain forests. How would a Bornean flying snake differ in population density, physiology, foraging behavior, reproductive behavior, demography (especially injury and mortality from falls), and dispersal from a similar-sized, but non-gliding, arboreal snake species, as closely related as possible and eating the same type of food, that lives in the Congo River basin or Amazonia? Repeating the same studies for frogs, reptiles, and squirrels would also be extremely valuable. Organizing such cross-continental comparative studies should be a priority for tropical ecology.

In this and past chapters, we have discussed the relationships between plants and vertebrate groups. In the next chapter, we will devote our attention solely to insects, the most species-rich group in the tropical rain forest.

Chapter 7

Insects: Diverse, Abundant, and Ecologically Important

Plants, birds, and mammals are the best-known and best-studied components of tropical rain forests, but they make up only a tiny fraction of the total number of species – probably less than 1%. Invertebrates are the dominant animals of the rain forest, contributing most of the species and the overwhelming majority of individuals. Of all invertebrates, the insects contribute most of the biomass and probably most of the species, although it must be admitted that we know very little about the rain forest diversity of such species-rich, non-insect groups as mites and nematodes.

In contrast to the plants, birds, and mammals, where most species have already been described and named by scientists, rain forest insects are much less known. Certain groups of large, popular insects, such as butterflies, dragonflies, large bees, and long-horn beetles, are reasonably well known, but smaller and less colorful insects, including many beetle families and moths, are poorly known. A million or so insects have been given scientific names (Adler & Foottit 2009), but most of these are from the temperate regions. Yet the majority of insect species live in the tropics, and most live in the rain forest. If estimates of 4 million insect species are correct, then only a quarter have been described so far. Sort through a leaf litter sample for tiny insects in any tropical rain forest and you are likely to catch at least one species new to science! Of the species that do have names, the name is usually all that is known. The ecology of tropical insects is largely unknown.

Some general patterns are apparent, however. First, insect faunas appear to be broadly similar in all the major rain forest areas that have been sampled, with the same insect orders and even families found in most rain forest areas. Both the relative abundance of different insect groups and the relative importance of different feeding guilds seem to be more or less the same. These similarities contrast with the major differences between the vertebrate faunas of the major rain forest regions discussed in previous chapters. Second, Neotropical forests seem to have more species than those in other regions, at least in the canopy. Third, there are conspicuous and ecologically important exceptions to the first two generalizations! We aim here to illustrate both the general patterns and some of the major exceptions.

Although many rain forest insect groups are very poorly known, there are some significant exceptions. Butterflies are well collected in some rain forests, and it is likely that more than 90% of the total species have been described, making them the best group to illustrate the general patterns of diversity in rain forest insects. It would be premature, however, to assume that all other insect groups show the same patterns, as described later. The social insects – ants, bees, and termites – are also relatively well studied and illustrate the overwhelming importance of insects in the ecology of tropical rain forests better than any other insect group, as well as providing striking exceptions to the rule of relative uniformity among rain forest insect faunas.

Butterflies

Butterflies are both the best-known and the best-loved group of insects, and most species live in tropical rain forests. A mere 500 ha (1250 acres) of lowland rain forest at Garza Cocha in Ecuador has more butterfly species – 676 – than the whole of North America (DeVries 2001). Species richness is lower in the Old World, with a small Neotropical country like Costa Rica having more species (1044) than either Peninsular Malaysia (777) or New Guinea (785). The Neotropics, with 5341 species, has roughly double the species of all of Africa (2729 species). Even so, one can still see more species in a day spent walking in an Asian or African rain forest than in a lifetime in more temperate climates. Tropical rain forests also support the biggest and most brightly colored butterflies as well as an amazing variety of smaller and less conspicuous species.

The insects we know as butterflies are only a fraction of the species in the huge order Lepidoptera, most of which are known as moths. The division between butterflies and moths is fairly arbitrary, and the precise boundaries differ between authors. The "true" butterflies, considered here, are most commonly divided into four families (Papilionidae, Pieridae, Lycaenidae, and Nymphalidae), although some authors recognize more. The relative contribution of each family to the total butterfly diversity is very similar in Asia, Africa, and the Neotropics, although several important subfamilies are largely or entirely restricted to one region. The Lycaenidae and Nymphalidae together account for around 80% of the total species, with just the lycaenids accounting for almost 50% of the total (Table 7.1). The Madagascan butterfly fauna, in contrast, is strikingly deficient in lycaenids (only 17% of the species) and disproportionately rich in nymphalids (67%), particularly in the genera *Heteropsis* and *Strabena* in the subfamily Satyrinae, commonly known as the browns. New Guinea's butterfly fauna is also somewhat different, with a higher proportion of pierids than other rain forest areas, largely in the genus *Delias*, the jezebels (Braby & Pierce 2007).

Papilionidae: the swallowtails

The Papilionidae are commonly known as swallowtails, because in many species – but by no means all – there is a long "tail" extending from the rear of each hindwing (Fig. 7.1). The family is distributed worldwide, though most species are tropical. Although they make up a relatively small proportion of the total

Table 7.1
Variation in the contribution of each butterfly family to species richness among
different faunas. The percentage of total faunal richness by each family is given in
parentheses.

Fauna	Papilionidae (swallowtails)	Pieridae (whites)	Nymphalidae (nymphs)	Lycaenidae (blues)	Total
All of Africa	80 (2.9)	145 (5.3)	1,107 (40.6)	1,397 (51.2)	2,729
Zaire	48 (3.7)	100 (7.6)	607 (46.5)	551 (42.2)	1,306
Kenya	27 (3.7)	87 (12.1)	335 (46.5)	271 (37.6)	720
Southern Africa	17 (2.3)	54 (7.2)	265 (35.6)	409 (54.9)	745
Madagascar	13 (4.9)	28 (10.7)	175 (66.8)	46 (17.5)	262
Australia	18 (6.5)	35 (12.6)	85 (30.6)	140 (50.3)	278
New Guinea	41 (5.2)	146 (18.6)	222 (28.3)	376 (47.9)	785
Malaysia	44 (5.8)	44 (5.8)	273 (35.9)	400 (52.5)	761
Costa Rica	42 (4.0)	71 (6.8)	438 (41.9)	493 (47.2)	1,044

From DeVries (2001).

Fig. 7.1 The small-striped swordtail, *Graphium policenes*, here on Bioko Island,
Equatorial Guinea, is a common swallowtail butterfly of African forests. (Courtesy
of Jessica Weinberg.)

(a)

(b)

Fig. 7.2 This "mud-puddling" behavior by male butterflies is thought to be a way of obtaining essential minerals. (a) Papilionids, Rajah Brooke's birdwing (*Trogonoptera brookiana*), in Ulu Temburong National Park, Brunei Darussalam, Borneo. (Courtesy of Max Dehling.) (b) Pierids (*Mylothris* sp.), on Bioko Island, Equatorial Guinea. (Courtesy of Jessica Weinberg.)

butterfly diversity, they are mostly large and colorful butterflies that are very popular with collectors. Papilionid diversity is greatest in the New World, but the largest and most spectacular species are found in the Old World.

The largest of all butterflies are the magnificent birdwings (*Trogonoptera*, *Troides*, and *Ornithoptera*) (Fig. 7.2a), which occur from Sri Lanka through

Southeast Asia to New Guinea and Australia. The world's largest butterfly is Queen Alexandra's birdwing (*Ornithoptera alexandrae*), found only in New Guinea, in which the relatively dull-colored black and yellow females can attain wingspans of up to 28 cm (11 inches). These butterflies usually fly high above the forest canopy; many early specimens are peppered with holes as a result of being brought down with a shotgun! The males are smaller but more colorful, with a bright yellow body and iridescent blue or green markings on the wings. Birdwing caterpillars feed on poisonous *Aristolochia* vines, incorporating the poison into their bodies to make both the caterpillar and the adult distasteful to potential predators.

This family also provides Africa's largest butterflies, the African giant swallowtail (*Papilio antimachus*), with a wingspan of up to 23 cm, and the slightly smaller giant blue swallowtail (*P. zalmoxis*), both of which fly high in the canopy of rain forests throughout Central and West Africa. The males of both species can be collected by their attraction to human urine, but the females are rarely seen and the caterpillars are unknown. The males of the giant blue swallowtail have a distinctive noniridescent blue color, which is produced by a completely different mechanism from that which produces the iridescent blues of most other blue butterflies (Kinoshita et al. 2008). The resulting appearance of *P. zalmoxis* in flight has been likened to fragments of blue sky.

Pieridae: the whites

The family Pieridae consists of small to medium-sized butterflies, most of which are white, yellow, or orange, leading to their common names, the whites and sulfurs: indeed, the yellow European species are probably the origin of the word "butter-fly." The cabbage butterflies are well-known temperate zone members of the group. By far the largest genus in the family, however, is *Delias*, known as the jezebels, with more than 200 species, mostly in montane habitats of New Guinea, but also extending into Southeast Asia (Braby & Pierce 2007). Unlike most other members of the family, the jezebels often have striking black or red patterns on a pale background. Throughout the tropics, pierids are a major component of the butterfly aggregations on exposed riverbanks and puddles along muddy rain forest tracks (Fig. 7.2b). These aggregations may contain several species of butterfly, but they are usually arrayed in groups of the same species. This behavior, called "mud-puddling," occurs almost exclusively with males. It is thought that they may be obtaining essential minerals, particularly sodium, which are then transferred to the females during mating (Molleman et al. 2005). This is advantageous because land plants have very low sodium content.

Lycaenidae: the blues

The Lycaenidae account for almost half of all butterfly species, although their typically small size makes them less conspicuous than the other families. They show an amazing diversity of shapes and colors. Many species are blue on the upper side and have common names such as blues, hairstreaks, coppers, and metal-marks. Many have one to three pairs of "tails" on the hindwings which, in extreme cases, such as the plane (*Bindahara phocides*) of tropical Asia, can be up to twice as long as the body. These tails are often associated with dark spots on the hindwings, and it has been suggested that the combination resembles eyes and antennae. This apparently deceives predators into striking at the false "head" and thus allows the butterfly to escape with only minor damage.

The lycaenids are even more diverse in terms of their life histories than in their appearances. Adults may feed on nectar, fruits, or the carcasses of animals, while the caterpillars range from herbivores to strict carnivores, feeding on insects, such as aphids or ant larvae. Many lycaenids have complex associations with ants (Braby 2000). In some cases these relationships are mutually beneficial; the caterpillars secrete a fluid rich in sugar and amino acids from their "honey glands" that attracts ants to act as bodyguards against predators and parasites. In other cases, however, the benefits are all in the butterfly's favor. One of the most extreme examples of such a relationship is the moth butterfly (*Liphyra brassolis*), which ranges from India through Southeast Asia to New Guinea and northern Australia. This species is among the largest of the lycaenids, with a wingspan of 8 cm (3 inches), and is easily mistaken for a moth. The caterpillars live in the arboreal nests of the weaver ants, *Oecophylla smaragdina*, and feed on the ant larvae. The moth then pupates in the nest, and the emerging adult is covered with sticky scales that clog the jaws, legs, and antennae of any ants that attack it on its way out of the nest. The female moth lays her eggs on the trunk and branches of trees where *Oecophylla* nest.

Among the 10 subfamilies of the Lycaenidae, the Riodininae are sometimes treated as a separate family, Riodinidae. This is a largely Neotropical group, although there is also a scattering of species throughout the Old World tropics. The common name, metalmarks, refers to the metallic appearance of raised areas on the wings of some species. This subfamily includes some of the most beautiful of all butterflies, but many species are rarely seen inhabitants of the forest canopy.

Nymphalidae: the nymphs

For diversity of adult form, no other butterfly family matches the Nymphalidae, commonly known as the nymphs (Fig. 7.3). They are referred to also as "brush-footed" butterflies because the front pair of legs are greatly reduced in size and often hairy or brush-like. They range in size from small to very large and include some of the most conspicuous species as well as some of the best-camouflaged species of butterflies. Although the family is represented in all rain forests, several of the 10–12 subfamilies and most of the genera have more restricted distributions. These differences, coupled with the conspicuousness of many nymphalids, make the butterfly faunas of the different rain forest regions distinctive to even the casual observer.

The largest nymphalids and probably the largest Neotropical butterflies are the owl butterflies (*Caligo* spp.), named for the big eye-shaped markings on the underside of each hindwing. These giant butterflies fly mainly at dawn and dusk, when they can easily be mistaken for bats. Owl butterflies are also notable for their mating behavior, which resembles that of some rain forest birds, such as the birds of paradise and the manakins (see Chapter 5). The male butterflies aggregate in "leks" at display sites along the forest edge, where they compete for perches while the females visit to choose a mate (Srygley & Penz 1999).

The most spectacular nymphalids are the morphos (*Morpho* spp.) (Fig. 7.3b), which are also confined to the Neotropics. The wings of most male morphos

(a)

(b)

Fig. 7.3 Some Neotropical nymphalid butterflies from Costa Rica illustrating their great diversity. (a) Glass-wing butterflies (*Ithomia patilla*). (Courtesy of Dale Morris.) (b) Morpho butterfly (*Morpho menelaus*). (Courtesy of Dale Morris.)

display a metallic blue of an intensity unmatched by any other butterfly. As with many butterfly colors, this blue is not created by pigments, but by optical effects in the microscopic layered structure of the scales. Such structural colors do not fade like pigments, and dead specimens retain the brilliance of the

Fig. 7.3 (*cont'd*) (c) Heliconia butterfly (*Heliconius hecale zuleika*). (Courtesy of Ethan Scott.)

living butterfly, a characteristic exploited to the morphos' cost by the makers of butterfly jewelry. The blue of morpho wings has continued to intrigue physicists because the color changes very little with viewing direction, which may be important in mutual recognition. Recent studies suggest that both interference and diffraction effects are involved (Kinoshita et al. 2008). In morphos, the undersides of the wings are brown, so the resting butterfly is almost invisible against tree bark. The alternate flashes of blue and brown in flight, as the wings open and close, may also make them hard for a predator to follow.

The males of many nymphalid species are territorial, flying from their perch to investigate passing butterflies. Male morphos, for instance, can often be attracted by waving a blue cloth of the appropriate hue. Another group of Neotropical nymphalids, the cracker butterflies (*Hamadryas* spp.), produce short bursts of loud cracking noises as they defend their territories, probably by buckling a stiff part of the wing (Yack et al. 2000). To a human observer these sounds can be audible 50 m (160 feet) away, but they are usually only produced when the butterflies are very close to each other (< 10 cm), suggesting that their own hearing is not as sensitive.

Many nymphalid butterflies escape from predators by camouflage. In the Indian leaf butterfly (*Kallima inachus*), the uppersides of the wings are brightly colored, but the undersides, which are all that is visible in the resting butterfly, are an almost perfect imitation of a dead leaf in shape, color, and pattern. There is even a "midrib" down the middle of each wing. The jungle queen (*Stichophthalma louisa*) in the montane rain forests of Vietnam takes camouflage even further, with both sides of the wing resembling dead leaves. When flying in a slow, zigzag course it can easily be mistaken for a falling leaf (Novotny et al. 1991).

In striking contrast, many other nymphalids have conspicuous colors and patterns. The tree nymphs (*Idea* spp.) of Asian rain forests, for example, are large and slow-flying, with grayish-white wings, appearing like a winged seed or piece of paper blowing in the wind, and making them easy prey for a hungry flycatcher. As with the birdwing butterflies mentioned above, however, these conspicuous nymphalids are often poisonous, or at least distasteful, as a result of both chemicals acquired from the plants on which their larvae feed and additional toxins consumed or synthesized by the adults. Such defenses are effective only if vertebrate predators learn to associate the unpleasant experience with the striking appearance of the butterfly. Predators cannot be expected to learn to avoid lots of different patterns, so unrelated unpalatable butterflies have often evolved very similar mimicry patterns. Experience with one member of the group teaches the predator to avoid all similar-patterned butterflies in future.

Palatable butterflies can also gain protection by mimicking unpalatable species, which has led to the evolution of complex multispecies "mimicry rings," including both unpalatable and palatable species, with the latter usually in a minority for obvious reasons (Fig. 7.4). Day-flying moths may also join these mimicry rings. A particularly striking example is the "tiger-striped" mimicry ring in Neotropical forests, consisting of butterflies and moths from several families, all of which are striped in yellow, orange, black, and brown. Up to five mimicry rings with different color patterns have been described from one Neotropical forest. Similar rings occur in rain forests elsewhere, mostly centered on unpalatable

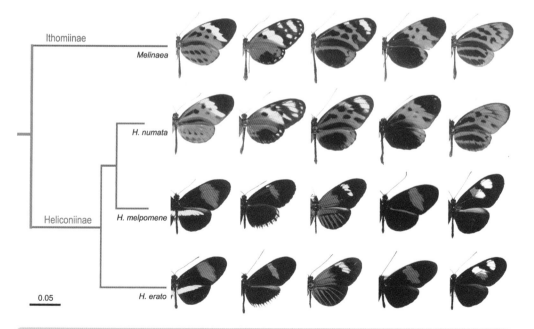

Fig. 7.4 Mimicry in unpalatable Neotropical nymphalid butterflies. *Heliconius melpomene* and *H. erato* are distantly related but have undergone parallel radiations into races with almost identical color patterns. *H. numata* is closely related to *H. melpomene* but each form is a precise mimic of a different species in the unrelated ithomiine genus *Melinaea*. (From Joron et al. 2006.) (Courtesy of Mathieu Joron.)

nymphalids. In some cases, the mimicry involves wing shape and flight behavior as well as pattern, although size seems to be unimportant.

Another group of nymphalids, the largely Neotropical heliconines (Fig. 7.3c), also feed on poisonous passionflower plants. They advertise their poisonous contents through warning patterns on both surfaces of their characteristically long and narrow wings. Female heliconines have an apparently unique method of feeding on pollen; they first accumulate it at the base of the proboscis while visiting flowers and then mix it with saliva (Eberhard et al. 2009). Amino acids from the pollen leach into the saliva and are ingested by the butterfly. The extra nitrogen acquired in this fashion increases egg production and permits an adult life span of several months, which is very long for butterflies. The nitrogen is also used to synthesize cyanide, making the butterfly even more toxic to birds. The lure of extra nitrogen for making eggs also explains the behavior of some species in a related group of Neotropical nymphalids, the Ithomiinae, in which the females feed on the droppings of the antbirds (Gotwald 1995). Up to a dozen may follow one swarm. These ant butterflies avoid becoming additional prey for the antbirds by accumulating toxins from their food plants, again advertising their presence by warning coloration.

Ants

Ants and termites, along with some species of wasps and bees, are distinguished from most other insects in that they form long-lived colonies with overlapping generations and have a division of labor between fertile queens and sterile workers. In some ways the whole colony acts as a single "superorganism," with an ecological impact at least as great as an individual vertebrate. Less than 2% of known insect species are social, but they account for a large fraction of insect abundance. In the rain forest of central Amazonia, it has been estimated that social insects (mostly ants and termites) account for around 80% of the total insect biomass and a third of the total animal biomass (Fittkau & Klinge 1973). The dominance of social insects in tropical rain forests would probably appear even greater if their role in ecological processes such as biomass harvesting, mineral nutrient cycling, and soil excavation were measured.

The most important group of social insects, in terms of numbers, biomass, and ecological impact, is undoubtedly the ants. Around 12,500 ant species have been described but the true total may exceed 25,000, with most of these in lowland tropical rain forests (Ward 2010). Ants are found in all layers and microhabitats of the forest and make use of a huge range of food resources. In comparison with termites, ants are a conspicuous part of the rain forest fauna because many species forage actively on the ground and on plant surfaces. The highly visible ant species are only a small fraction of the total, however; most species are small and secretive, including a newly discovered Amazonian species, *Martialis heureka*, which appears to be the sister lineage group of all other ants (Rabeling et al. 2008). Some species forage on the forest floor, some below ground, and others live in the forest canopy above. Although certain species, such as Neotropical leaf-cutting ants (in the genera *Atta* and *Acromyrmex*) and the Malaysian giant ant (*Camponotus gigas*), nest in the ground and forage mostly in the canopy, many more species spend their entire lives in one stratum, forcing scientists interested in sampling the entire ant fauna to use a variety of techniques for their research. In the few cases where this has been done, the total number of ant species found is staggering: for example, 524 ant species in just 6 ha (15 acres) of lowland rain forest in Sabah, Malaysia (Brühl et al. 1998). This compares with about 700 species in the whole of the United States and Canada.

Ants are a widely dispersed family of insects, facilitated by the ability, in most species, of a single, winged female to start a new colony. Only Hawaii and other highly isolated Pacific islands lack native ants. As a result of this wide dispersal, there are many similarities between the ant faunas of the major rain forest areas. There are similar numbers of species in tropical Asia, Africa, and the Neotropics and three of the largest ant genera – the *Camponotus*, *Pheidole*, and *Crematogaster* – are found in all rain forests, except in the remote Pacific. There are, however, many exceptions to this basic uniformity and around half of all ant genera are restricted to one of the major regions (Fisher 2010). Several of these exceptions are of great ecological significance.

Army ants

Most of the ant diversity on the rain forest floor is made up of small and inconspicuous species living in small colonies (with fewer than 150 workers) in decaying twigs and other short-lived nest sites, and foraging 1 m or so from their nests (Byrne 1994). Many of these ants are thought to be primarily predators on tiny litter invertebrates or scavengers, but some also harvest seeds from bird droppings and may have a large impact on the regeneration of small-seeded plant species.

At the other extreme, there are ground-dwelling ants that form huge colonies and forage over vast areas. The most spectacular of these species are the swarm-raiding army ants of Africa and the Neotropics (Fig. 7.5a,b). Until recently, it was thought that the army ant syndrome had originated independently in the New and Old World species: one of the classic examples of convergent evolution. Recent molecular studies, however, have shown that all the true army ants had a common ancestor (Brady 2003). This ancestor was initially suggested to have lived in the mid-Cretaceous, which would be consistent with an origin on Gondwana before its final break-up (see Chapter 1), but more recent dating suggests a later divergence, which would imply a more complex history (Kronauer 2009).

True army ants in the Old and New Worlds share a number of important characteristics. The first of these is the group raiding and retrieval of animal prey that gives rise to their common name. This collective foraging gives them access to a far wider range of prey, such as large invertebrates and small vertebrates, than is available to a solitary forager. It also allows them to attack colonies of other social insects. The second, which is probably a consequence of the first, is that the colonies are nomadic, moving on when the supply of prey is exhausted. The third is that the queen ants are highly modified in comparison with all other ant species. Army ant colonies can be enormous but they include only a single fertile queen. Army ant queens must therefore produce extraordinary numbers of eggs throughout their lifetime: estimated to be as many as 3–4 million per month in the African species, *Dorylus wilverthi* (Gotwald 1995). Not surprisingly, these queens do not look much like ordinary ants. They are much larger than their workers – *Dorylus* queens can exceed 50 mm (2 inches) in length and are

(a)

(b)

Fig. 7.5 (a) Army ants (*Eciton burchelli*) attacking a scorpion in Costa Rica. (Courtesy of Dale Morris.) (b) African driver ant soldier (*Dorylus* sp.), with worker driver ants below, in Kibale Forest, Uganda. (Courtesy of Tim Laman.)

the largest ants known – and have a greatly enlarged abdomen. In most ants, new colonies are founded by a single, winged female, but in army ants, the queens are wingless and new colonies are produced by division, or "budding" off of parts of the colony, as in the honeybees. This explains both the existence of separate radiations in the Old and New World, and the absence of army ants from Madagascar, since dispersal across water is unlikely.

The most dramatic development of the army ant syndrome is found in a few species of *Eciton* in the Neotropics and of *Dorylus* (often known as the "driver ants") in Africa, with colonies consisting of hundreds of thousands or millions of workers, which forage on the surface in huge swarms. In *Eciton* (Fig. 7.5a), a swarm of ants moves forward on a broad front, sweeping up everything in their path or driving more mobile species before them. In *Dorylus* (Fig. 7.5b), the snake-like feeder columns expand at the front into a broad crown. Although the threat these swarms pose to large vertebrates has often been exaggerated, they will consume any prey they can overcome: mostly invertebrates, often other social insects, but also the occasional small and helpless vertebrate, such as bird nestlings. Prey items too large for an individual ant to carry back to the nest – which can be 200 m (650 feet) from the raid front – are cut up or transported by teams of ants and slung beneath their bodies. The widespread Neotropical *Eciton burchelli* has a specialist porter caste of large workers, which shuttle at high speed between the nest and the swarm front, while African *Dorylus wilverthi* relies on teams of smaller, slower workers for the same task (Franks et al. 2001). Much of the more mobile prey escapes the ants, only to be snapped up by birds and other vertebrates that follow these surface raids for this very reason (see Chapter 5).

Although army ants, including species of *Dorylus* and *Aenictus*, occur in Southeast Asian rain forests, these form much smaller and much less conspicuous colonies. *Dorylus laevigatus*, for instance, is found throughout Southeast Asia, but lives a largely subterranean existence, except for small, short-lived raids on the surface at night (Berghoff et al. 2003). One colony contained an estimated 325,000 workers, compared with millions in some African species. There is no Asian equivalent of the spectacular and devastating swarm raids of Africa and the Neotropics, and there are no records of antbird followers. Perhaps Asian rain forests do not provide the high density of insects needed to support these huge colonies? Even in Africa and the Neotropics, most army ant species do not move on the surface, but conduct their raids entirely underground or beneath the litter layer. In rain forests in the Neotropics up to 20 species of army ants can coexist at one site, suggesting that most species must have specialized diets. Such army ants are found in all the major rain forest regions, except Madagascar, but have not attracted the same attention as the more conspicuous surface-raiding species, although they may be of equal ecological importance.

In addition to the "true" army ants, in the subfamilies Aenictinae (Africa to Australia), Dorylinae (Africa and Asia), and Ecitoninae (America), elements of army ant behavior are seen in species scattered in other subfamilies (Kronauer 2009). On Madagascar, which lacks true army ants, some species of *Cerapachys* (Cerapachyinae) have developed some of the characteristics of army ants (Fisher 2010). In Southeast Asia, *Leptogenys distinguenda* (Ponerinae) is a generalist, group-raiding, nomadic predator that could reasonably be termed an army ant. Their colonies are smaller than those of the true army ants, however, and the differences between the queen and her workers are much less pronounced. Two Asian species of *Pheidologeton* (Myrmicinae) are also swarm-raiders with massive colonies, but have winged queens and do not change their nest sites as frequently as the true army ants. This may, in part, be a result of their extremely broad diet that includes insects, earthworms, snails, spiders, and other invertebrates, and the corpses of vertebrates, as well as a variety of seeds and fruits (Moffett 1987). These ants have a very complex division of labor and a more than 500-fold range in weight between the largest and smallest workers. Both *Leptogenys* and

Pheidologeton have been reported to drive insects and small vertebrates ahead of their swarms, but there are no reports of birds or other vertebrates foraging at the swarm front.

Leaf-cutter ants

In contrast to the generally wide distribution of ant groups, leaf-cutter ants are unique, conspicuous, ecologically important, and found only in the New World. Leaf-cutter ants are fungus-growing ants in the tribe Attini, known as attines; specifically, the 50 or so species in the genera *Acromyrmex* and *Atta*. One of the most characteristic sights in Neotropical rain forests is the columns of leaf-cutter ants carrying pieces of leaves or flowers back to their underground nests (Fig. 7.6a,b). Each ant holds its leaf section above its head, giving rise to the nickname of umbrella or parasol ants. In mature rain forests, leaf-cutter ants consume a relatively small proportion of the total leaf production, although they have a significant impact on their preferred plant species, which may lose up to 40% of their leaves (Wirth et al. 2003). These ants reach much higher densities in young secondary forests and forest edges, reflecting a preference for the leaves of pioneer vegetation (Meyer et al. 2009).

No other ants are completely herbivorous, presumably because the ant digestive system, unlike that of termites, cannot cope with a diet that is so high in cellulose and so low in protein and sugars. Herbivory for the attines is made possible by a unique and obligatory symbiosis with fungi. Molecular evidence suggests that this relationship initiated once, more than 50 million years ago, when a member of the fungus family Lepiotaceae (Agaricales, Basidiomycotina)

(a)

Fig. 7.6 (a) A leaf-cutter ant (*Acromyrmex octospinosus echinatior*) cutting a leaf in Santa Rosa, Costa Rica. (Courtesy of Dale Morris.)

(b)

(c)

Fig. 7.6 (*cont'd*) (b) Leaf-cutter ants carrying leaf fragments back to their nest. (Courtesy of Bruce Webber.) (c) Leaf-cutter ants from Panama cultivating their nest fungus. (Courtesy of Max Dehling.)

was domesticated by an attine ancestor, but that new fungal cultivars have subsequently been domesticated from the wild and exchanged between ant species (Schulz & Brady 2008; Mikheyev et al. 2010). Several primitive genera of attines form small colonies of a few hundred workers that cultivate fungi on the droppings of herbivorous insects and other dead plant matter. The two genera of leaf-cutters, in contrast, form enormous colonies with, in some species, several million workers. These workers cut fresh leaves, flowers, and fruits with their mandibles, traveling more than 100 m (325 feet) – and several hours – from their nest and climbing up to 30 m into the canopy. Back at the nest, this fresh material, along with fallen flowers, leaves, and stipules gathered from the forest floor, is taken into underground chambers, where other ants process it into a soft pulp on which the fungus is planted (Fig. 7.6c). Molecular phylogenies show that leaf-cutting attines originated only 8–12 million years ago and that they acquired their current fungal cultivar lineage only within the last 2–4 million years (Mikheyev et al. 2010).

The gardens are not a fungus monoculture and support a diverse community of additional microbes (Scott et al. 2010). The ants keep the garden free of harmful microbial "weeds" by both physical and chemical means, but a specialist garden parasite, the ascomycete fungus *Escovopsis*, can rapidly overgrow even ant-tended gardens. Phylogenetic analyses suggest that *Escovopsis* originated in the early stages of fungus cultivation by ants. In most attines, this fungus is kept in check by a third partner in the symbiosis, a filamentous bacterium (Actinomycete) in the genus *Pseudonocardia*, which produces antibiotics that are effective against *Escovopsis*. This bacterium is cultured by the ants on specialized body surfaces where it produces a conspicuous white bloom. When a young queen leaves the colony to found a new one, she carries the garden fungus in her mouth and the "weed-killing" bacterium growing on her cuticle. In the genus *Atta*, however, the use of bacterial antibiotics has apparently been abandoned in favor of broad-spectrum antimicrobial secretions produced by the ants themselves (Fernández-Marín et al. 2009).

The leaf-cutter ants feed on protein- and sugar-rich knobs (gongylidia) that bud off from the fungus. For the ant larvae, the fungus is the only food, but leaf-cutter workers also drink sap directly from the cut edges of the leaves that they harvest. The fungus not only acts as an external digestive system, giving the ants access to cellulose-rich plant material, but also overcomes the plant's chemical defenses. Leaf-cutters avoid many plant species, but they still collect a greater variety of plants than is known for any other herbivore. Experiments with unknown plants and fungicide-treated leaves show that the worker ants learn within 1–3 days to reject plants that are harmful to the fungus, but not themselves, and that this behavior is retained for up to 18 weeks after the harmful plant is removed (Saverschek et al. 2010). Nitrogen-fixing bacteria in the fungus gardens supply the extra nitrogen that enables these ants to survive on a plant-based diet (Pinto-Tomás et al. 2009). Large colonies of *Atta* leaf-cutting ants are so well defended that they are virtually immune to natural enemies. However, at least one species of army ant, the largely subterranean *Nomamyrmex esenbeckii*, attacks and can overwhelm even mature colonies (Powell & Clark 2004).

Canopy ants

It is a long walk from a nest at ground level up into the forest canopy; only relatively large and fast-moving species like the leaf-cutter ants do this regularly. In the Neotropics, the 2.5 cm (1 inch) long "bullet ants" (*Paraponera clavata*) nest at the base of trees and ascend the trunk to forage in the canopy, with an average round trip of over 100 m. This widespread ant is one of the most noxious to humans, with both a painful bite and a wasp-like sting that can cause a person's hand to swell for days. The Malaysian giant ant, *Camponotus gigas*, is even larger (up to 3 cm) but is largely nocturnal, is far less aggressive, and has no sting. Like the bullet ant, this species nests underground but forages mostly in the canopy, with each colony occupying a huge three-dimensional territory (Pfeiffer & Linsenmair 2001).

Most canopy ants, in contrast, spend their whole lives in the canopy (Tanaka et al. 2010) and some species can even glide back to tree trunks after a fall (Yanoviak 2010). Many species nest in tree hollows or accumulations of leaf litter, while the weaver ants, *Oecophylla*, which are common from Africa (but not Madagascar) to tropical Asia and Australia, make nests by joining leaves together with silk produced by their larvae, which the ants use like living shuttles (Fig. 7.7). Other ant species construct nests from "carton," a cardboard-like mixture of soil, chewed-up wood pulp, and ant secretions. As many as 61 ant species have been collected by "fogging" a single subcanopy tree in Sabah, Borneo, with insecticide (Floren & Linsenmair 2000). Most of these species were rare, however, and there were only one to three species common in each tree, with 2–12 additional somewhat common species. In tropical plantations, several aggressive and abundant ant species establish nonoverlapping territories, so the canopy is divided into an "ant mosaic." Each dominant species has a characteristic set of subdominant species, and both dominants and subdominants influence the non-ant fauna within the territory. Ant mosaics also occur in the canopy of at least some undisturbed rain forests, but investigations into their biology and significance are only just beginning (Davidson et al. 2007).

Many species of canopy ants are known to be predators on other insects, but it is not obvious how the most abundant invertebrates of the forest canopy, in terms of both numbers and biomass, can live as carnivores on less abundant prey.

Fig. 7.7 Weaver ants (*Oecophylla smaragdina*) in Australia. (Courtesy of Harald Schuetz.)

Studies of the nitrogen isotope ratios of canopy ants have shown that many of the most abundant species get very little of their nitrogen from predation (Davidson et al. 2003). It has therefore been suggested that the principal food of the commonest species is plant carbohydrates, including nectar from extrafloral nectaries (see below) and "honeydew" – the sugary sap excreted by plant-sucking homopteran insects, such as scale insects, aphids, and treehoppers. This diet is also consistent with the observation that most canopy ants are active only during the day, when such foods are being actively produced (Tanaka et al. 2010). The most abundant species of canopy ants actively farm homopteran insects, protecting them from predators and parasites. In Australian rain forests, for example, all colonies of the weaver ant *Oecophylla smaragdina* tend homopterans, often constructing sheltering "pavilions" out of leaves woven together with larval silk in the same way they make their nests (Blüthgen & Fiedler 2002).

Plant-sucking insects are undersampled by the insecticidal fogging used to collect canopy insects, but the abundance of their associated ants suggest that the ant–homopteran partnership may be responsible for more consumption of plant biomass than all other invertebrates and vertebrates together (Davidson et al. 2003). In a sense, these canopy ants and their plant-sucking homopterans may be the major "herbivores" in the rain forest. There is increasing evidence that these herbivorous ants use symbiotic nitrogen-fixing bacteria in their guts to supplement their nitrogen-poor diet and this may have been the key innovation that enabled these basically carnivorous insects to radiate into the forest canopy (Russell et al. 2009).

Ant gardens and ant-house epiphytes

Ant gardens (Fig. 7.8) have received most study in the Neotropics, and until recently were considered to be rare or absent in other tropical rain forests. In the Neotropics, the carton nests of canopy ants are often associated with a limited number of epiphytic plant species. These so-called ant gardens are not simply chance associations. The plants are there because the ants have "planted" their seeds in the nest carton. In lowland Amazonia, 15 epiphytic plant species in seven families are found only or largely in ant gardens and at least two ant species are obligate gardeners that depend on the plants for the support they give to their nests, which would otherwise disintegrate in the wet season (Youngsteadt et al. 2008, 2009). This complex, multi-species relationship is clearly mutualistic. In some rain forest areas the ant-garden ants are the most abundant arboreal insects and their nests are the most important substrate for epiphytes. The seeds of ant-garden epiphytes have specific chemical cues that attract the ants and induce them to carry the seeds back to their nests. At the same time, these seeds are avoided by other ants, including seed-eaters.

As well as seed dispersal, the plants benefit from the nutrient-rich carton, which often contains vertebrate feces, and they probably also benefit by the protection the ants give them from both insects and vertebrates that might eat them. The ants, in turn, benefit from the supply of fruits and elaiosomes (oily structures on many ant-dispersed seeds), as well as the structural support that the plant roots give the nest. The plants include bromeliads, figs, cacti, aroids (Araceae), and gesneriads. Several unrelated ant species are involved, but many Neotropical ant gardens are occupied by two coexisting species, the larger and fiercely aggressive

Fig. 7.8 This ant garden in Peru houses the ants *Camponotus femoratus* and *Crematogaster levior*, and the epiphytic plants *Peperomia macrostachya* and *Codonanthe uleana*. These gardens are established when ants embed seeds of ant-garden epiphytes into their arboreal carton nests (Youngsteadt et al. 2009). (Courtesy of Elsa Youngsteadt.)

Camponotus femoratus and the smaller *Crematogaster levior* (Orivel & Leroy 2010). The exact nature of the relationship between these ant species – termed para-biosis – is unclear, but they share foraging trails and nest in different chambers of the same garden. Ant gardens were considered to be much less common in Old World rain forests, but a recent survey found a great variety of ant-garden systems in Southeast Asia (Kaufmann & Maschwitz 2006), suggesting both that the difference between the New and Old World has been exaggerated and that a pantropical comparison could be very interesting.

In Asia and New Guinea, much more attention has focused on so-called ant-house epiphytes, where the plant itself forms some type of hollow structure that is occupied by ants (Benzing 1990). These ant houses are occupied by a diverse group of canopy-dwelling ants, with *Crematogaster* species being most common. The ant-house plants form structures that have clearly evolved to be occupied by ants and do not have any other obvious function to the plant. Dramatic examples are found in the genera *Myrmecodia* (Fig. 7.9) and *Hydnophytum* (Rubiaceae), which range from Southeast Asia to New Guinea, Australia, and Fiji. In these genera, the plant stems form a tuber, which is honeycombed with tunnels leading from the surface to chambers inside. Ants occupy these chambers to raise their young, but also store droppings, debris, and dead ants in refuse chambers. Ants derive obvious benefit from the relationship by having a strong, secure nest; the benefit to the plant comes from the ability to absorb essential

Fig. 7.9 An ant-house plant in Papua New Guinea, *Myrmecodia* sp., cut open to show the honeycomb structure of the tuber, which is inhabited by ants in the genus *Anonychomyrma*. (Courtesy of Milan Janda.)

mineral nutrients from the decaying animal material in the refuse chambers (Watkins et al. 2008) and the protection afforded by the ants, which readily attack insects encountered on the plants. Species of *Dischidia*, in the milkweed family, form very different ant houses. *Dischidia* plants grow as epiphytic vines with thick purplish leaves that press against tree stems like overturned bowls. Ants establish homes in the hollow space formed between the leaf and the tree stem. In the case of two ant-house ferns in Borneo, *Lecanopteris* sp. and *Platycerium* sp., the *Crematogaster* ants that inhabit them defend not just their epiphytic hosts against herbivores, but also the leaves of the emergent dipterocarps on which they grow (Tanaka et al. 2009). Curiously, these highly aggressive ants tolerate large populations of a cockroach species, *Pseudoanaplectinia yumotoi*, in their houses, and the cockroaches are never found without the ants (Inui et al. 2009).

In Southeast Asia and New Guinea, ant houses are extremely common in heath forests on nutrient-poor sandy soils and sometimes in regenerating forests. One survey of a 25×9 m (80×30 feet) site recorded over 500 ant-house plants (Benzing 1990). Ants occupied virtually every one of these plants. Ant-house epiphytes are present in the New World as well, including the bromeliads *Tillandsia caput-medusae* and *T. butzii*, the potato-ferns (*Solanopteris* spp.), and a few species of orchids, but they seem to be considerably less diverse and abundant. This may reflect the dominance of bromeliads in the exposed, dry habitats favored by ant-house epiphytes in the Old World, and the fact that many of these bromeliads obtain nutrition from debris collected in their leaf rosettes and so have no need for ants (Kaufman & Maschwitz 2006). A recent study of an understorey epiphytic fern, *Antrophyum lanceolatum*, however, showed that even without special structures, plants with nests of *Pheidole flavens* ants among their rhizomes derived half their nitrogen budget from this source (Watkins et al. 2008).

More plant-ants and ant-plants

The impact of leaf-cutter ants on the plants they harvest for their fungus gardens is entirely negative, but many ant species are involved in relationships with plants in which both partners benefit (Beattie & Hughes 2002). The ant gardens and ant epiphytes mentioned above are examples of this. Many other rain forest plants have extrafloral nectaries – that is, nectar-producing structures that are not inside flowers – that are visited by ants. The proportion of trees and shrubs reported to have such nectaries varies greatly between rain forest sites, from as low as 12% (Malaysia) to as high as 40% (Cameroon), but differences in definitions, methodology, and sample sizes mean that these different percentages should be interpreted with caution (Blüthgen & Reifenrath 2003). The ants in turn defend the plant against herbivorous insects. The nectaries are sited so as to attract ant defenders to the most vulnerable parts of the plant, such as young, expanding leaves, softer stems, and flower buds. Some plants also produce solid "food bodies" for the same purpose. These food bodies are protein-, lipid-, or carbohydrate-rich nodules produced on the edges of young leaves that are eagerly collected by ants.

These relationships of nectaries and food bodies are relatively unspecialized and opportunistic: they involve many species of plants and many species of ants. There are also a much smaller number of more specialized relationships, which in some cases are obligatory for both partners. Although the details of these relationships vary tremendously, the common feature is that the plant provides a nesting space for the ants, as well as food in exchange for protection. Two of the most conspicuous and best-studied examples have evolved independently in unrelated but ecologically similar tree genera: *Macaranga* (Euphorbiaceae) in Southeast Asia (Fig. 7.10) and *Cecropia* (Cecropiaceae) in the Neotropics. Both genera consist of

Fig. 7.10 Macaranga tree in Brunei with *Crematogaster* ants living inside its hollow stems. (Courtesy of Wong Khoon-Meng.)

small, fast-growing, large-leaved pioneer trees that grow naturally in tree-fall gaps and often become very abundant along logging roads, as well as in abandoned fields and other human-made clearings.

The 280 or so species of *Macaranga* range from West Africa through tropical Asia to Fiji. Many species have casual relationships in which ants are attracted to extrafloral nectaries and food bodies, but 26 species from the wetter parts of Southeast Asia have developed obligatory relationships, usually with ant species in the genus *Crematogaster* (Fiala et al. 1989; Ueda et al. 2010) (Fig. 7.10). In these *Macaranga* species, the ants nest in hollow stems and feed on food bodies produced by the host plant, as well as honeydew from sap-sucking scale insects in the genus *Coccus* that they keep inside the stems. Several *Macaranga* species have a bluish, waxy surface covering over the stem that is very slippery for most insects, but presents no problems to their specific ant partners (Federle et al. 1997).

The costs to the plant can be high: in *Macaranga triloba*, food body production accounts for 5% of the daily biomass production (Heil et al. 1997). In return, the plant receives protection from herbivorous insects and some pathogenic fungi. The ants also bite off any part of a foreign plant that comes into contact with their host, thus preventing overgrowth by climbers, which are very common in the well-lit habitats that most *Macaranga* species require. Ant-free individuals of these *Macaranga* species do not survive long, while the ants are never found without their *Macaranga* hosts. Recent DNA studies have shown that the plants and ants have diversified together over 16–20 million years, but the scale insects apparently joined the association later and do not form specific associations with particular ant or plant species (Ueda et al. 2010).

Like the Asian *Macaranga* species, many of the more than 100 species of *Cecropia* in Central and South America have hollow stem internodes and produce food bodies on specialized structures at the base of the leaf stalk. The ants in this case are usually members of the Neotropical genus *Azteca*, and they forage entirely on their host tree. Despite the availability of food bodies, as well as honeydew from hemipterans, these ants also retain the predatory behavior of their free-living relatives (Dejean et al. 2009). The *Azteca* ants provide the same services to their *Cecropia* hosts in the Neotropics as their *Crematogaster* counterparts do for *Macaranga* in the Old World tropics.

Termites

At first sight, termites seem very similar to ants, and the two are sometimes confused by observers. Both are small social insects with distinct castes, nonflying workers, and nests of varying complexity. In most other ways, however, these two groups of insects are very different. Ants are related to bees and wasps, and some species have stings. Their immature stages are helpless larvae and pupae, and the entire colony is female, except for the relatively brief period when the short-lived males are produced to mate with the young queens. In contrast, termites are related to cockroaches. The immature termites progressively resemble the adults and there is no pupal stage. The workers are both male and female. Termite colonies are founded by a long-lived "king" as well as a queen. Most ants are carnivores, although many other types of tissues are consumed, while the diet of termites is dominated by cellulose.

Fig. 7.11 *Cubitermes* termite mounds, Korup National Park, Cameroon. The local people protect these mounds because they are believed to house spirits of their dead ancestors. (Courtesy of Bruce Webber.)

Termites are very abundant in tropical rain forests but much less conspicuous than ants. A few species, such as the Southeast Asian processional termites, *Hospitalitermes*, travel in ant-like columns in the open air, carrying their resources like leaf-cutter ants. In contrast, most termites remain concealed in their nests and construct long-lived covered trails when they must cross open areas. The nests of many species, however, are much more conspicuous than their inhabitants. Some termites build nests on the soil surface, such as the mushroom-shaped nests of *Cubitermes* in Africa (Fig. 7.11), while others, including many wood-feeding *Nasutitermes*, construct large, ball-shaped nests attached to trees at various heights. Many species, however, make nests inside dead wood or below the ground, which are invisible from the outside.

Estimates of total termite abundance at rain forest sites mostly exceed a thousand individuals per square meter, with a total biomass of at least 2,000 kg/ha (Martius 1994). To put this in perspective, this is greater than the total biomass of primates at most rain forest sites. Termites are the dominant decomposers in lowland tropical rain forests. They consume up to a third of the annual litter fall and, although they digest only part of this, their activities break up woody litter into fragments that are more easily attacked by other decomposers such as beetles, fungi, and worms. The activities of soil-feeding termites also have large, but little understood, effects on soil structure and chemistry. The nests tend to concentrate nutrients, and nest material is used as a fertilizer for crops in several parts of the tropics. Bacteria in termite guts also fix nitrogen and produce methane, a potential greenhouse gas. The termites are also prey for a huge range of more or less specialized predators, from army ants to Neotropical anteaters (Myrmecophagidae), African and Asian pangolins (*Manis* spp.), and the bizarre short-nosed echidna (*Tachyglossus aculeatus*) of New Guinea and Australia, an egg-laying monotreme related to the platypus.

Termite diversity

There are approximately 3000 known termite species, most of which are confined to the tropics and subtropics (Engel et al. 2009a). Fifty or more species can coexist in one rain forest site. The richest termite faunas are in West Africa, followed by South America, Southeast Asia, New Guinea, Madagascar, and Australia (Davies et al. 2003). This pattern is different from most other groups of rain forest organisms, for which the Neotropics is usually the richest area, perhaps because termites are less sensitive to drying. Although the termites originated in the Mesozoic and several genera, such as *Nasutitermes*, are pantropical, there are large differences in the composition of termite faunas in the various rain forest regions. Termite queens cannot fly far on their own, although some species may be transported long distances after being picked up by air currents. In general, the biggest differences between regions are in the soil-nesting groups. More than a century after it was sterilized by a volcanic eruption, the island of Krakatau, between Java and Sumatra, still has no soil-nesting termites (Gathorne-Hardy et al. 2002). Soil-feeding species are particularly poor at crossing water gaps and the low diversity of soil-feeding termites in Southeast Asia, Madagascar, and Australia seems to reflect the isolation of these regions from the major radiations of soil-feeder diversity in Africa and South America (Davies et al. 2003). By contrast, termite groups that feed inside dead wood are very widespread, since they can be "rafted" across ocean gaps in floating wood.

Rain forest termites belong to three (of a global total of seven) families. The dry-wood termites (Kalotermitidae) and damp-wood termites (Rhinotermitidae) feed almost exclusively on wood, which they digest with the aid of symbiotic flagellate protozoa and bacteria (Ohkuma 2008). The dry-wood termites are pantropical, but seem to be rare in most rain forests, where they are largely confined to dead wood in the canopy. In Madagascar, however, they are extraordinarily diverse – probably because of the rafting ability mentioned above – and also attack wet wood on the forest floor (Eggleton & Davies 2003). The damp-wood termites feed mostly on moist and partly decomposed wood on the forest floor and appear to be more common in most areas. Neither group of these so-called "lower termites" is seen by most visitors to the rain forest because they feed on the wood within which they nest.

Members of the third termite family, the Termitidae, are known as "higher termites" because of their more complex social organization involving a true sterile worker caste. They appear to have radiated only within the last 40 million years (Engel et al. 2009a). They lack the flagellate protozoans on which the lower termites depend for cellulose digestion, relying instead on their own enzymes and a diverse community of bacterial symbionts in their guts. With their huge colonies and diverse diets, the Termitidae overwhelmingly dominate in all tropical forests, but the major subfamilies have very different distributions. Soldiers of the largest subfamily, the Nasutitermitinae (the "nasute" or long-nosed termites), can be recognized by a long projection, or "nasus," on their head and

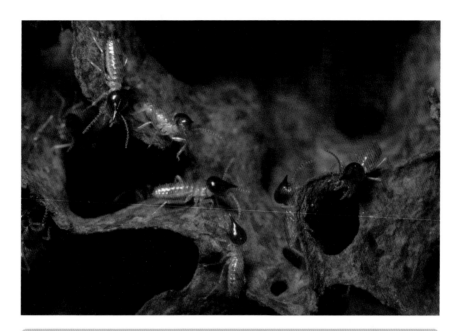

Fig. 7.12 Nozzle-headed nasute termite soldiers in Costa Rica. (Courtesy of Dale Morris.)

greatly reduced, nonfunctional mandibles (Fig. 7.12). When threatened, they can eject from this snout a sticky and irritating fluid composed largely of terpenes – a chemical defense more effective than the powerful mandibles of lower termite soldiers. The Nasutitermitinae – particularly the genus *Nasutitermes* – dominate Neotropical forests in terms of both abundance and number of species. They are also very important in the Asian region and New Guinea, but are relatively less abundant in Africa and Madagascar. Most species in this subfamily are wood-feeders but some have other specialties. In the Neotropics, species of *Syntermes* forage at night, making circular cuts several millimeters in diameter in dead leaves and carrying the pieces back to their underground nests. In the Amazonian rain forest, 20–50% of the leaves in the litter layer show evidence of feeding by these termites (Martius 1994). In Southeast Asia, columns of *Hospitalitermes* individuals forage high in the forest canopy for lichens, bryophytes, and algae, which are carried back to the nest as food balls (Miura et al. 1998).

Other Nasutitermitinae, such as the Neotropical *Subulitermes*, are soil-feeders. Soil-feeders dominate the rain forest termite community but are very difficult to study and this general designation probably hides a great deal of variation in what is actually eaten and where. In addition to fragments of dead plant material and humus, the guts of soil-feeders contain fine roots and fungal mycelium, as well as soil minerals. Soil-feeding termites must pass huge quantities of soil through their guts to obtain enough digestible organic matter and they probably have profound effects on soil chemistry and structure. The other pantropical subfamily of higher termites, the Termitinae, includes both soil- and wood-feeding species, and is important in all regions, although relatively less so in the Neotropics. Different branches of the subfamily have penetrated the soil-feeding niche in Africa and Asia.

The other two subfamilies of the Termitidae have more restricted distributions. The largely soil-feeding Apicotermitinae are most diverse and abundant in West Africa, but this subfamily is also well represented in the Neotropics and a few species are found in most of the Oriental region. It has not reached Madagascar, New Guinea, or Australia. There is evidence that the coexistence of many species within the soldierless *Anoplotermes* group in the Neotropics is aided by their specialization on organic matter with different degrees of degradation (Bourguignon et al. 2009).

The remaining subfamily, the Macrotermitinae ("big termites"), is confined to Africa, Madagascar, and Asia, and has not reached the Neotropics, New Guinea, or Australia. The 350 or so members of this subfamily are distinguished by the fact that they cultivate a fungus in order to degrade the cell walls in their food materials and thus increase the digestibility of the cellulose. The termites grow the fungus – usually a species of *Termitomyces* (Basidiomycotina) – on specialized structures known as "combs" built from their feces, largely the undigested cell walls of plant material, and then feed on both unripe mushrooms, known as "nodules," and the older parts of the comb after fungal degradation. Molecular studies have shown that this fungus-farming mutualism arose only once, in Africa, from where there has been a single migration to Madagascar (Fig. 7.13) and several independent migrations to Asia (Aanen et al. 2002). This association allows the Macrotermitinae to make extremely efficient use of a wide range

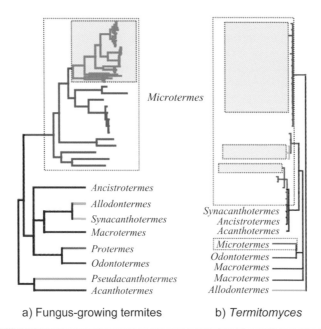

a) Fungus-growing termites   b) *Termitomyces*

Fig. 7.13 Madagascar was colonized 13 million years ago by a single species of *Microtermes* that gave rise to all the island's fungus-growing termites (in the shaded box) (Nobre & Aanen 2010). Unlike most Macrotermitinae, this clade carries a strain of the Termitomyces fungus when dispersing, but the fungus phylogeny shows that some of the Madagascan termites have subsequently switched strains, suggesting that the fungus disperses much better than its hosts. (Courtesy of Tânia Nobre.)

of relatively undecomposed plant materials. The fungus, in turn, receives pre-conditioned plant material as food in the form of termite feces, an equable microenvironment for growth, and control of potential competitors. The Macro-termitinae are most abundant in drier forests and become less important with increasing rainfall, perhaps because continuous dampness favors free-living fungi and reduces the benefits derived from the symbiosis.

It seems a curious coincidence that the fungus-growing leaf-cutter ants are confined to the New World and the fungus-growing termites to the Old World, but despite the obvious parallels, the diets of the two groups show little over-lap, and there seems to be no reason why they could not coexist in the same forest. One interesting contrast between the fungus-farming systems of the two groups is that, while young queens of leaf-cutter ants carry the fungus with them when leaving the parental colony, most fungus-growing termites rely on collecting fungal spores from around the new nest. Two groups of termites do carry the fungus with them, however, and it is surely no coincidence that it was a species from one of these groups that colonized Madagascar 13 million years ago and gave rise to all the island's fungus-growing termites (Nobre et al. 2010; Nobre & Aanen 2010) (Fig. 7.13).

The importance of termites in soil processes means that the biogeographical differences in the composition of rain forest termite communities described above may translate into real differences between regions in ecosystem processes, such as decomposition, energy flow, and nutrient cycling. With regard to ecosystem function, the most important differences are likely to be the low diversity of soil-feeding termites in Southeast Asia, Madagascar, New Guinea, and Australia, and the absence of fungus-growing Macrotermitinae from the Neotropics, New Guinea, and Australia (Davies et al. 2003). However, while there is some evi-dence for differences in the importance of termites between regions, with much lower energy flow through termites in Borneo than Africa (Eggleton et al. 1999), this will not necessarily lead to differences in ecosystem processes, since other invertebrates and free-living microbes may compensate for the missing termites.

Bees

Although butterflies are the best-known group of insects, the honeybee, *Apis mellifera*, is almost certainly the best-known single species. Honeybees and the other highly social bees, notably the 600 stingless bee species, make up only a small fraction of the total of 20,000 or so bee species, most of which are solitary – that is, they nest and forage as single individuals (Michener 2007). In contrast to most other insect groups, bee diversity is not highest in the rain forest, but appears to reach a maximum in dry, warm temperate regions, such as the southwestern United States. The social bees, however, are concentrated in the wet tropics, and their numerical dominance there may be a major reason for the lower diversity of solitary species.

Bees originated around 140 million years ago in the Cretaceous, evolving from wasps that became specialized on pollen instead of animal prey as a source of protein-rich food for their larvae. Their later diversification proceeded in parallel with the speciation of flowering plants over the same period. While bees came to depend on plants for both protein-rich pollen and energy-rich nectar, many plant species became dependent on bees for cross-pollination. The

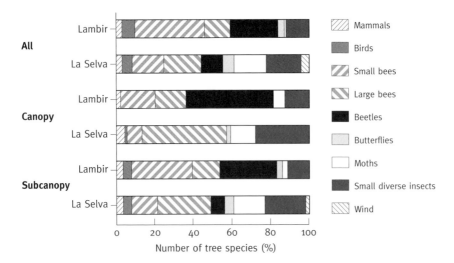

Fig. 7.14 Bees are important pollinators at two rain forest sites: Lambir in Borneo and La Selva in Costa Rica. Shown are the percentages of all trees, canopy trees, and subcanopy trees pollinated by major classes of pollinators. (From Turner 2001b.)

evolutionary histories of bees and flowers have thus been intertwined from near the origins of both groups. Today, bees are the most important group of pollinators in all tropical rain forests (Fig. 7.14). In a lowland dipterocarp forest in Sarawak, Borneo, for instance, 32% of the 270 plant species studied were pollinated by social bees and an additional 18% by nonsocial bees (Momose et al. 1998). At the La Selva field station in Costa Rica, bees were the main pollinators of at least 42% of the tree species studied and were frequent visitors to many others (Kress & Beach 1994).

Rain forest bees range in size from the 1.5 mm-long workers of some stingless bees to the giant mason bee, *Megachile pluto*, from the North Moluccas, in which the females reach a length of 39 mm (1.6 inches) (Michener 2007). Most are black or brown, but many are boldly striped, while the orchid bees of the Neotropics rival the flowers they visit in their shiny green and blue colors. Most bee species are solitary, some form communal nests, while others show a gradient of increasing cooperation and specialization, culminating in the permanent highly social colonies of the honeybees and stingless bees, which consist of a single queen and her worker daughters.

Despite their ancient origin and powers of flight, there are some distinct differences between the bee faunas of different rain forest regions. The biggest differences, as is so often the case, are between the Old and New Worlds. Neotropical rain forests have the most diverse bee faunas, possibly because of the absence (until recently) of the honeybee genus, *Apis*, with its ability to recruit huge numbers of foragers and thus dominate large patches of flowers. Indeed, Southeast Asian rain forests, which support two or three species of honeybee at any one locality, have only half as many bee species as Neotropical sites (Roubik 1989). Another factor in this difference, however, may be the phenomenon of community-wide mass flowering at multiyear intervals in Southeast Asian rain forests, which favors large perennial colonies that are able to forage over very large areas, and then either migrate or store enough resources to get them through the lean periods (Corlett 2004).

Large bees

Large bees are not only a conspicuous element of the rain forest fauna but are also very important as pollinators of a variety of large flowers, both in the canopy and the understorey. Bees 20 mm or more in length are found in several groups, including the orchid bees and honeybees, which are considered separately below. In the Neotropics, the most important of the large bee pollinators are in the family Apidae, including the large, hairy, fast-flying solitary bees in the genus *Centris* (Michener 2007). Most species of *Centris* use oil, rather than nectar, as a supplementary larval food. This oil is collected by the female bees from the flowers of various plants, particularly in the largely New World family Malpighiaceae, which produce it as a floral reward instead of nectar. Although both oil-collecting bees and oil-producing flowers occur in other parts of the world, it is only in the Neotropics that they are an important component of rain forests. The association between the Neotropical Malpighiaceae, which are mostly lianas, and oil-foraging bees appears to be an ancient one, stretching back to the late Cretaceous or early Tertiary (Renner & Schaefer 2010). Interestingly, the ability to produce oil was lost in all the 6 or 7 clades that invaded the Old World tropics (mostly Africa) and these have other pollinators.

Another group of large-bodied bees (< 30 mm long), the carpenter bees, in the pantropical genus *Xylocopa*, have been described as important pollinators of rain forest trees in the Malay Peninsula. In Sarawak, however, carpenter bees foraged mostly in open habitats and along the forest edge, and were rare in the undisturbed forest (Momose et al. 1998). The common name of the *Xylocopa* bees refers to the female bee's excavation of its nest in solid wood or in the hollow internodes of bamboo, for which it is provided with exceptionally powerful jaws. Species of *Megachile* (Megachilidae), including the giant mason bee mentioned above, are another prominent group of relatively large bees in Asian forests. Overall, however, large solitary bees appear to be less important as rain forest pollinators in Southeast Asia than in the Neotropics, perhaps because such bees are particularly vulnerable to long-term fluctuations in the availability of floral resources (Corlett 2004).

Orchid bees

The common name of the orchid bees (Apidae, Euglossini) (Fig. 7.15a) refers to the role of the male bees in pollinating many of the larger flowers of orchids of the Neotropics, but could equally well reflect the brilliant colors of many of these bees themselves, which include metallic blues, greens, reds, or coppers. These bees and the plants that depend on them are confined to the Neotropics, with most species found in rain forests (Cameron 2004). They range in size from 8.5 mm to almost 30 mm in length, and their scientific name refers to their characteristic long tongues. Although in the same family as the highly social stingless bees and honeybees, most orchid bees are solitary.

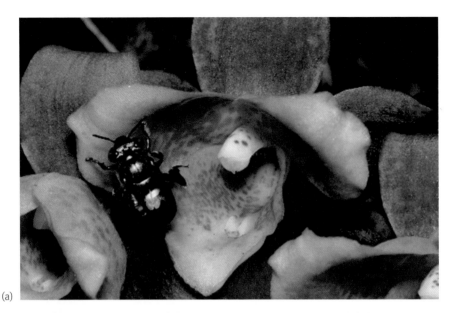

(a)

(b)

Fig. 7.15 (a) Euglossine, or orchid, bee visiting a flower of the orchid *Lycaste brevispatha* in Costa Rica. (Courtesy of Dan Perlman.) (b) Stingless bees guarding a nest entrance in Santa Rosa, Costa Rica. (Courtesy of Dale Morris.)

Both male and female euglossines visit flowers to collect nectar and pollen like other bees, but the males alone also visit the flowers of certain species of orchids and other plants, as well as decaying wood and fruits, to gather volatile chemicals – fragrances. More than 700 Neotropical orchid species in numerous genera provide these fragrances as their only reward for flower visitors

(c)

(d)

Fig. 7.15 (*cont'd*) (c) African honeybee (*Apis mellifera*) visiting flowers on Pico Basile, Bioko Island, Equatorial Guinea. (Courtesy of Jessica Weinberg.) (d) Giant honeybee (*Apis dorsata*) nest on a tree in Bangladesh. (Courtesy of Tim Laman.)

(Pemberton 2010). The complex shapes of the orchid flowers manipulate the bees so that the pollinia (waxy masses of pollen) of different species are usually precisely placed on different parts of the bee's body. Male euglossines have fiber-filled pouches on their legs to collect and store fragrances. Each bee species collects fragrances from multiple sources, but the final mixture is distinct in each species. It has been suggested that these complex fragrance blends, which are actively released during courtship displays, function like the species-specific pheromones produced by many other insects to attract mates (Zimmerman et al. 2006). The effort that male euglossines put into collecting fragrances suggests that the females – which mate only once – may also use the fragrance to assess the quality of the males (Zimmerman et al. 2009).

Stingless bees

Most individual bees seen in tropical rain forests are stingless bees (tribe Meliponini) (Fig. 7.15b) and they are probably the most important single group of flower visitors and pollinators. These bees form permanent colonies with a social system very similar to that of the better-known honeybees. The largest species, the Neotropical *Melipona fuliginosa*, is slightly bigger than the common honeybee, and will kill honeybees in nest raids. Most stingless bees are smaller than honeybees and many are less than 5 mm (0.2 inches) in length. Colonies range in size from fewer than 100 individuals in some Neotropical *Melipona* species to more than 100,000 individuals in some species of *Trigona*. Most species nest in preexisting cavities in tree trunks or the ground. Although they lack functional stings, the nests of some are defended aggressively by worker bees. They crawl over human intruders, biting and pulling at hairs. Some species known as "fire bees" eject a caustic chemical that can irritate skin.

   Stingless bees are an ancient group and occur in all major rain forest regions, but they are most diverse in Central and South America, relatively impoverished in Madagascar and New Guinea, and absent from oceanic islands (Rasmussen & Cameron 2010). Many species can coexist at a single rain forest site, with up to 64 species co-occurring at sites in the Neotropics and up to 22 in Southeast Asia. Coexisting species appear to reduce competition by adopting different foraging strategies. Some species forage individually on scattered floral resources, while others can rapidly recruit hundreds or even thousands of coworkers to a large flowering tree. Recruitment ability varies widely among stingless bees, but in some species it appears to be as effective as in the honeybees discussed below. Some stingless bees are known to leave a trail of scent marks as a guide to followers, but in other species some form of three-dimensional communication is apparently used for recruitment (Nieh 2004). Some stingless bees defend clumps of flowers against other bees, while others depend on locating them rapidly. A few Neotropical species, including *Trigona necrophaga* in Central America, have abandoned the key bee innovation of feeding pollen to their larvae, and instead use animal protein collected from the carcasses of animals. Some species even keep scale insects in their nest as a source of nectar and wax (Camargo & Pedro 2002).

Honeybees

The true honeybees, in the genus *Apis*, occur naturally only in the Old World tropics, with three exceptions. Two cavity-nesting species, *A. mellifera* and *A. cerana*, extend north into temperate Eurasia, and the giant Himalayan honeybee, *A. laboriosa*, is found in the temperate zone of the Himalayas. The common domesticated honeybee, *A. mellifera*, also occurs naturally throughout tropical Africa (Fig. 7.15c) and has now been widely introduced into other regions of the world for honey production. There are at least nine *Apis* species in the Asian tropics, occurring as far east as Sulawesi and the Philippines. Prior to the introduction of the domestic honeybee, there were no honeybees in New Guinea, Australia, or the Neotropics. The recent – surprising – discovery of a Middle Miocene fossil *Apis* in Nevada has shown, however, that their absence from the New World reflects post-Miocene extinction, rather than a failure to get there (Engel et al. 2009b).

In addition to a group of at least five medium-sized, cavity-nesting honeybee species, which includes the familiar *A. mellifera*, there are two much smaller species (7 mm long), *A. florea* and *A. andreniformis*, and two very large species (19 mm long), *A. dorsata* and *A. laboriosa* (Michener 2007). Southeast Asian rain forests typically support one species from each group. Instead of using the hollow trees favored by the medium-sized honeybees, the nests of both the small and the large species consist of a single exposed vertical comb of hexagonal wax cells. In the small species, these nests are hidden low down among dense foliage, but the huge nests of *A. dorsata* (Fig. 7.15d) and *A. laboriosa*, sometimes 1 m^2 (10 sq. feet) or more in area, hang conspicuously from rock faces or the branches of emergent trees, depending for defense on their inaccessibility and the formidable stings of their occupants. These nests are often aggregated, with up to 100 in a single huge tree.

Tropical honeybees move their nest sites more readily than do their temperate counterparts. This habit is taken to the extreme in *A. dorsata*, which makes long-distance migrations of 100 km (60 miles) or more. In the seasonal areas of tropical Asia, such as Sri Lanka, these migrations are annual and take advantage of the predictable seasonal patterns of flowering. In the lowland dipterocarp forests of Sarawak, in contrast, *A. dorsata* colonies appear only during the mass flowering episodes that occur at irregular, multiyear intervals (Itioka et al. 2001). Only 11% of the 305 plant species studied at Lambir Hills, Sarawak, are pollinated by *Apis* bees, but these species, which are mostly large canopy trees, include a high proportion of the total available floral resources during mass-flowering episodes. *A. dorsata* bees (Fig. 7.15d) become active before dawn and may continue flying after sunset. Several of the tree species they pollinate have open flowers at these times, when no other social bees are foraging, suggesting that the relationship between mass flowering and honeybee migration in Southeast Asian rain forests is an ancient one.

The success of *A. dorsata* and other honeybees in the rain forest depends to a large extent on the efficiency with which individual worker scouts recruit large numbers of their sisters to newly discovered sources of food. The "waggle dances" with which *A. mellifera* communicates the direction of a food source relative to the sun, as well as its distance and quality, are well known (Dyer 2002). Similar dances have been observed in several other *Apis* species and are assumed to occur in all. In *A. florea*, the dances are carried out on the expanded horizontal base of the comb, while *A. dorsata* dances on the vertical comb itself. The ability to exploit large flowering trees by rapid recruitment, coupled with large colony sizes and wide flight ranges, makes *Apis* bees very important components of Old World rain forest ecosystems.

Honeybees have been introduced into many other regions of the world, altering the organization of rain forest insect communities and pollination relationships.

European strains of *A. mellifera* have been deliberately transported all over the world in the last few centuries, but are poorly adapted to the lowland tropics and have invaded rain forests only on islands, such as Hawaii and Mauritius, and in Australia, where they are now the most common visitors to some mass-flowering rain forest trees (Williams & Adam 1997). It was not until queens of an African strain of *A. mellifera* were imported to Brazil in 1956 to increase honey production and subsequently escaped that honeybees became established in Neotropical rain forests. Fifty years later, these "Africanized" bees have spread throughout South and Central America. The bees are abundant in the mosaic of forest fragments and agricultural habitats created by human activities, but rare in most undisturbed tracts of primary forest (Dick et al. 2003; de Oliveira & Cunha 2005). They differ from European strains of honeybees in several important ways, including their ability to nest in a greater variety of smaller cavities, their willingness to emigrate when the local resources are exhausted, their preference for pollen over nectar, and their much more aggressive defense of nest sites (Winston 1992; Harrison et al. 2006). This latter characteristic, which presumably evolved in response to nest predation by vertebrates in Africa (including, for several millennia, humans), has led to a number of human deaths and the popular name "killer bees."

Apis mellifera is not the only honeybee species that has been deliberately introduced outside its natural range. The closely related and widely domesticated Asian hive bee, *A. cerana*, is now established on many islands east of Sulawesi and has recently colonized New Guinea. Occasional colonies of this and other honeybee species have been found aboard ships in Australian ports, and there is a clear risk of further introductions both there and in the Neotropics.

Introduced honeybees have been blamed for a dramatic decline in the native solitary bees of Hawaii, and research is urgently needed to determine whether a similar decline in bee diversity is occurring in Neotropical or Australian rain forests. Africanized honey bees do compete with native bees for specific floral resources in the Neotropics, but in the few communities that have been studied, this has not resulted in the expected population declines, because the native bees compensate by switching to other flowers (Roubik 2009; Roubik & Villanueva-Gutiérrez 2009). The impact of introduced honeybees may be greater, however, on specialized native species, which cannot easily switch food sources, or in low diversity vegetation, where there are few alternatives.

Conclusions and future research directions

This chapter has touched only briefly on a few of the most well-known groups of insects: the butterflies and social insects. Other important groups such as the beetles, flies, and moths, which together account together for two-thirds of all the insects described, must await further investigation before cross-continental comparisons can be made. Although the butterfly communities of the various regions each include unique and visually distinctive groups of species, these differences between the regions do not appear to have any major ecological importance. In contrast, there are at least three major differences between the social insects of the different rain forest regions that have considerable ecological significance. First, leaf-cutter ants in the Neotropics selectively harvest large amounts of biomass on which they cultivate fungi in underground chambers.

Second, the aggressive foraging of army ants in the Neotropics and Africa removes animal life from local areas of the forest floor and creates foraging opportunities for many other species. And third, honeybee species in the Asian and African tropics forage as a group and dominate many flowering plants, excluding other insects from those resources. The major differences between the termite communities of different regions are probably also of considerable ecological significance, but more studies are needed to identify which differences are important and why.

The recent introduction of Africanized honeybees into the American tropics is of grave conservation concern, but it is also a chance to examine the interaction between species of two divergent biogeographical regions. Although no one could reasonably suggest carrying out such an experiment for scientific reasons, now that honeybees are in both the New World and the Australian region, we can study their impact on other insects and native plant species. Massive changes in the insect fauna of the New World have been predicted as a result of the introduction of honeybees, but these have not (yet?) been observed (Roubik 2009). Honeybee densities are greatest in areas disturbed by selective logging, farming, and other human disturbances, while honeybees are sometimes present but always less abundant in primary rain forest. If this is always the case, we may not be able to distinguish the specific impact of honeybees on active insect communities from the more general impact of direct human activities. It has also been shown that honeybees may have a positive impact on the survival of forest trees in highly disturbed landscapes by providing pollination services to isolated trees that the native pollinators do not visit (Dick et al. 2003).

Honeybee densities are currently being manipulated in many Asian and African forests through the collection of honey from bee colonies. When honey is recovered carefully from the colony, the colony can recover. However, if the harvesting is done carelessly and the queen is killed, the colony may not recover. Destructive harvesting of honey is a particular threat to *Apis dorsata* (Oldroyd & Nanork 2009). In areas where honeybee populations have decreased due to the over-harvesting of honey, it would be valuable to determine whether fruit set has declined in certain plant species through lack of pollination or, alternatively, perhaps other insect species have increased in abundance and are carrying out pollination activities.

The ecological significance of leaf-cutter ants could potentially be investigated in a comparative manner by clipping the leaves of African or Asian trees in a similar way to the damage inflicted on a Neotropical forest by these ants. While such an experiment is almost certainly impractical, it is at least worth contemplating. An alternative and more practical approach would be to experimentally eliminate leaf-cutter ants from an area and monitor the impact on the surrounding forest. It would be expected that the trees most heavily harvested by the ants would respond positively, while the numerous subterranean animals that live in association with the leaf-cutter ants would suffer. This manipulation of leaf-cutter ant colonies is already being carried out by farmers who poison colonies that are harvesting their plantation trees. The farmers know from experience that their orange trees do not do well when ants clip off their leaves and flowers.

One of the most challenging groups to investigate, and one of great ecological significance, is the termites. With their subterranean foraging behavior and nests, they are not readily suited to observation and manipulation. One group that can be observed and investigated is the processional termites of Southeast Asia. These termites can be manipulated by damaging trails and removing entire colonies. Using such manipulations, the role of at least these highly specialized termites in forest processes could potentially be investigated.

Insects remain one of the great, unknown forces in tropical rain forests. The uneven distribution of honeybees, army ants, termite families, and leaf-cutter ants highlight the differences among rain forests. But other groups of invertebrates, such as soil spiders, seed-eating beetles, and canopy thrips, might be found to have an enormous role in ecological processes once they are more completely investigated and compared across continents.

Chapter 8

Island Rain Forests

When the first Polynesian voyagers arrived in the Hawaiian Islands in their double-hulled sailing canoes around 300 AD, they found rain forests totally unlike those their ancestors had left behind in Southeast Asia more than two thousand years earlier. The forests were wet and evergreen, but there the resemblance ended. There were no mammals, except for a single species of insectivorous bat, and there were no frogs, snakes, or lizards. There were birds aplenty, but no parrots, pigeons, or doves. Indeed, only 13 families of birds were represented in the forest avifauna, and only six of these were passerine songbirds. Insects too were abundant and diverse, but there were no ants or termites, almost the defining insects of the rain forest, and there were also no social bees, swallowtail butterflies, mosquitoes, cockroaches, or mantids. There were also no earthworms to eat the fallen leaves or scorpions to hunt through the leaf litter. The vegetation had the green exuberance of the rain forest, but with far fewer plant species than even the least diverse continental site. Many important large, pantropical, rain forest tree families were completely absent, including the Annonaceae (soursop family), Meliaceae (mahogany family), and Myristicaceae (nutmeg family). Also missing were herb families such as the Araceae (aroid family) and Zingiberaceae (ginger family). Other large plant families, such as the Lauraceae (laurel family) and Sapotaceae (sapodilla family), were represented by only one or two species. There were no figs (*Ficus*), the pre-eminent keystone species producing food for birds and mammals, no *Piper* (black pepper family, so common in New World forests), no gymnosperms, only three species of orchids despite thousands in rain forests elsewhere, and only a single genus of palms, in contrast to their diversity in continental rain forests.

The explanation for these unique Hawaiian rain forests is the same as that for the late arrival of humans, who had long since spread over almost all other habitable parts of the Earth's surface: extreme isolation. The Hawaiian Islands form the most isolated archipelago on Earth, located as they are at the center of the Pacific Ocean, 3765 km (2340 miles) from the nearest major land mass (Fig. 8.1). They rose out of the sea as a result of a series of volcanic eruptions as the Pacific tectonic plate moved westnorthwest over a fixed geological hotspot,

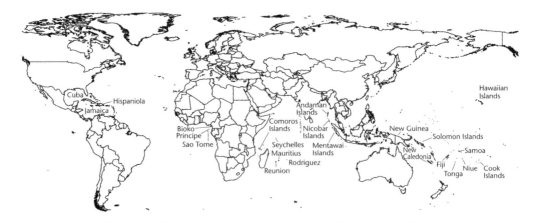

Fig. 8.1 Map showing the locations of the major islands and island groups with tropical rain forest that are mentioned in this chapter.

forming a linear sequence of islands of different ages (Neall & Trewick 2008). The oldest of the high islands, Kaua'i, has existed for only 5 million years, but eroded remnants of older islands extend for 1600 km to the northwest, and the archipelago as a whole is much older.

The ancestors of the plants and animals that inhabited the Hawaiian Islands before the arrival of the Polynesians all had to cross thousands of kilometers of ocean. It has been estimated that the rate of successful colonization needed to account for all native species of land-dwelling plants and animals in Hawaii was only one species every 35,000 years. The rate could have been even lower if species moved along the archipelago as new islands formed and old islands eroded. The likelihood of dispersal across the vast distances differs greatly between families and species, so the resulting biota was highly unbalanced, with relatively well-dispersed groups, such as ferns or mosses, much better represented than poorly dispersed trees or vertebrates.

Rain forest will grow on any tropical island that has enough soil and adequate, well-distributed rainfall (Fig. 8.1). However, the rainfall over much of the tropical oceans is too low to support rain forest. On many islands, therefore, such forest is confined to those areas that receive extra, orographic rainfall as a result of the rapid uplift of the prevailing winds as they blow over mountains. In general, rainfall is greater on the windward side of an island than the leeward side and also increases with altitude above sea level. Thus, the annual rainfall over open sea around the Hawaiian Islands is less than 700 mm a year, and rain forest was confined to the windward (north and east) sides of the six high islands, with drier vegetation types on leeward sides of the high islands and all over the low islands.

Pacific islands

The Hawaiian Islands are an extreme case, but isolation has influenced the floras and faunas of rain forests on all islands that have never been connected to a major land mass. Rain forest animals are often unable or unwilling to cross even small open areas, never mind a large expanse of sea. Rain forest plants often have large, poorly dispersed seeds that are generally too heavy to be dispersed by wind, often do not float, and most are killed by seawater. The resulting attenuation of the biota with distance from the mainland is well illustrated by the islands east of New Guinea.

The huge island of New Guinea, which has had dry land connections with Australia but not Asia, has a unique but diverse mammalian fauna of marsupials, monotremes, rats, and bats (Flannery 1995a). Marsupials and monotremes are almost entirely confined to the islands that were connected to New Guinea during periods of low sea level. Native rats extend as far east as the Solomon Islands, which never had a dry land connection and must have been colonized by animals carried on floating vegetation (Flannery 1995b) (Fig. 8.2). The Solomons also have a very diverse bat fauna descended from ancestors that reached the island by flying over water. A few large bat species, mostly flying foxes in the genus *Pteropus*, extended far out into the Pacific, where they were the only land mammals. Fruit bats reached as far as the Cook Islands, but no bats reached the Hawaiian Islands from the west: they received the ancestor of their single, insectivorous, bat species, the Hawaiian hoary bat (*Lasiurus cinereus semotus*), from North America.

Frogs, snakes, and lizards reached as far east as Fiji, presumably rafting there on tree trunks and other floating debris. Several species of lizard now extend further east, although it is possible that human movement assisted this range extension. Among the rain forest birds, the hornbills, pittas, sunbirds, and flower-peckers reached their eastern limit in the Solomons, but – rather surprisingly for

Fig. 8.2 The Isabel naked-tailed rat is a large tree-dwelling species confined to the 3000-km² island of Santa Isabel in the Solomon Islands, where it is the only native non-flying mammal. (Courtesy of Pavel German.)

such heavy-bodied birds – megapodes got as far east as Tonga, Samoa, and Niue (Steadman 2006). Parrots, pigeons, and doves occupied most islands in the tropical Pacific except Hawaii, which derived at least half of its bird fauna from the New World. It is interesting to speculate on how different the community structure in the Hawaiian Islands might have been if these islands had been reached by either fruit pigeons or *Pteropus* fruit bats, the major inter-island seed dispersal agents elsewhere in the Pacific.

Rain forest floras show a similar, but more gradual, decline in diversity with distance from the mainland, with the greatest reduction among the canopy trees, which tend to have large, poorly dispersed seeds and fruits. A single hectare of rain forest in New Guinea can support more than 200 species of trees and woody climbers (Wright et al. 1997). Rain forests in the Solomon Islands have fewer species than New Guinea but a similar range of families. Fijian rain forests are surprisingly diverse, with 124 tree species in a single hectare (Keppel et al. 2010), but several rain forest families are missing. Many rain forest genera reach their eastern limits in Samoa, but figs have almost the same distribution as the pigeons and doves that disperse them. By the time the Hawaiian rain forests are reached, the flora is substantially reduced. Hawaiian rain forests are mostly dominated by a single species in the myrtle family, 'Ohi'a lehua (*Metrosideros polymorpha*) (Fig. 8.3b), which has tiny, wind-blown seeds, with koa, *Acacia koa*, a leguminous species, co-dominant in some areas. Although the rain forest flora has largely reached Hawaii from the west, there is also a substantial tropical American component.

The large (16,750 km^2) island of New Caledonia must be considered separately because, uniquely in the tropical southwest Pacific, it is not volcanic in origin, but is a fragment of the ancient supercontinent of Gondwana (see Chapter 1). It separated from Australia at least 65 million years ago, possibly much earlier, and has been isolated since then. The ancestors of its diverse rain forest flora could have reached New Caledonia before the dry land connection was severed, but the fauna seems to be the result of more recent colonization over water, suggesting that the island may have been entirely flooded by sea after separation from Australia (Leigh et al. 2007). Thus, as with similarly isolated volcanic islands, prior to human occupation there were no amphibians, bats were the only mammals, reptiles were represented by a diversity of skinks and geckoes, and there were relatively few families of land birds.

Evolution on islands

If a selective reduction in biological diversity were the only consequence of isolation, island rain forests would be merely an ecological curiosity. However, the absence of whole groups of typical rain forest organisms and the low diversity of others creates unusual opportunities for those species that do become established. In some cases, the new colonizers spread into a much wider range of habitats than their mainland ancestors. Thus, the Hawaiian 'Ohi'a lehua is not only the dominant canopy tree in rain forests of all ages from sea level to over 2000 m altitude, but also invades dry young lava fields and wet high-altitude bogs, changing its growth characteristics to suit the local environment. Despite a growing number of studies, the relative importance of genetic differences and extreme phenotypic plasticity of 'Ohi'a lehua in this ecological range is still unclear

(Harbaugh et al. 2009). Similarly, the Hawaiian hoary bat today makes use of almost all natural and human-made Hawaiian habitats, from rain forests to urban areas, whereas in continental areas species of bats are typically found only in particular habitats.

In other cases, in contrast, the colonizers on isolated islands evolved in their new home into large numbers of closely related species, each of which occupied a different, previously vacant niche. As the number of colonizers declines with distance from the mainland, the opportunities for such evolutionary radiations may increase. In the Solomon Islands, two genera of giant rats (*Solomys* and *Uromys*) evolved into forms similar to some of the arboreal and terrestrial marsupials of New Guinea rain forests (Flannery 1995b) (Fig. 8.2). In Fiji, which had no terrestrial mammals at all, recent fossil discoveries include a giant frog, 25 cm in length, a 1.5 m iguana, a giant tortoise, a 2 m land crocodile, flightless megapodes, and a giant flightless pigeon, 80 cm tall (Worthy et al. 1999). New Caledonia also had a giant megapode, *Sylviornis neocaledoniae*, which stood 1.2–1.6 m tall and weighed around 40 kg (Steadman 2006). Flightless birds, some of them ecologically similar to herbivorous mammals, evolved on almost all Pacific Islands, particularly among the chicken-like rails (Rallidae) (Kirchman & Steadman 2007).

Birds

The most spectacular radiations occurred on the most isolated large islands: those of the Hawaiian archipelago. Most conspicuous to the newly arrived Polynesians would have been the birds. The 112 known non-marine bird species in Hawaii evolved from fewer than a quarter as many colonizers. Moreover, more than half these species (61) belonged to a single tribe, the honeycreepers (Drepanidini) (Fig. 8.3a,b), which are derived from a single colonizing species of seed-eating finch (*Carpodacus* sp., Fringillidae). DNA studies suggest that this colonization event may have occurred only 2 to 5 million years ago (Fleischer & McIntosh 2001), yet the honeycreepers have radiated into an amazing range of shapes, sizes, colors, habits, and habitats. If Darwin had visited Hawaii instead of the Galapagos, the honeycreepers would have provided an even clearer example of adaptive radiation than the Galapagos finches. Molecular studies have also shown that, as with other island radiations, the degree of genetic divergence is less than expected given the morphological diversity, reflecting their relatively recent evolution.

Some nectar-feeding honeycreepers, such as the orange-red I'iwi (*Vestiaris coccinea*) (Fig. 8.3b), evolved long, curved bills to feed from the similarly curved flowers of the endemic lobelias. Others, such as the scarlet 'Apanane (*Himatione sanguinea*), had short bills and fed on the more accessible nectar of the 'Oh'ia lehua. Insectivorous species evolved short, thin bills for picking insects from foliage or, as in the yellow-green Nukupu'u (*Hemignathus lucidus*), thin, strongly down-curved bills used for probing bark crevices. The green 'Akeke'e (*Loxops caeruleirostris*) had a peculiar scissor-like bill, which it used to force open closely spaced leaves to get at insects hidden within. The bicolored 'O'u (*Psittorostra psittacea*) had a parrot-like bill, which it used to eat fruits and seeds. Other species had even stouter bills, strong enough to crack the hardest seeds.

(a)

(b)

Fig. 8.3 Honeycreepers of the Hawaiian Islands. (a) Some specimens of extinct honeycreeper species. (Courtesy of Dennis Hansen.) (b) An I'iwi (*Vestiaris coccinia*) visiting the flower of an 'Ohi'a lehua (*Metrosideros polymorpha*) plant. (Courtesy of Jack Jeffrey.)

The honeycreepers were the most diverse and numerous bird radiation in the Hawaiian Islands, but there were also several other, smaller ones. At least a dozen flightless species of chicken-like rails (*Porzana*) evolved from two or more colonizers, and six species of Hawaiian honeyeaters (Mohoidae) (Fig. 8.4)

Fig. 8.4 The Hawai'i 'O'o (*Moho nobilis*) was part of a radiation of endemic Hawaiian honeyeaters. It was heavily hunted by both native Hawaiians and Europeans and the last known sighting was in 1934 on the slopes of Mauna Loa. Painting by John Gerrard Keulemans (1842–1912).

evolved from one original species. Six species of thrush (*Myadestes*) were probably derived from a single colonist, as were at least three crows (*Corvus*), four species of long-legged, bird-feeding owls (*Grallistrix*), and at least two species of flightless ibis (*Apteribis*). Another radiation produced at least four species of flightless moa-nalos (Anatidae: Thambetochenini), the size of large swans, from normal ducks. These bizarre birds seem to have browsed plants in the forest understorey and grasses in open habitats in a manner comparable to mammalian grazers (James & Burney 1997). Weirdest of all was another recently described flightless duck, *Talpanas lippa* (literally, "nearly blind mole-duck"), with tiny eyes, which may have occupied a kiwi-like niche on Kauai, using its sensitive beak to search for prey among the forest litter layer (Iwaniuk et al. 2009).

Insects

While the bird radiations were the most visible, they were by no means the largest in terms of numbers of species. That honor belongs to the tiny Hawaiian fruit flies in the genus *Drosophila*, which may number as many as 1000 species, with most evidence suggesting they were all derived from a single colonist species (Edwards et al. 2007). These flies show a huge range of ecological, morphological, and behavioral diversity, greater than that shown by all the non-Hawaiian species in the genus (Fig. 8.5). They have evolved an amazing diversity of courtship rituals, some of which are reminiscent of those performed by birds of paradise. Most species breed on decaying plant material, but 11 species are parasitic on the cocoons of spiders.

Fig. 8.5 A small part of the diversity of the genus *Drosophila* in Hawaii. This photograph shows the diversity of wing patterns in one subgroup of the "picture wing" group of *Drosophila* (Edwards et al. 2007). (Courtesy of Kevin Edwards.)

There were many other spectacular insect radiations on the Hawaiian Islands, and it has been estimated that fewer than 400 separate colonizations can account for all of the more than 5000 insect species (Leigh et al. 2007). In addition to the fruit flies, at least eight other genera contained more than 100 species. More than a third of the known Hawaiian moth species belong to the single endemic genus *Hyposmocoma* (Cosmopterigidae), which is so ecologically diverse that the adults were originally placed in 14 separate genera. More than 350 *Hyposmocoma* species have been described and it is estimated that there are many more. The larvae include grazers, scavengers, woodborers, and even aquatic species, and they occupy almost all habitats, from rain forests to dry lava flows (Rubinoff 2008). One species even kills and eats snails.

Flightless forms have evolved in almost all insect families that arrived with wings and there have been many other unusual adaptations to the island environment and simplified fauna. Perhaps the most bizarre is the evolution of carnivory in caterpillars of the moth genus *Eupithecia*. Uniquely among the Lepidoptera, these caterpillars are ambush predators: when potential prey touches the rear end of the motionless caterpillar, it suddenly loops back to seize and devour it. The Hawaiian crickets were probably derived from as few as four flightless ancestors, which may have arrived as eggs on floating vegetation. These species have radiated to produce more than twice as many species as the continental United States. The Hawaiian yellow-faced bees (*Hylaeus*: Colletidae) included at least 60 species that were found in every available habitat and probably pollinated a large proportion of the native flora.

Other invertebrates

Major radiations also occurred among the Hawaiian spiders, particularly in the genus *Tetragnatha*, the long-jawed spiders, where some of the 100 or so species abandoned the ancestral web-building habit in favor of actively pursuing prey (Gillespie 1999).

Other major radiations involved the Hawaiian land snails, which, despite the failure of most tropical snail families to reach the islands, totaled more than 750 species (Cowie 1995). The endemic family Amastridae included more than 300 species of generally ground-dwelling snails, while the endemic subfamily Achatinellinae produced tree snails with an extraordinary diversity of shell colors and patterns, probably from a single colonization. Most species were confined to a single island, and only 2 to 4 species occurred on other islands in the Pacific.

Plants

Several flowering plant colonists also gave rise to spectacular evolutionary radiations in the Hawaiian Islands. One eighth of the native flora – 126 species – is made up of species in the family Campanulaceae (the bellflower family), all descended from a single ancestral species that arrived in the archipelago 13 million years ago, before any of the present-day high islands existed (Givnish et al. 2009). Most species in the Campanulaceae are temperate-zone herbs with dry, capsular fruits, but most species in Hawaii are woody and have fleshy fruits. There are species adapted to all major habitats, but most are found in rain forests. The largest genus, *Cyanea*, has an amazing range of growth forms, leaf sizes and shapes, and floral morphology (Fig. 8.6a,b,c). The smallest species has a maximum height of around 1 m, while the largest is a palm-like tree up to 14 m tall. The leaves in different species range widely in size and include both simple leaves and doubly compound, fern-like fronds. The floral diversity seems to have co-evolved with the pollinating honeycreepers and honeyeaters, resulting in a huge range of corolla tube lengths, degrees of curvature, and colors. In some species, the leaves of seedlings are spiny, a characteristic which may, in the absence of grazing mammals, have evolved to protect the young foliage from giant flightless geese and moa-nalos, which grazed in the understorey.

Indian Ocean islands

The Pacific is unique for its large number of tropical islands, but the effects of isolation and ecological opportunity on rain forests biotas were similar in other oceans. In the northeast Indian Ocean, the Andaman and Nicobar Islands form an arc stretching 550 km between Myanmar and Sumatra (Fig. 8.1). These are true oceanic islands that were uplifted in the Tertiary and have never been connected to the Asiatic mainland. They were largely covered in rain forests, but these forests lack many characteristic plant and animal groups of their mainland counterparts (Corlett 2009a). There are no primates, ungulates, or carnivores, and no babblers, the most diverse group of Southeast Asian rain forest birds but known to be weak flyers. Dipterocarp trees, which are another poorly dispersed group, are present in the Andamans, but not the Nicobars.

The Mentawai Islands of Indonesia are a southern continuation of the same island arc, but, unlike the Andamans and Nicobars, they appear to have been connected to Sumatra at some time within the last million years. Thus, the ancestors of their rain forest biotas need not have dispersed across the current 85 km water gap. Although the fauna is somewhat impoverished relative to the Sundaland rain forests, these islands support a number of endemics, including four species of primates.

Mid-way between Africa and the northern tip of Madagascar are the four Comoro Islands. These were formed volcanically and are much younger than Madagascar. In common with many other oceanic islands, their rain forests are relatively species-poor, and the vertebrate fauna consisted only of bats, birds, and reptiles. To the east of Madagascar are the volcanic Mascarene islands of Mauritius, Rodriguez, and Reunion, whose faunas included giant, turkey-sized flightless pigeons – the dodo on Mauritius (Fig. 8.7) and the solitaire on Reunion – as well as flightless rails, bats, giant tortoises, and a variety of other reptiles (Cheke & Hume 2008).

North of the Mascarenes is the Seychelles archipelago, which includes 41 granite islands that originated not as volcanoes but as a fragment of Gondwana that separated from India an estimated 65 million years ago. The nearest major land mass is Madagascar, 930 km away. As expected for such isolated islands, the rain forests of the Seychelles are relatively species-poor, but they have many endemic plants. These include a giant rain forest dipterocarp with wingless fruits, *Vateriopsis seychellarum*, which seems unlikely to have dispersed across the ocean and may be an ancient survivor from Gondwana. The same origin has been

(a)

(c)

(b)

Fig. 8.6 Species in the Hawaiian plant Genus *Cyanea* have an amazing range of different growth forms. (a) Degener's cyanea (*Cyanea degeneriana*). (Courtesy of Jack Jeffrey.) (b) Shipman's cyanea (*Cyanea shipmanii*), with fern-like fronds. (Courtesy of Jack Jeffrey.) (c) Flowers of the ridge rollandia (*Cyanea longiflora*). (Courtesy of Jack Jeffrey.)

Fig. 8.7 A 17th century Dutch painting of a dodo: a giant, flightless pigeon endemic to the Indian Ocean island of Mauritius.

suggested for several other poorly dispersed groups, including an endemic genus of land snails (*Pachnodus*, Enidae), an endemic family of frogs (Sooglossidae), and three endemic genera of caecilians (legless, worm-like amphibians; Caeciliidae) (Zhang & Wake 2009), but rare oceanic dispersal events have been suggested for others.

Atlantic islands

The Atlantic, excluding the Caribbean, has relatively few oceanic islands, and only four of these had extensive tropical rain forests (Fig. 8.1). These four, however, provide an excellent illustration of the effects of isolation on rain forest biotas (Jones 1994). Bioko (2027 km^2), Principe (139 km^2), São Tomé (857 km^2), and Annobon (17 km^2) are mountainous volcanic islands that are, respectively, 32, 220, 225, and 340 km from the coast of West Africa. Bioko, the largest, was connected to the mainland during periods of low sea level and has only been an island for around 10,000 years. The other three are true oceanic islands that have never been connected to the African mainland or to each other. All four were covered largely in tropical rain forest. Bioko has a rich, continental rain forest biota, including 10 species of primates. It has few endemic species, but the island has been isolated long enough for distinct races or subspecies to have evolved in many vertebrates, including at least five of the primates.

The other three islands have never had any large mammals. All three have bats, birds, and lizards, while Principe and São Tomé also have shrews, snakes, and frogs. Levels of endemism in all groups are high. São Tomé, in particular, has three endemic genera and 15 endemic species of land birds (Dallimer et al. 2009), plus an endemic genus of plants, an endemic species of shrew, several endemic species of reptiles, an endemic genus of amphibians, and an endemic family of snails with a single species (Jones 1994). Annobon is more isolated but much smaller, so it has fewer species and fewer endemics.

Caribbean islands

The islands of the Caribbean were almost entirely forested until the last few centuries, although not all this was rain forest. These islands had a more complex geological history than most of those considered so far, and the origins of some elements of their biota are still disputed. The majority of the flora and fauna seems to have dispersed to the islands over water, but there may also have been dry land connections or island stepping-stones in the past between at least some of the islands and South and Central America (Ricklefs & Bermingham 2008). Although these connections, if they existed, were broken more than 30 million years ago, there is fossil evidence that the ancestors of some of the recent vertebrate biota had arrived by this time. The mammalian fauna consisted of bats, rodents, and primates – the same groups that reached Madagascar – as well as the uniquely Neotropical sloths. The three known species of primates, which occurred on the large islands of Cuba, Hispaniola, and Jamaica, may have had a common ancestor, but the sloths provide the clearest example of adaptive radiation. In the absence of potential competitors, species ranging in size from huge ground sloths to smaller arboreal forms occurred on Cuba and Hispaniola (White & MacPhee 2001). Bats and rodents were the most diverse mammalian groups and were present on islands that were too small to support the larger mammals.

There were also spectacular radiations in the *Anolis* lizards (Iguanidae or Polychrotidae, depending on the system of classification) (Fig. 8.8), the *Sphaerodactylus* geckos (Gekkonidae), and the *Eleutherodactlyus* frogs (Leptodactylidae). A total of 151 species of Caribbean island *Anolis* are currently recognized, with 63 species on Cuba, the largest island, and at least one on most islands larger than 0.25 km^2 (Losos 2009). There are also at least 79 species of *Sphaerodactylus* (including the world's smallest lizards, at *c.* 16 mm long) and 139 species of *Eleutherodactlyus*, although by no means all of these are found in rain forests. In Caribbean rain forests, the diverse and abundant *Anolis* lizards dominate the daytime insectivore niche, which is occupied largely by birds in continental rain forests, while the *Eleutherodactlyus* frogs dominate the equivalent niche at night.

Natural disasters

Getting there was the biggest problem faced by the ancestors of island rain forest organisms, but not the only one. As well as vacant niches and under-utilized resources, tropical island environments also have a dark side: a tendency to natural disasters, from which the surrounding ocean and their small size makes escape impossible. Not surprisingly, in view of their mostly volcanic origin, there are still active or recently active volcanoes in many tropical island archipelagos. A few years ago, an eruption of the long-dormant La Soufriere volcano on the Caribbean island of St Vincent destroyed much of the remaining forest. Tsunamis, caused by earthquakes and other undersea disturbances, are also

Fig. 8.8 Male Plymouth anole (*Anolis lividus*) displaying its dewlap on the Caribbean island of Montserrat. (Courtesy of Jonathan Losos.)

a threat to low islands, particularly in the tectonically active Pacific. The most frequent natural catastrophe on most tropical islands, however, is a direct hit by a tropical cyclone.

Cyclones, also known as hurricanes or typhoons, bring both extreme wind speeds and concentrated rainfall. Tropical islands and their forests are particularly vulnerable, because their small areas have little influence on the cyclone's intensity. In contrast, the energy of a cyclone diminishes rapidly after it has made landfall on a continent. Repeated cyclone damage is therefore a fact of life on many islands in the tropical Pacific, the Indian Ocean, and the Caribbean. Trees are uprooted or snapped off by wind only in the most exposed sites, but major canopy damage and near total defoliation are more widespread (Lugo 2008). Animals are killed both directly and indirectly through reductions in food supply. Fruits, flowers, and leaves become unavailable for weeks or months until the damaged stems sprout new growth, but insects survive and are easier for insectivorous species to find. The native species on cyclone-prone islands must be adapted to such events, but the same cannot be assumed for new colonizers, and there may well have been species that survived the perilous journey to a remote island, only to succumb to the first severe cyclone.

Human impacts

This account of island rain forests has been written largely in the past tense. The reason for this is that many of the species discussed are now extinct: the melodious 'o'o'a'a no longer sings from the 'ohi'a trees of the Hawaiian rain forest, and the moa-nalo no longer browses in the understorey. These and countless other species were victims of a very unnatural disaster. As humans spread beyond the edges of the continental shelves, they came into contact with island faunas that had no experience of terrestrial predators and no resistance to continental diseases. For people, these were truly island paradises: no dangerous predators or diseases, and animals that could be gathered for food as easily as plucking fruits. The result was an expanding front of extinctions, eventually encompassing more than a thousand species of birds and hundreds of species of reptiles, as well as many small mammals and countless insects, snails and other invertebrates.

The impact of human arrival must have depended to some extent on the social and technological sophistication of the humans involved, but even hunter-gatherers arriving on unpopulated islands drove large vertebrates to extinction. Most damaging, however, were peoples, like the Polynesians colonizers of the Pacific, who grew crops and raised domestic animals. The great voyaging canoes that reached the Hawaiian Islands brought not only people but also their crop plants, domestic animals, and assorted stowaways. The Polynesian settlers decimated populations of native seabirds and hunted many forest birds to extinction for food and their colorful feathers (Boyer 2008). They cleared almost all the lowland rain forest for growing crops, and their pigs and chickens, along with stowaway geckos and skinks, spread into the forest that was left. Rats (*Rattus exulans*) may have been accidental or deliberate introductions, since they were sometimes used as food. Their exponential population increase on islands free of competitors and predators may have devastated native animals and large-seeded plants, such as palms, well ahead of the more slowly expanding human population (Athens 2009). On the smaller islands of the Pacific, the Polynesian arrival signaled the almost instant collapse of the native vertebrate community, while on larger islands the impact was more gradual, although the end was the same.

When Captain James Cook accidentally "discovered" Hawaii in 1778, on his third voyage of Pacific exploration, the islands had already been transformed by human impact. The lowland rain forest had almost entirely gone, as had half of the native bird fauna, including the moa-nalos and most other large, ground-nesting species (Boyer 2008). Post-Polynesian impacts on the rain forest, which was largely confined to upland areas, have included further clearance, commercial harvesting of timber, and the introduction of a huge array of additional exotic plants and animals. No species was deliberately introduced to the rain forest, and most species that have become established in the islands are confined to the disturbed lowlands. Enough, however, have recently invaded the rain forest to cause massive damage to the native flora and fauna.

There are still rain forests in the Hawaiian Islands, but they are now very different from those seen by the first Polynesian voyagers 1500 years ago. Exotic pigs root through the understorey, destroying native plants, feeding on exotic earthworms, and dispersing the seeds of exotic plants, such as the strawberry guava (*Psidium cattleianum*) and the banana poka (*Passiflora mollissima*) (Fig. 8.9a,b) (Nogueira-Filho et al. 2009). The black rat (*Rattus rattus*) climbs into the trees, eating seeds, snails, and insects, as well as the eggs and nestlings of the surviving native birds. Flocks of Japanese white-eyes move through the canopy, feeding on insects, fruits, and nectar. European honeybees visit the flowers of native trees, while introduced ants and yellowjacket wasps (*Vespula pensylvanica*)

(a)

(b)

Fig. 8.9 The banana poka (*Passiflora mollissima*), a climbing vine, is an invasive species in the Hawaiian Islands. (a) Banana poka flower. (Courtesy of Jack Jeffrey.) (b) Banana poka vines overgrowing native vegetation. (Courtesy of Jack Jeffrey.)

hunt both native and exotic invertebrates (Wilson et al. 2009). The predatory rosy wolfsnail (*Euglandina rosea*) stalks the few native snails that have survived the onslaught of the rats. Native moths are attacked by exotic parasites that were introduced to control pests of lowland crops, and bird-biting mosquitoes (*Culex quinquefasciatus*) transmit avian pox and malaria to susceptible native birds.

The same story, with varying intensity, has been repeated on other rain forest islands. In some cases, separating natural from human-caused extinctions is made difficult by gaps in the fossil and archeological records. An incredible 88% (67 of 76) of the known nonflying land mammal species of the Caribbean islands have become extinct in the last 10,000–20,000 years (Morgan & Woods 1986). Many extinct species, particularly rodents, are common in Amerindian kitchen middens, showing that they survived into the period of human occupation, but for others, including the primates and most of the sloths, the extent of overlap with people is less clear (MacPhee et al. 2007). In view of the record of human impacts on island biotas elsewhere, however, many ecologists are inclined to see a human hand behind most Holocene extinctions.

The arrival of European explorers had most impact on those few rain forest islands that had not already been settled by other people. The best-documented examples are the Mascarene Islands of Mauritius, Reunion, and Rodrigues in the Indian Ocean (Cheke & Hume 2008). Although known already to Arab navigators, the first European visitors in the 16th century found the islands uninhabited, covered in forest, and with their native biotas apparently intact. The Mauritian dodo (Fig. 8.7), which has become synonymous with extinction, was only one of many endemic species to succumb to hunting, deforestation, and introduced species over the next four hundred years. Today, the rain forest on Mauritius has been reduced to a few upland remnants, covering less than 5% of the total land area and heavily invaded by exotic plants, where nest predation by introduced black rats and long-tailed macaques threatens the survival of the remaining native forest birds.

Worldwide, around 75% of terrestrial vertebrate extinctions in the last 500 years have been on islands, although by no means all in rain forest (Sax & Gaines 2008). At least 13 endemic bird species have been lost from the Caribbean alone in this period, and 48 more are threatened with extinction (McGinley 2008). Nineteen endemic mammal species have also gone extinct from the Caribbean in the past 500 years (McGinley 2008). Island plants are also particularly vulnerable to extinction (Caujapé-Castells et al. 2010). The continued vulnerability of island biotas is highlighted by the fact that the Hawaiian Islands account for 75% of documented plant and animal extinctions in the United States and more than a third of the current endangered species lists for plants and birds, even though they make up less than 0.2% of the total land area of the United States (Stein et al. 2000).

Although tropical island rain forests have suffered a huge range of adverse impacts, the small areas that remain are usually more intact than drier vegetation types on the same islands, and are often the last strongholds for what survives of the endemic flora and fauna. Moreover, although the total area of island rain forests worldwide is small, and the forests on each island are much less species-rich than their continental counterparts, the high rate of endemism means that island rain forests together support a significant proportion of all known rain forest species. The extinction crisis in island rain forests is far from over.

Conclusions and future research directions

Rain forests on islands represent both crucibles of evolution and hotspots of recent extinction. They have provided some of the best examples of the process of speciation, and they have greatly influenced the thinking of evolutionary biologists. Islands have also been central to biogeographic theory ever since Robert MacArthur and E.O. Wilson developed a theory to explain the numbers of species they support (MacArthur & Wilson 1967). Lessons from island biogeography have also been applied to conservation planning on continents, with surviving patches of natural habitat viewed as "islands" in a "sea" of agricultural and urban development. This analogy should not be taken too far, however, since human-made habitat islands differ in many important ways from true islands.

Human impact has already driven thousands of rain forest species to extinction, with a high proportion of the documented extinctions taking place on islands. Island species are especially vulnerable due to the small areas they occupy, as well as their lack of resistance to introduced predators, grazers, competitors, and diseases. Many more will go extinct in the near future unless current conservation efforts at habitat preservation and species management are successful. Not only will species be lost, but our ability to investigate the process of evolution will also be diminished.

Biologists and governments need to become actively involved in protecting island species because this is where there are so many vulnerable species; that is, species with limited range, few populations, and declining numbers of individuals. The richness of unique species on islands and their vulnerability to extinction has been highlighted by international conservation organizations, such as the Hotspot approach of Conservation International (Mittermeier et al. 2005). Research is needed not only to describe the ecology of these species, but also to map their ranges, monitor population sizes, and document the threats to species.

One critical area of research is documenting the impact of exotic species of plants and animals on the native flora and fauna. Such documentation will assist in providing arguments for the control of invasive species. When invasive species are controlled, programs to restore native species need to be implemented. Areas formerly occupied by rain forest can be replanted with native tree species, and native animal species can be gradually reintroduced. The restored rain forests can be monitored and compared with undisturbed forests to determine how effective the new forests are in maintaining native species and the ecosystems functions of forests. If such programs are successful, they provide a positive demonstration to the public that the remaining diversity of island life can be protected.

Chapter 9

The Future of Tropical Rain Forests

In this book, we have examined the unique features of rain forests in many regions of the world. Although we have focused on the intact rain forests, in almost every case these rain forests have been damaged by human activities – in the worst cases, they are in the process of being utterly destroyed. In this final chapter, we will consider the threats to rain forests and strategies to protect them. Whenever possible, we will try to highlight threats and possible solutions that are unique to each rain forest area.

The major threats to tropical rain forests are the clearance and fragmentation of the forest, largely for agriculture, and the overexploitation of plant and animal species in the areas that remain. Invasions by exotic species are also a growing threat, particularly on oceanic islands, and the impact of global climate change is likely to increase in significance over the next few decades. Many threatened rain forest species face two or more of these threats, speeding their way to destruction and hindering efforts to protect them. Often, these threats develop so rapidly and on such a large scale that efforts to save species and representative examples of rain forests are difficult. The threats also build upon one another in a spiral of destruction. For example, logging companies build roads to take logs out of the forest, and then timber workers and local people use this road network to hunt animals for the camp mess hall. Later, landless people move in on the logging roads and establish farms and ranches, cutting down the remaining trees and burning them to create a nutrient-rich ash to fertilize their crops. When such agriculture is practiced near logged forests, the fire can easily spread from the fields into the forest – particularly during dry periods – creating thousands of fires over a large area, devastating forests, and producing a persistent smoky haze over an entire region.

Different forests, different threats

Just as rain forests differ both within and between the major rain forest regions, so does the amount of rain forest remaining and the nature of the threats to its

Tropical Rain Forests: An Ecological and Biogeographical Comparison, Second edition.
© Richard T. Corlett and Richard B. Primack. Published 2011 by Blackwell Publishing Ltd.

continued survival. In many parts of the world, tropical rain forests have already been largely destroyed by human activity, with lush forests converted into cropland or, all too often, into barren grasslands and scrub. Yet despite the extent of the destruction that has occurred already, there are still large areas of rain forest in several parts of the tropics. To protect these remaining rain forests, it is important to understand not only the biological differences between each forest as detailed in the preceding chapters, but to also account for the very different economic and social factors threatening forests in each area of the world. Economic globalization is having an impact, with Malaysian logging companies operating in Papua New Guinea and Africa, and Amazonian soybean oil in direct competition with Indonesian palm oil, but the major threats still differ greatly between regions.

Population growth and poverty in Africa

The rain forest countries of Africa have among the highest population growth rates in the world, as a result of women bearing an average of 4.5–6.5 children each, and they have among the lowest per capita incomes, with most people living on less than US$1 a day (Table 9.1). These problems are combined with high child mortalities and low adult life expectancies. Threats to rain forests in

Table 9.1
Some statistics relevant for the future of rain forests in five countries. Note the different dates to which they apply.

	Brazil	DRC	Indonesia	PNG	Madagascar
Area of forest (thousands km^2) (2010)[1]	5,121	1,541	909	286	121
Percentage forest cover (2010)[1]	62	68	52	63	22
Percentage of intact forest landscapes (c. 2000)[2]	32	29	20	35	8
Annual change in forest cover (%) (2005–2010)[1]	−0.4	−0.2	−0.7	−0.5	−0.5
Annual log production (mil m^3) (2008)[3]	25	0.3	34	3	0.1
Number of cattle (millions) (2007)[4]	200	1	11	0.1	10
Human population (millions) (2005)[5]	186	59	219	6	18
Population density (per sq. km) (2005)[5]	22	25	119	13	32
Human population growth rate (%) (2005)[5]	1.0	2.8	1.2	2.4	2.7
Projected human population in 2050[5]	219	148	288	13	43
Fertility (children per woman) (2005)[5]	1.9	6.1	2.2	4.1	4.8
Mortality before age 5 (per thousand) (2005)[5]	29	198	32	69	100
Life expectancy (2005)[5]	72	48	71	61	60
Per capita GDP (PPP) (US$) (2008)[6]	10,200	300	3,900	2,200	1,000

[1] FAO Global Forest Resources Assessment 2010. Includes all forest types; in the case of Madagascar most of this is not tropical rain forest.
[2] Potapov et al. (2008).
[3] International Tropical Timber Organization.
[4] FAOSTAT.
[5] United Nations Population Division.
[6] International Monetary Fund. GDP, gross domestic product; PPP, purchasing power parity.
DRC, Democratic Republic of Congo; PNG, Papua New Guinea.

these countries stem largely from the poverty of this rising population dependent upon subsistence agriculture and herding.

The rain forests of West Africa have already largely gone, transformed by fire, cattle grazing, agriculture, and the gathering of firewood. The surviving fragments are scattered within a matrix of plantations and farmland that is being invaded by savanna species that have expanded south with deforestation (Davis & Philips 2009). Vast areas of rain forest still remain in Central Africa, but these now face the same threats as access improves. Throughout the accessible forests of the region, subsistence and commercial hunting of wildlife for "bushmeat" is far more intense than in other regions (Fa & Brown 2009), with meat from wild game an important source of protein for rural populations and sometimes preferred to that from domestic animals even in urban areas. Over the last decade, the Democratic Republic of Congo (DRC) has also seen the deadliest human conflict since World War II, the Second Congo War, with devastating impacts on people and wildlife that still continue today (Nellemann et al. 2010).

Less than 1% of the forest in the Congo Basin was cleared between 1990 and 2000, but this clearance was fine-scale and pervasive, leaving relatively little forest totally free of human impacts (Hansen et al. 2008a). With a projected human population of 187 million by the year 2050, the future of the DRC's rain forests does not look good. The greatest threat in the immediate future is the dramatic expansion of road building in the Congo Basin to facilitate logging and mining (Blake et al. 2008; Nellemann et al. 2010) (Fig. 9.1). Logging is still

Fig. 9.1 Expansion of mining and the associated roads and settlements is a major threat to the rain forests of the Congo Basin. This photograph shows deforestation at a gold mining site called Paradizo, near Pili-Pili, in the Ituri District of DRC. (Courtesy of Dan Fahey.)

highly selective, minimizing local direct impacts, but very extensive. Large-scale agro-industrial clearing has been largely absent so far, accounting for the relatively low deforestation rate, but there are plans for a massive expansion of oil palm production by foreign companies to meet rising global demand.

Madagascar on the brink

Madagascar faces these same problems of poverty and population growth as Africa, but its biota is in more immediate peril because so much of its forest has already been lost (Harper et al. 2007). Estimates of the original forest extent vary widely, but more than half the humid forest present in the 1950s has now gone and the surviving forest is badly fragmented. High human population densities occur throughout most of the island, including adjacent to the rain forest areas, where land continues to be cleared for rice and cattle pasture. Logging is a relatively small but rapidly growing problem, reflecting the high value of rain forest hard-woods, such as rosewood (at least 25 *Dalbergia* spp.) and ebony (*Diospyros* spp.) on the global market (EIA & Global Witness 2010). Hunting is also a growing conservation issue in some areas (Golden 2009). In the absence of the ungulates or large rodents that are favored in other tropical forests, the hunters focus on the lemurs and carnivores. Endemic reptiles and amphibians are also widely collected for the international pet trade (Andreone et al. 2008). Madagascar had a relatively effective protected area system, but political instability in 2009 resulted in an increase in commercial hunting and illegal logging within rain forest reserves (Barrett & Ratsimbazafy 2009; Bohannon 2010).

Logging and cash crops in Asian rain forests

Southeast Asia has the highest rates of forest loss and degradation in the tropics and there is little immediate prospect of improvement. There are no large intact areas of forest left on the Asian mainland and the biggest remaining areas, on Borneo, Sumatra, and Sulawesi, are under serious threat. This region has a higher population density than most of the tropics, but until recently the population in the wetter, eastern half of the Asian tropics was concentrated in a few areas, such as central and eastern Java, where conditions favored intensive rice cultivation. In the last decades of the 20th century, however, the Indonesian government relocated more than 6 million people to the rain forest-rich provinces of Kalimantan (Borneo), Sumatra, Sulawesi, and Papua (on the island of New Guinea; see below), with predictable impacts on forests and wildlife. Although poverty and population growth still play an important role in the destruction of the region's rain forests, the high rates seen today are mainly due to a combination of the region's large logging industry, much of it illegal, and the conversion of forest to cash crops, particularly oil palm (Laurance et al. 2010a). In the 15 years between 1990 and 2005, an astounding 42% of the lowland forests of Sumatra and Kalimantan was cleared (Hansen et al. 2009) (Fig. 9.2). Of particular concern is the logging, drainage, clearance, abandonment, and burning of the region's extensive peat swamp forests, creating regional "haze" episodes and releasing huge amounts of carbon dioxide into the atmosphere – as much as 3% of total global emissions (van der Werf et al. 2009; Hooijer et al. 2010).

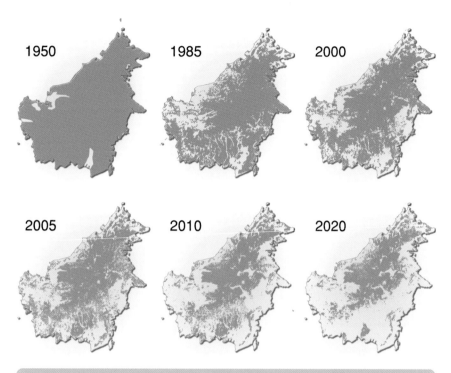

Fig. 9.2 Deforestation on the island of Borneo 1950–2005, with projections to 2020. Map by Hugo Ahlenius, UNEP/GRID-Arendal Maps and Graphics Library (http://maps.grida.no/go/graphic/extent-of-deforestation-in-borneo-1950–2005-and-projection-towards-2020).

Roads, ranching, and hydrocarbons in the Amazon

In comparison with many other major rain forest countries, Brazil is an economic success story: relatively wealthy with low fertility, low child mortality, and high life expectancy. Yet Brazil's rain forest faces many of the same threats as in Indonesia: clearance for agriculture, logging, and an influx of settlers, but with the added impact of a huge cattle industry. The increasing deforestation in the Amazon region during the 1980s and 1990s was greatly facilitated by the development of a new road system, financed by multibillion dollar loans from the World Bank and other international lenders. These new roads, coupled with government policies intended to encourage immigration and economic development in the Amazon region, resulted in a rapid growth in the population of the region, from 2.5 million in 1960 to 20 million in 2000. Deforestation, logging, and forest fires are all concentrated along the new roads (Adeney et al. 2009; Laurance et al. 2009) (Fig. 9.3). Cattle ranching is now the biggest cause of deforestation, but landless migrants, small farmers, and large-scale farming of soybeans are also important, with soy apparently displacing cattle ranching into the interior (Fearnside 2008; Barona et al. 2010). Industrial logging is expanding rapidly and now affects an area as large as that which is cleared, greatly

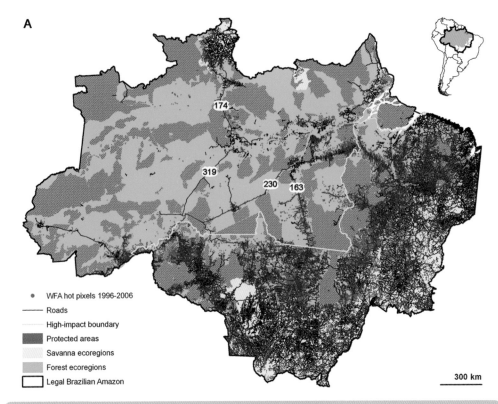

Fig. 9.3 Roads, fires, and protected areas in the Brazilian Amazon, 1996–2006 (Adeney et al. 2009). Deforestation fires declined exponentially with increasing distance from roads. Fewer fires occurred in protected than unprotected areas and the difference was greatest near roads. (Courtesy of Stuart Pimm.)

increasing fire risk. Timber, soybeans, and beef in turn justify further expansion of the road system. In the western Amazon region, vast areas of hyperdiverse rain forest are threatened by a rapid expansion of oil and gas exploration (Finer et al. 2008; Finer & Orta-Martínez 2010).

Papua New Guinea: paradise lost?

The western half of the island of New Guinea is the Indonesian province of Papua (previously Irian Jaya), whose future prospects are linked to those of that complex nation. Until recently, in contrast, the rain forests of Papua New Guinea, which occupies the eastern half of the island, appeared to have a brighter future. The human population density in the lowlands is relatively low, rugged topography over much of the country reduces accessibility, and a unique system of clan control of forest lands creates barriers to large-scale logging or conversion to cash crops. In the last decade or so, however, things have started to change, with an exponential increase in population and a dramatic increase in logging (Shearman et al. 2009; Laurance 2010; Novotny 2010). As in Africa, subsistence agriculture still dominates deforestation, with commercial plantations playing a relatively minor role, and overhunting is a growing problem. Continuing high birth rates (averaging 3.8 children per woman) pose a problem for the future, with the human population of Papua New Guinea expected to grow from its present 6 million to 11 million by 2050.

Australia: paradise regained?

Australia is the only developed country in the world with a significant area of tropical rain forest, so it is not surprising that it now has the world's best-protected rain forests. After a long period of exploitation, followed by an epic struggle between competing interest groups in the 1980s, Australia has now protected most of its remaining tropical rain forests, including all of the larger blocks, in the 8944 km^2 Wet Tropics of Queensland World Heritage Area (Valentine & Hill 2008). Despite the absence of such spectacular wildlife as primates, elephants, and hornbills, the protected rain forests have been a hugely successful tourist attraction (Fig. 9.4). Packaged with the Great Barrier Reef in "rain forest and reef" tours, the rain forest attracts large numbers of Australian and international tourists. The region still has many real environmental problems, but most of these seem relatively minor in comparison with the massive threats to rain forests elsewhere. One exception is the threat from climate change, since the concentration of the Australian endemic rain forest vertebrates in upland areas makes them especially vulnerable to global warming (Wet Tropics Management Authority 2008).

Fig. 9.4 Observation tower (front left) on the Mamu rainforest canopy walkway in the Queensland Wet Tropics World Heritage Area, Australia. (Courtesy of the Queensland Parks and Wildlife Service.)

The major threats

In the preceding section, we described in brief some of the major threats that face each rain forest region. In the following section, we will elaborate on each of these threats and show how they affect rain forest biodiversity.

Clearance for agriculture and ranching

Few rain forest plant and animal species can survive complete removal of the forest. Newly created open habitats are invaded by non-forest species, often from outside the region. These non-forest communities have a much lower diversity of species than the forests they replace, except in regions with extensive natural savannas, like West Africa (Davis & Philips 2009). On a global scale, the majority of rain forest destruction may still result from small-scale cultivation of crops by poor farmers, often forced to remote forest lands by poverty. In an increasing proportion of the tropics, however, clearance by peasant farmers to meet subsistence needs is now dwarfed by clearance by large landowners and commercial interests, to create pasture for cattle ranching or to plant cash crops, such as oil palm (Fig. 9.5) and soybeans. Cattle ranching and soybean cultivation are particularly important in tropical America, while plantations of tree crops are the major cause of deforestation in much of Southeast Asia and are increasing elsewhere. Commercial agriculture displaces poor farmers and justifies the expansion of roads. It is generally worse for biodiversity than clearance by peasant farmers, because large areas are maintained under a uniform crop cover. Tree crops, in general, are somewhat better than pasture or soybeans, because their structure and microclimate is more similar to that of forest, but even then, only the more tolerant rain forest species can survive in these artificial habitats. Oil palm plantations (Fig. 9.5) appear to be particularly hostile to native forest species, perhaps because of their intensive management and unusually simple structure (Brühl & Eltz 2010; Sheldon et al. 2010).

Plans for a massive expansion in the production of biofuels, including bioethanol (made from the sugar and starch components of crops and, in the future, cellulose) and biodiesel (made from vegetable oils), pose a growing new threat to tropical rain forests. Oil palm is the most productive of the major first-generation biofuel crops, but is currently the most damaging, incurring huge net carbon debts when it replaces rain forest, and even more if grown on peat, in addition to the biodiversity impacts mentioned above. Indirect land-use changes triggered by biofuel expansion may be just as damaging. The Brazilian government is pushing for a large increase in the production of sugarcane ethanol and soybean biodiesel, with most plantations expected to replace rangelands, rather than forest. This, however, may push the rangeland frontier further into the Amazon region, with cattle ranches replacing forest (Lapola et al. 2010). The interactions between biofuel production, deforestation, food production, and climate change are complex and sensitive to both changing markets and new technologies. However, the sheer scale of current plans suggests that the biofuel boom will place massive additional strains on tropical lowland forests.

A recent estimate from satellite imagery suggests that 2.4% of the world's "humid tropical forest" (tropical rain forest in the broadest sense) was lost between 2000

Fig. 9.5 Oil palm plantation replacing tropical rain forest in Kedah, Malaysia. (Courtesy of Tim Laman.)

and 2005 (Hansen et al. 2008b). More than 60% of that loss occurred in the Neotropics, with Brazil alone accounting for almost half. Another third occurred in Asia, with Indonesia second to Brazil in the absolute rate of forest loss. Africa contributed only 5.4% to the total area lost, reflecting the current absence of industrial-scale agricultural clearance. Strikingly, 55% of all forest loss occurred within only 6% of the total area, with these "hotspots" including the "arc of deforestation" in the south and southeast of the Brazilian Amazon, much of Malaysia, and Sumatra and parts of Kalimantan in Indonesia. The impacts on rain forest biodiversity, however, are not simply proportional to the absolute area lost. As we have seen throughout the book, each rain forest region is unique and the impacts of deforestation increase as the area remaining declines. These global figures may therefore hide massive impacts in places such as the Chocó region of Columbia, West Africa, Madagascar, continental Southeast Asia, and the Philippines, where so little rain forest remains that even low rates of deforestation risk destroying the last habitats for many species.

Fragmentation

In addition to outright clearance, tropical forests are being threatened by fragmentation. When rain forest is cleared for cultivation, a patchwork of forest fragments is often left behind. In many cases, these are sites that escape immediate clearance because they are too steep, swampy, or infertile for cultivation. Patterns of land ownership or legal restrictions may also prevent complete clearance, and forest patches may also be deliberately retained as sources of timber and firewood, as shelters for cattle, and as protection for the local water supply. These rain forest fragments are often isolated from one another by a highly modified landscape.

Numerous observational and experimental studies have been carried out, particularly in the Amazon rain forest (Laurance et al. 2010b), to determine the effects of rain forest fragmentation. These studies have shown that when a habitat is fragmented, movement between fragments is reduced. Many bird, mammal, and insect species of the forest interior will not cross even narrow open areas. Even a road can be a barrier for many animal species, including birds, bats, and other mammals (Laurance et al. 2009). When animal movements are reduced by habitat fragmentation, plants with fleshy fruits that depend on these animals for dispersal are also affected. As species go extinct within individual fragments through natural succession or random chance, new species will be unable to arrive due to barriers to colonization, and the number of species present in the habitat fragment will decline.

Forest fragmentation also changes the microenvironment at the fragment edge, and this can have a significant impact on species composition. Some of the more important edge effects include changes in light, temperature, wind, humidity, and incidence of fire (Fig. 9.6). In studies of Amazonian forest

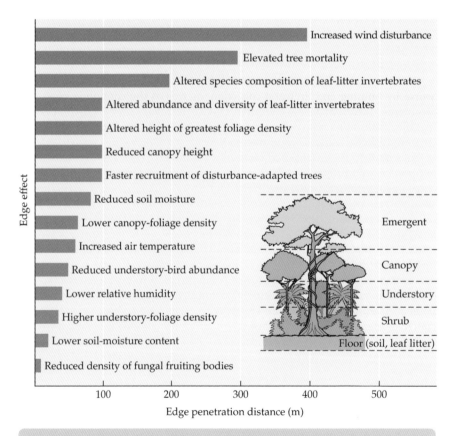

Fig. 9.6 Edge effects in the Amazon rain forest. The bars indicate how far into the forest fragment the specified effect occurs. (From Primack 2010, after Laurance et al. 2002.)

fragments, microclimate effects were evident up to 100 m (300 feet) into the forest interior. These changes in temperature, humidity, and light levels exclude sensitive forest organisms and promote invasion by non-forest species, including aliens. The smaller the fragment, the higher the edge to interior ratio, and the smallest fragments (< 5 ha) are all edge with no interior habitat at all. Habitat fragments are also susceptible to fires spreading from nearby cultivated fields into the forest interior.

Studies of fragments have also revealed the crucial importance of the non-forest matrix around the fragments in determining how many and which species survive (Ewers & Didham 2006). Species that are able to live in and move across the matrix will increase in abundance in small, isolated fragments, while other species decline. The more similar the matrix habitats are to the original rain forest, the weaker are the edge effects in forest fragments, and the more species can survive. Pasture or low-growing crops are the worst matrix types, while tree crops or secondary forest regrowth can act as a buffer and protect the fragment from the external microclimate. These observations suggest ways in which the conservation value of fragments could be increased by planting buffer zones or encouraging regrowth in the surrounding area.

The overall effect of all these fragment processes is that all forest fragments tend to lose species, with both the rate of loss and the precise species that are lost depending on fragment size, fragment isolation from other forest areas, and the nature of the matrix. Species loss from rain forest fragments of the size most commonly left in agricultural landscapes – typically 1–500 ha (2–1200 acres) in area – is rapid, with the most sensitive species disappearing within a few years. In real, rather than experimental, fragments, the situation is made worse by hunting and firewood collection, as well as the continued erosion of fragments by fires and new clearance activity. The good news is that a surprising diversity of the more tolerant rain forest species can persist even in tiny fragments (< 5 ha; e.g. Arroyo-Rodriguez et al. 2009), particularly if they are protected from further exploitation and damage. The bad news is that a surprising number of the more sensitive species are lost even from "fragments" so large (> 10 km^2) that they can seem endless to the casual visitor (e.g. Brühl et al. 2003).

Logging

An estimated 20% of the total area of humid tropical forest was subject to logging operations between 2000 and 2005 alone (Asner et al. 2009) (Fig. 9.7a,b,c). Most of this was highly selective, with only one or a few high-value tree species harvested per hectare. Logging intensities can be much higher, however, in Southeast Asian dipterocarp forests, because dozens of similar dipterocarp species can be grouped into a small number of categories for marketing (see Chapter 2). Because the logs are straight and of considerable length, and the wood is often light in weight, dipterocarps are in great demand by the timber industry to be used for inexpensive construction and plywood manufacture. Of the top 10 timber-exporting countries in the world, only two are tropical countries, and both of these are in Asia: Indonesia and Malaysia. In contrast, annual timber production in tropical Africa is much lower, due to low extraction rates and a smaller area being logged.

(a)

(b)

Fig. 9.7 Logging contributes to rain forest destruction in every part of the tropics.
(a) Logging truck coming out of a rain forest in Nenasi, Pahang, Malaysia. (Courtesy
of M. Sugumaran.) (b) Small-scale wood cutting in the Ituri District of the DRC.
(Courtesy of Dan Fahey.)

(c)

Fig. 9.7 (*cont'd*) (c) Logging yard in Libreville, Gabon, showing the massive scale of the logging industry. (Courtesy of Olivier Langrand, Conservation International.)

This pattern is now changing with the rapid exhaustion of accessible forests in Asia and increasing investment by transnational (often Malaysian) logging companies in Africa, the Neotropics, and Papua New Guinea. In addition, the growing domestic and Asian markets for rain forest timber are far less fussy about the species, size, and quality of logs than the traditional markets in Europe, North America, and Japan. This means that more species are saleable and at a smaller size, so the initial logging intensity is greater and accessible areas are often re-logged later for smaller timber and less valuable species. Throughout the tropics, however, logging in remote areas is still mostly "mining" of huge ancient trees in virgin forest, since only the most valuable species can justify the large investment in machinery and infrastructure required. Some luxury timbers, such as Madagascan rosewoods (*Dalbergia* spp.; Patel 2007) and the Neotropical big-leaf mahogany (*Swietenia macrophylla*), are extraordinarily valuable – up to US$1800 per cubic meter at the point of export for mahogany in 2009 (Grogan et al. 2010) – and their harvest thus correspondingly difficult to control.

This logging is "selective" only in terms of the logs actually removed from the site; the process of finding, cutting, preparing, and extracting these logs is devastating and leaves the whole forest looking like a war zone. Often forests are degraded over large areas just to extract a few trees per hectare, with most of the trees knocked down and the soil exposed and compacted by heavy machinery. Satellite data from the Brazilian Amazon shows that even low (5–10%) levels of canopy damage can cause long-lasting changes in forest phenology, reducing dry-season canopy greenness (Koltunov et al. 2009). Despite the damage, numerous studies have now shown that most wildlife can survive a single cycle of selective logging (Berry et al. 2010). Recovery is slow, but it is possible. The major impact is not the extraction of timber from the site, but the opening up of the forest by the construction of roads and other infrastructure. Improved access brings in hunters and encourages recurrent cycles of logging. Landless farmers often move into the area and cut down the remaining trees for agriculture. Recently logged forests are also far more likely to burn than those that have not been logged or were logged long ago.

Fire

Under normal conditions, tropical rain forests do not burn. The presence of charcoal buried in the soil beneath many rain forest areas shows that fires are possible under extreme conditions of drought, but in the absence of other human impacts such conditions were clearly very rare. Closed-canopy rain forests are remarkably resistant to drought, remaining immune to fires even after months without rain. Because fires were so rare, rain forest trees have not evolved resistance to fire and most species have very thin bark. Natural rain forest fires must therefore have been rare, but catastrophic, events.

Fires are no longer either natural or rare in the rain forest, but they are still catastrophic (Barlow & Silveira 2009; Cochrane 2009). Fire is the most convenient tool for clearing rain forest and preventing regrowth, but it is not an easy tool to control. Out-of-control fires burned an estimated 200,000 square kilometers (77,000 square miles) of forest in Southeast Asia and the Neotropics during the unusually dry conditions caused by the 1997–8 El Niño event. These fires primarily affect logged forests, because the canopy has been opened and logging waste increases the fuel supply. Forest fragments are also vulnerable, because structural changes at the edges increase the fuel load while exposure to wind and sunlight reduces humidity. The fires in turn kill trees and leave huge amounts of flammable dead wood, greatly increasing the risk of recurrence. Rain forest recovery after a single fire is a slow process, aided by the seed sources provided by surviving trees and unburned "islands" of forest. Recurrent fires rapidly remove these seed sources and encourage the invasion of grasses and vines, further increasing flammability. The result is a landscape that becomes vulnerable to fire after weeks, rather than months, without rain. In much of the rain forest region of Southeast Asia, for example, the fires that were an exceptional occurrence a few decades ago, and occurred every few years in the 1980s and 1990s, are now becoming an annual event (Harrison et al. 2009) (Fig. 9.8). People refer to this period as the "haze season," a concept that did not exist 20 years ago.

Hunting

Even where rain forests still appear intact, many rain forest animals are being threatened with extinction. This current round of extinctions is the latest episode in the long-term elimination of vulnerable species from the landscape. As people arrived for the first time in each rain forest region, they drove the large native species to extinction, largely by hunting. In the Americas, New Guinea, Madagascar, and the Pacific Islands, the arrival of people signaled the decline and eventual extinction of numerous animal species (see Chapters 1 and 8). The

Fig. 9.8 Smoke from forest fires in Borneo on August 19, 2002, during a moderate El Niño episode. (NASA.)

extinction of large animals appears to have been much less common in Africa, where animals evolved with people. The picture in tropical Asia is less clear, but there have been several suspicious extinctions since modern humans arrived 50,000 years ago and several large vertebrate species that still survive in the region, such as orangutans, suffered huge reductions in their distributional ranges (Corlett 2010).

In all rain forests areas, people have hunted and harvested the food and other resources they need in order to survive (Fig. 9.9). As long as human populations were small and the methods of collection unsophisticated, this hunting and harvesting could be sustainable, at least after the most vulnerable species had gone. Animal populations could be lowered in size, particularly near villages, but some animals would survive, often at distance from human settlements. In recent decades, human populations in rain forest areas have increased, and their methods of harvesting wildlife have become dramatically more efficient. In addition, previously isolated human populations, which hunted for their own use, are now connected to regional and international markets whose demand is insatiable. Hunting has been transformed from a subsistence activity into a commercial enterprise. This has led to an almost complete depletion of larger animals from many rain forests, leaving strangely "empty" forests, with animal densities reduced by 90% or more. People only appreciate how empty a typical modern rain forest is when they visit a rain forest that is extremely remote or that has been vigorously protected, such as Tikal National Park in Guatemala or Barro Colorado Island in Panama. At these locations animals are at a high density, and they are readily seen because they are not afraid of people.

Fig. 9.9 Partly butchered forest elephant in the Evinayong province of Rio Muni, Equatorial Guinea. (Courtesy of Jessica Weinberg.)

In traditional societies, restrictions were often imposed that prevent over-exploitation of natural resources. In other cases, traditional taboos, with supernatural sanctions, have limited exploitation of certain species (e.g. Jones et al. 2008). In much of the world today, however, traditional social constraints are no longer effective and resources are exploited opportunistically. Where they are available, guns are now used instead of blowpipes, spears, or arrows for hunting in the tropical rain forests, especially for primates and other arboreal animals. Hunting with leg-hold cable snares is particularly common in Africa, due to their low cost and the abundance of large mammals foraging on the ground (Fa & Brown 2009). Snares are also used for ungulates in Asia, but they are rare in other rain forest regions where arboreal species are often more important to hunters. The small game that dominated the catch of traditional subsistence hunters has been replaced by larger-bodied primates, ungulates, and rodents, which find a ready market as "bushmeat" in towns and logging camps. In the Asian tropics, besides the exploitation of animals for meat, the demand for animal parts for use in traditional medicines fuels a huge regional trade (Corlett 2007b). Globally, besides a surprisingly large legal trade (Nijman 2010), billions of dollars are involved in the illegal trade of wildlife or wildlife products from tropical forest habitats. This trade has many of the same characteristics, the same practices, and sometimes the same players, as the illegal trade in drugs.

Throughout this book, we have emphasized the interrelationships between plants and animals in the rain forest. These interactions mean that the hunting of forest vertebrates has impacts far beyond the survival of the species that are harvested (Wright et al. 2007). Hunters favor larger species and it is these that are responsible for dispersing the seeds of many rain forest plants. Plants with small fruits and small seeds will find a wide range of dispersal agents in even the most degraded landscapes, but larger, bigger-seeded fruits are consumed by fewer dispersers, with the largest depending on the few species of large birds and mammals that are most vulnerable to hunting. And it is not only seed dispersal that is affected. Vertebrates that consume seeds influence the competitive balance between plant species, so their elimination will change the plant community. In the Neotropics, the hunting of large rodents favors the survival of the large seeds they consume. Selective hunting of carnivores can lead to proliferation of their prey, while the elimination of the prey threatens any carnivores that are not hunted themselves. These various interactions between species are complex and little understood, so the long-term impacts of hunting and the animal trade are impossible to predict.

Invasive exotics

A further threat to degraded and fragmented rain forests is invasion by exotic species, to the detriment of native species. So far, exotic species have been a problem mainly on oceanic islands (see Chapter 8). It appears that the low diversity of island biotas and the complete absence of some major groups of continental organisms make island rain forests particularly vulnerable to invasion. However, there are now an increasing number of cases where species from one continent have invaded disturbed and fragmented rain forests on other continents. The invasion of "Africanized" Old World honeybees into Amazonian forest fragments is a striking example (see Chapter 7), as are the invasions of several Neotropical pioneer trees and shrubs (e.g. *Cecropia* spp., *Clidemia hirta*, and *Piper aduncum*) into Old World forest fragments. Invasive aliens often dominate the regrowth on abandoned lands and it is not yet clear if they will eventually be displaced by native species. After a brief initial period when control may be possible, species introductions, like extinctions, are forever, so this is a trend that can only get worse.

Air pollution

Air pollution is not on most lists of threats to tropical rain forests, but all rain forest organisms are exposed today to an atmosphere that differs significantly in its composition from any that their ancestors would have experienced. The climatic impacts of the rising concentrations of greenhouse gases are considered later, but other changes have been largely overlooked, in part because there are so few non-urban studies of air pollution in the tropics. Industrialization in much of Southeast Asia and parts of the Neotropics is a major source of pollutants, as are intensive agriculture and widespread biomass burning. Atmospheric nitrogen deposition is the best-recognized threat, because the changes have been so large, so rapid, and so pervasive (Fig. 9.10), but it is not clear what the impact will be in forests that are not normally considered to be nitrogen-limited (Bobbink et al. 2010). Ozone, sulfur dioxide, and particulates are additional threats in some areas (Corlett 2009a). Air quality is declining almost everywhere in the tropics so this is another problem that can only get worse.

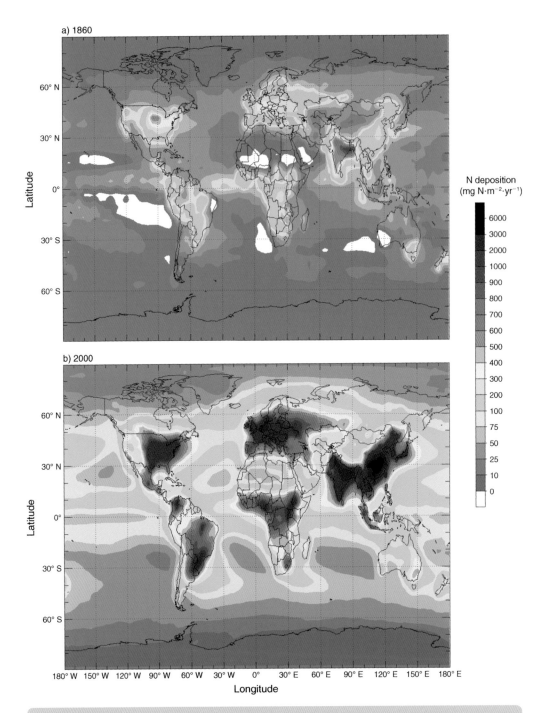

a) 1860

b) 2000

N deposition
(mg N·m⁻²·yr⁻¹)

Fig. 9.10 Air pollution is a commonly overlooked threat to tropical rain forests. This figure shows the modeled total nitrogen deposition from the atmosphere for 1860 and 2000 (Bobbink et al. 2010). Major anthropogenic sources include biomass burning, fossil fuels, and intensive agriculture.

The forces behind the threats

The threats described above can be seen as symptoms of a smaller number of underlying problems, including population growth, poverty, inequity, wars, and corruption in the rain forest countries, and excessive consumption in the developed world.

Population growth

The threats to rain forests are all ultimately caused by an ever-increasing use of the world's natural resources by an expanding human population, particularly in species-rich tropical countries. Until the 19th century, the rate of human population growth was relatively slow, with the birth rate only slightly exceeding the mortality rate, but the human population has exploded since, going from 1.2 billion in 1850, to 2 billion in 1930, and reaching 6.8 billion in 2010 (Fig. 9.11). World population is expected to increase until around 2050, reaching a maximum of 9–10 billion, with most of this increase occurring in tropical

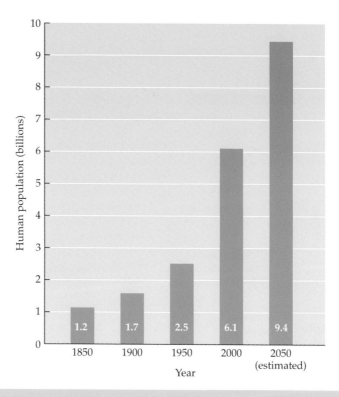

Fig. 9.11 The exponential growth of the global human population over the last 160 years is the ultimate factor behind most threats to tropical rain forests. This growth rate is now declining, but the world population is still expected to reach at least 9–10 billion before it begins to stabilize. (From Primack 2010. Data from the US Census Bureau, www.census.gov.)

countries (Fig. 9.10a). The period of explosive global population growth is now over, with the global growth rate almost half of its peak in 1963, but growth remains very high in some rain forest countries, including all those in tropical Africa, as well as Madagascar, the Philippines, and Papua New Guinea. Even in countries such as Brazil and Thailand, where fertility rates have fallen below long-term replacement (2.1–2.15 children per woman), the youthful age profile of the population will insure that it keeps growing rapidly for several more decades before it begins to stabilize.

The number of people is not the only factor in rain forest destruction. The environmental impact of a human population can be expressed as the product of three factors: the population size, the per capita consumption of resources, and the impact per unit of consumption (Ehrlich & Goulder 2007). Even as rates of population growth decline, per capita consumption continues to rise as poor people aspire to the middle-class lifestyle they see every night on their televisions. This results in impacts far beyond those needed for mere subsistence, as rural people hunt more wildlife, harvest more forest products, and grow more crops to pay for things the readers of this book take for granted. Trends in the third factor, the impact per unit of consumption, are more difficult to assess. Positive trends, such as the increase in agricultural yields per unit area, tend to be offset by increased negative impacts from energy consumption and the excessive use of pesticides and fertilizers.

Poverty and inequality

In many tropical countries there is extreme inequality in the distribution of wealth, with the majority of the resources (money, good farmland, livestock, timber resources, etc.) owned by a small percentage of the population. Brazil, for instance, has one of the least equal distributions of income in the world, with the richest 20% of the population receiving two-thirds of the total national income, while the poorest 20% receive less than 3%. Poverty has a complex relationship with rain forest destruction. Poor people in rural areas are more likely to use rain forest resources to meet subsistence needs and to pay for a chance of a better life for their children. But poor people lack both the financial capital and the political influence needed for the commercial logging and large-scale forest clearance that are now the greatest threats in most of the tropics. Moreover, it is rich people in urban areas that provide the biggest market for the illegal trade in wildlife and other rain forest luxuries.

Wars and political instability

More than half the wars of the 20th century were in forested areas and many of those in the last 50 years were fought at least partly in tropical rain forests: in Vietnam and Cambodia, in the Democratic Republic of Congo and the Central African Republic, in Colombia and several countries in Central America (Donovan et al. 2007). An even more striking correlation is between warfare and biodiversity hotspots, with more than 80% of the major armed conflicts between 1950 and 2000 taking place in hotspots that cover 16% of the Earth's land area (Hanson et al. 2009). The reasons for these trends are not clear, but probably include the remoteness and inaccessibility of many biodiversity-rich forested land-scapes and the potential for natural resource exploitation to finance conflicts. "Conflict timber" was particularly important in funding the devastating conflicts in Liberia, the Democratic Republic of Congo (DRC), and Cambodia (Price et al. 2007). The consequences for wildlife have been varied, but generally negative, with the collapse of protected area systems, the accelerated exploitation of timber, the proliferation of weapons leading to increased hunting, and the displacement of refugees into forested areas. The eastern DRC illustrates all these problems, with heavily armed militias making tens of millions of dollars annually from the illegal trade in minerals, charcoal, timber, and bushmeat (Nellemann et al. 2010).

Corruption and bad governance

Damage to biodiversity can be limited by the enforcement of laws that protect threatened habitats and species, that regulate the exploitation of non-endangered species, and that limit the adverse environmental impacts of new developments. This approach has been used most effectively to protect the remaining rain forests in Australia, but has also worked to a varying degree in many less wealthy parts of the tropics. The reasons why it is not work-ing everywhere are many and varied, but one widespread problem, which is repeated in many rain forest countries, is the poor quality of governance at the national, provincial, and local level (Yu et al. 2010). Good governance will not save the rain forests, but it may be a necessary condition for the rain forests to be saved.

The term governance covers all aspects of the way a country is governed, includ-ing government policies, laws, and regulations, and the ways in which these are implemented in practice. Good governance includes such characteristics as the rule of law, transparency, efficiency, effectiveness, and accountability. Bad governance is the lack of these characteristics. Bribery and corruption are often the most visible symptoms of bad governance, but inefficiency and ineffective-ness can be just as damaging. In relation to rain forest habitats and species, the major problem is sometimes outdated laws – usually a relict of colonial admin-istration – but even good laws are often ineffectively enforced by weak forestry and conservation departments. Poorly paid staff can be bribed or simply ignored. In many rain forest countries, decisions on forest exploitation are made by a small group of powerful people in, or closely linked to, the government, who see rain forests as a short-term source of personal revenue, rather than a resource to be managed in the long-term national interest. At the local level, major landowners and powerful local families are effectively above the law and can have a major influence on conservation-related decisions, while at the forest frontier there may no law at all.

Poor governance encourages an exploitive "mining" mentality and discourages long-term investment. Why should a peasant farmer invest labor in developing a permanent farm that may be taken away at any moment? Why should a timber company invest scarce capital in sustainable forest management when its patron in the capital city may be removed in the next election or coup? Good governance encourages a longer-term view of costs and benefits, but only when all participants have confidence that the bad days will not return.

Roads and development

Roads may be the single biggest threat to the remaining tropical forests. The impact of highway construction has already been mentioned for the Amazon region of Brazil, where 95% of all deforestation and fires occur within 50 km of a road (Fig. 9.3), but roads are an increasing threat in Africa, Asia, and New Guinea (Laurance et al. 2009). The expansion of road systems is driven largely by the exploitation of natural resources, particularly timber, minerals, oil, gas, and hydropower sources, and by the desire of cattle ranchers and industrial agribusinesses to export their products. These roads then provide easy access for hunters, farmers, loggers, and miners, both legal and illegal. The "official roads" are supplemented by an unofficial network of logging roads. Paved roads, which provide year-round access, cause more damage than unpaved roads, which become inaccessible at wetter times of the year.

Roads and money encourage the development of urban settlements – both planned and spontaneous – away from the traditional coastal or riverside locations. These new towns attract more people into the region and provide an additional market for bushmeat, timber, firewood, and charcoal, further increasing pressure on the rain forest. In the longer term, however, the growth of urban areas in rain forest countries may be positive for conservation as migration to urban centers reduces rural population densities and thus the pressure on marginal land. These potential conservation benefits of rural–urban migration form a major element of the relatively optimistic scenario for tropical forest biodiversity presented by Wright and Muller-Landau (2006), in which secondary forests on abandoned agricultural land reduce the expected number of extinctions. Unfortunately, while declining rural populations may protect and promote forest in areas where smallholders are the major threat, an increasing proportion of tropical deforestation is carried out by ranchers and agri-businesses that can maintain and expand the deforested area despite the decline in the rural population (Fearnside 2008). The same applies to the extraction of forest resources, as nonresident professionals with a "mining" mentality replace subsistence resource users in depopulated areas (Parry et al. 2010).

The responsibility of consumer nations

The stark inequities that exist in most rain forest countries are played out on an international scale as well. Just as a few individuals consume the lion's share of the resources in many tropical countries, some nations in the world consume a disproportionate share of the world's energy, minerals, wood products, and food, and therefore have a disproportionate impact on the environment. Developed countries fuel rain forest loss not only through their consumption of products, but also by financing the companies directly involved in rain forest exploitation. Financial institutions in the developed world could – but rarely do – exert a great deal of influence on the practices of logging and plantation companies. The same applies to the potential influence of international institutions, such as the World Bank and the International Monetary Fund, at the government level. Good governance and "the environment" now feature prominently on the websites and publications of these institutions, but it is too soon to judge how much these issues influence important decisions. There is an additional problem with the economic reforms that these institutions promote. The effect is often to open up the economy to foreign investment, without any strengthening of the government's capacity to reduce the resulting environmental damage.

We must be careful, however, not to exaggerate the influence of the developed world on rain forest exploitation. Although North America, Europe, and Japan continue to be important markets for rain forest-related products, China is the biggest importer of tropical timber – much of it illegally exported from the country of origin (Laurance 2008) – and soybeans, while India is the biggest importer of palm oil. The same story applies to sources of investment, with changing players, increasingly from developing countries. These alternative markets and alternative sources of investment have inevitably reduced the leverage that consumers, bankers, and governments in the developed world have with the producers. A Malaysian company logging forests in Gabon for export to China will not be very concerned about the environmental sensitivities of the American, Japanese, or European public. The good news is that both China and India are showing increasing sensitivity to criticisms of their environmental impacts abroad.

Global climate change

There is one human impact that affects all forests, tropical or temperate, no matter where they are located: global climate change that stems from human activities (Figs. 9.12, 9.13). The influence of anthropogenic climate change on tropical rain forests is of global concern not only because of their rich biodiversity and irreplaceable ecological services, but also because these forests are a major element in the global carbon cycle, with the potential for both positive and negative feedbacks. Deforestation – mostly in the tropics – is the second biggest source of anthropogenic carbon dioxide, after fossil fuel combustion, contributing an estimated 12% of total global emissions (van der Werf et al. 2009). On the other hand, the remaining tropical rain forests are currently a significant carbon sink (Phillips et al. 2008; Lewis et al. 2009), although this seems unlikely to continue in the face of significant climate change.

For a given greenhouse gas scenario, the different global climate models make similar predictions about future temperatures in the tropics: around 3–4°C warmer than today by 2100 for a range of scenarios (IPCC 2007) (Fig. 9.13). This unanimity on future temperature trends does not carry over into the projections for total rainfall and rainfall seasonality, however, which continue to be imprecise in much of the tropics, as shown by the different predictions made by each of the models. The models also have difficulty predicting annual climate variability and the future frequency and intensity of El Niño events. Further, the models remain uncertain concerning potential feedbacks between climate change and the land and ocean carbon sinks. It is also worth noting that the current rate of increase in atmospheric carbon dioxide concentrations (more than 2 ppm per year) is greater than was assumed in even the IPCC's worst-case scenario.

An increasing amount of evidence suggests that tropical rain forest communities are highly vulnerable to increases in temperature of the magnitude predicted by all models. At sites outside the tropics, species that cannot tolerate

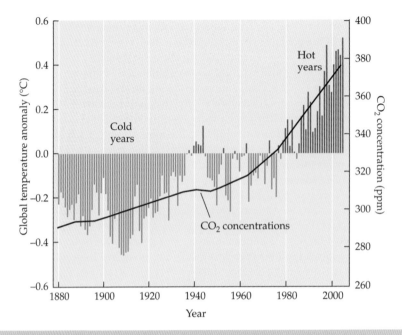

Fig. 9.12 Over the last 130 years, atmospheric carbon dioxide concentrations have increased dramatically as a result of human activities. Global temperatures, here shown as differences from the 1961–1990 average, have also risen over the same period, with most scientists now agreeing that the two phenomena are causally linked. (From Primack 2010, after Karl 2006.)

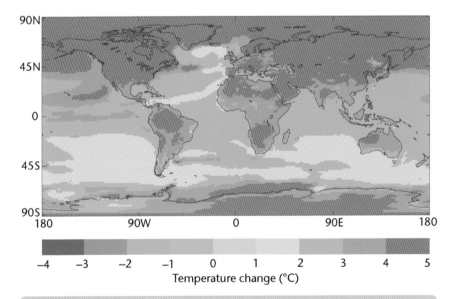

Fig. 9.13 Predicted changes in annual average surface air temperature from 1960–1990 to 2070–2100 according to the Met Office Hadley Centre global environment model, HadGEM1. (Crown Copyright 2009, the Met Office, UK.)

the increase in temperature will disappear and be replaced by species that can. We already see evidence of this wherever the records are long enough: the populations of numerous species of plants and animals have moved polewards and/or upwards in elevation over recent decades, tracking their optimum temperatures. However, within about 10 degrees either side of the equator, there is no temperature gradient, so latitudinal movement is not a feasible way of adapting to climate change. Altitudinal gradients are still just as steep, so movement uphill is still feasible. In keeping with this expectation, some moth species on Mount Kinabalu in Sabah, Borneo, have already moved upwards in elevation by an average of 67 m over the last 42 years (Chen et al. 2009).

In the tropical lowlands, however, there is no source of species adapted to warmer conditions, so species lost as a result of warming will not be replaced: a concept that has been termed "lowland biotic attrition" (Colwell et al. 2008; Feeley & Silman 2010). If this theory is correct, diversity can only decline, but by how much and how fast is not clear. The climates projected for 2100 do not exist today anywhere on Earth, but, as we have seen in earlier chapters, many present-day rain forest species originated in warmer times several million years ago and could perhaps survive such a climate again. On the other hand, three million years of relatively cool temperatures may have eliminated any tolerance of greater heat, along with the ability to evolve such tolerance within 100 years.

Most of the available evidence is on the side of the pessimists, who fear that climate change will bring about forest destruction and the extinction of species. Tropical forest ectotherms (those animals whose body temperature is controlled by the environment, such as insects, frogs, and reptiles) live in a relatively constant thermal environment and appear to tolerate a very narrow range of temperatures. It is likely that many of these species currently live very close to their optimum temperatures and will experience lethal temperatures after a few degrees of warming (Deutsch et al. 2008; Huey et al. 2009). Outside the tropics, in contrast, ectotherms must tolerate a wide seasonal range of temperatures, often live well below their optimums at present, and thus could in many cases actually benefit from warming. Among vertebrates, amphibians, reptiles, breeding birds, and bats, but not rodents, have smaller altitudinal ranges in the tropics than elsewhere, in keeping with the expectation of narrower temperature tolerances (McCain 2009). Endothermic mammals and birds actively regulate their body temperatures, but can suffer from heat stroke and other problems under excessively warm conditions. Plants are also ectotherms, at the mercy of external temperatures, and tree growth in tropical rain forests declines significantly with small ($< 1°C$) increases in temperature (Clark et al. 2010).

Uncertainties over rainfall projections make it hard to predict the overall impact of anthropogenic climate change on lowland tropical rain forests. This is exemplified by the wide spread of predictions for the Amazon basin, from extensive forest dieback to little change (Meir & Woodward 2010). Tropical rain forests appear to be relatively resistant to occasional droughts, but unusually dry years significantly decrease growth and increase carbon losses (Clark et al. 2010). Such carbon losses will change the regional carbon balance and further accelerate climate change. Moreover, any increase in drought frequency or severity will interact with other human impacts, such as fragmentation and logging, to increase the frequency and intensity of fires, releasing yet more carbon into the atmosphere.

Saving the many rain forests

The threats to the survival of tropical rain forests that we have outlined above have all intensified over recent decades, despite a massive pantropical expenditure of funds on conservation, both by national governments and by international agencies, conservation organizations, and foundations of various types. There are multiple reasons for this failure, but the major one is probably that, while the goals of conservation are biological, the means to achieve the goals are social, economic and political (Polasky 2008). The science of biological conservation has developed rapidly, but the complexities of social and political conditions in the tropics and the rapid pace of change have made it very difficult to produce large and long-lasting improvements on the ground. The best hope for the future of tropical rain forests probably lies in the development of local and regional conservation awareness and capabilities, with international support provided through partnerships. Here we outline the major general requirements for the successful conservation of tropical rain forests, but their means of implementation will undoubtedly need to be adjusted to local social, economic, and political circumstances.

National parks and other protected areas

This book has been about the ecology of more-or-less intact rain forest communities: rain forests with giant trees, primates, carnivores, herbivores, birds, bats, and insects. With a few exceptions, such rain forests survive today only in regions with very few human inhabitants or in national parks and similar areas set aside for their protection. An extra billion people in the tropics over the next 15 years, coupled with the expanding exploitation of rain forest resources, will insure that isolation alone will not provide protection much longer. The single most important strategy for protecting intact rain forest communities is therefore to establish protected areas and then to effectively manage them. The last 30 years has seen a succession of debates about how to design and manage protected areas, starting with the SLOSS (Single Large or Several Small) debate of the 1970s and 1980, and followed by a "use it or lose it" debate on sustainable exploitation, a debate on links between conservation and rural development, and an ongoing debate about the role of indigenous peoples and other local communities in conservation. In recent years, however, there has been a growing acceptance that there is no one-size-fits-all solution and that conservation success is pluralistic.

Existing rain forest protected areas range in size from the 39,000 km^2 (15,000 mi^2) Tumucumaque National Park in Brazil and the 36,000 km^2 (14,000 mi^2) Salonga National Park in the Democratic Republic of Congo, down to the 1.6 km^2 (0.6 mi^2) Bukit Timah Nature Reserve in Singapore and a number of even smaller reserves. "Megareserves" of > 10,000 km^2 (4000 mi^2) are probably needed to conserve intact rain forest communities for the long term, including top predators, megaherbivores, and specialists on rare habitats (Peres 2005; Laurance 2005). New reserves of this size are still possible in Amazonia, Central Africa, and (at least in theory) New Guinea, but not in lowland tropical Asia. At the other extreme, as discussed in the section on fragmentation above, a surprising diversity of the more tolerant rain forest species can persist for decades even in tiny fragments (< 5 ha) if they are well protected. The 1.6 km^2 Bukit Timah Nature Reserve has lost around half of its vertebrate fauna in the last 150 years, but still retains a hyperdiverse plant and invertebrate fauna. A simple rule is that reserves should be as large as practically possible, but while small reserves are no substitute for large reserves, they can and do protect unique species that would not survive in an otherwise unprotected landscape. In Borneo, for example, the massive Heart of Borneo project seeks to protect an area large enough to support wide-ranging large vertebrates, such as the orangutan, but excludes the last remnants of forest in the coastal lowlands of north and west Borneo, which may be the botanically richest rain forests in the Old World (Ashton 2010).

Rain forest protected areas also vary in who they are protected by. In most countries, national governments play the leading role in rain forest preservation, but state and local governments, local communities, conservation organizations, and even private individuals are important in some areas. While national governments are potentially the most powerful protectors, they are not always effective in projecting that power at the local level, so national reserves are not necessarily the best protected. Protected areas also differ widely in what they are protected against, both in theory and in practice. Most permit recreational and scientific visitors, but some also allow the sustainable extraction of natural resources, such as timber or Brazil nuts, and others allow the continuation of traditional exploitation by indigenous people. Indigenous reserves are particularly important in the Brazilian Amazon, where they cover 20% of the region (Fig. 9.14). The largest, the 130,000 km^2 (50,000 mi^2) reserve of the Kayapó and Upper Xingu people, is also the largest protected forest area in the tropics. The long-term sustainability of exploitation by indigenous peoples continues to be debated, but these reserves are currently an important barrier to deforestation (Ricketts et al. 2010).

The effectiveness of protected areas in limiting deforestation and over-exploitation also varies hugely, with most offering at least some degree of protection. The vegetation, at least, is in significantly better condition inside parks than outside (Joppa et al. 2008; Adeney et al. 2009; Gaveau et al. 2009). This is true even when the protective effect of the remote location of many parks is allowed for. The same surveys, however, also show that many tropical rain forest protected areas have major problems. Protection appears to be particularly weak in Indonesia, where many parks are subject to almost uncontrolled logging and hunting, as well as variable levels of deforestation (Curran et al. 2004; Gaveau et al. 2009). Similar problems have been reported from rain forest parks across the tropics. While unmanaged "paper parks," protected only by their designation, can be better than nothing (Martin & Blackburn 2009), improving protection on the ground is a high priority in most rain forest countries. This requires, at minimum, a trained and motivated staff equipped with vehicles and the basic equipment that most rain forest researchers take for granted, including binoculars, cameras, communication equipment, and GPS. Access to satellite images is also essential if large areas are to be managed effectively. None of this will be of help when logging or clearance has strong political backing, so investment at the front line must be backed up by reform at higher levels.

Experience throughout the tropics has shown the importance of local support when parks are established. The costs of establishing new protected areas in inhabited regions are typically borne largely by the local people, who lose access

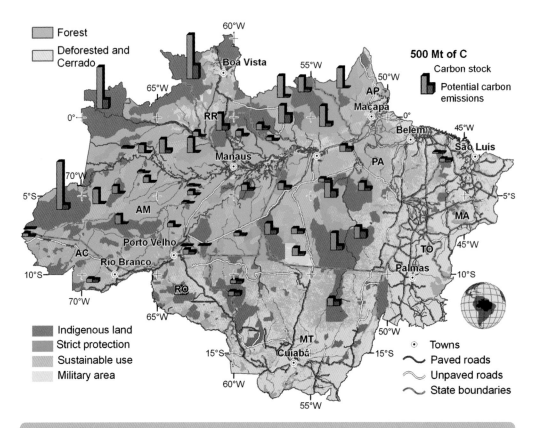

Fig. 9.14 Indigenous reserves and other protected areas in the Brazilian Amazon, showing their estimated carbon stocks as well as the potential carbon emissions from deforestation through 2050 if they were not present (Ricketts et al. 2010). (Courtesy of Taylor Ricketts.)

to resources within the park boundaries, and who may be displaced from their homes and farms (Whitten & Balmford 2006). In theory these costs may be offset by benefits, both financial, from jobs and tourism, and ecological, such as clean water and reduced soil erosion. In practice, however, the financial benefits often go largely to outsiders and the ecological benefits may not be obvious enough to compensate for the losses, although they are valued by local people (Sodhi et al. 2010). Not surprisingly, the establishment of parks and other protected areas has often bred resentment among local communities, making the task of park management much more difficult. The impact on local communities needs to be considered at the planning stage, with active steps taken to insure that the cost of the park, in terms of lost income and resources, is not paid by those local people least able to afford it.

While we have emphasized here the protection of the remaining areas of more-or-less intact rain forest communities, it is also important that protection is given to forests that are recovering from logging and hunting, and that are regrowing after clearance, with or without active restoration efforts (see below). Indeed, the inclusion of logged primary forests and areas of secondary regrowth is the only way that areas of adequate size for the support of large vertebrates can be protected in many parts of the tropics. Logging, by itself, has fewer adverse consequences for biodiversity than is often assumed and the worst impacts appear to be reversible (Berry et al. 2010). The jury is still out on the conservation value of secondary regrowth in isolation, but there is no doubt about its ability to buffer and connect fragmented primary forests. Overall, it makes sense to have a two-pronged conservation strategy: on the one hand, proactively minimizing the initial penetration of threats into the last great wilderness areas, while on the other reactively acting to minimize extinctions in the most disturbed and fragmented landscapes (Boakes et al. 2010).

Regulating exploitation

Even though they are important, parks by themselves will not be enough. However successful we are in expanding the present coverage of protected areas and insuring their proper management, they will inevitably be too few, too small, and not sufficiently representative to preserve all of the rain forest biodiversity. Most rain forest regions will continue to have a larger area of forest that is not inside parks. Whether this is in legally designated production forests and extractive reserves, or just blank areas on the map, regulating the exploitation of rain forest outside the parks can make a major contribution to the protection of rain forest diversity. It is also important to remember that even the regulated exploitation of timber and wildlife, as long as it maintains forest cover, protects much more biodiversity than clearance for pasture or crops (Berry et al. 2010). The aim of regulating exploitation outside parks should not therefore be to prevent any exploitation, but rather to insure that extraction is done with the least loss of species and habitats. It is important to insure that onerous legal restrictions on what can be done do not drive people and companies into far more damaging illegal exploitation.

   Reducing the adverse impacts of the logging industry is – or should be – a priority in most rain forest countries, not only because of the massive direct damage logging does to the rain forest, but also because of its role in catalyzing further degradation and deforestation. When very little forest is left, as in Thailand, a total ban on logging is simpler and more effective than attempts at regulation, although this tends to simply displace logging activity across international borders. Most rain forest countries, in contrast, need the income from logging and the employment that it and the wood-processing industry can provide. As a result, a great deal of research effort has gone into devising improved methods of logging that cause less environmental damage. These methods, known collectively as reduced impact logging or RIL, involve the application of a series of guidelines designed to minimize damage to soils and the next generation of commercial trees, as well as nontarget species of plants and animals (Putz et al. 2008). RIL includes such practices as: training of workers; protection of forest on steep slopes and along streams; careful planning of the road network to reduce soil damage; a pre-logging inventory to set cutting limits; directional felling of trees to avoid injuring neighboring trees; and pulling trees out of the forest using cables to minimize vegetation damage and soil disturbance. A number of studies have now shown that the application of RIL can potentially benefit everybody, reducing not just environmental damage (Fig. 9.15), but also

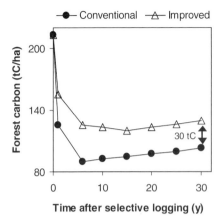

Fig. 9.15 The adoption of reduced impact logging (RIL) can substantially reduce environmental damage in tropical rain forests. This graph shows estimates of forest carbon in a hectare of lowland dipterocarp forest before and after conventional and improved (RIL) logging (Putz et al. 2008). (Courtesy of Francis Putz.)

the financial costs of logging. Why then do destructive logging practices persist almost everywhere in the tropics?

One major reason why RIL is not more widely adopted is that the long-term benefits do not accrue to the logger, who is very rarely the owner of the forest. Currently, most rain forest logging is either illegal or involves only a short-term concession, so the logger derives no financial benefit from protecting soils and future generations of trees. The logger's interest is to take out as many logs as quickly and cheaply as possible. Some aspects of RIL, such as the training of workers, careful planning of roads, and directional felling of trees, make sense in any logging operation, but others, such as the exclusion of steep slopes and streamside forests, merely cut profits. The forest owner – the state, in most rain forest countries – would undoubtedly benefit from the strict application of RIL guidelines, but enforcing them on logging companies requires a well-trained, adequately paid, and highly motivated team of forest officers, and few rain forest countries have this. This is not an excuse for abandoning all attempts at controlling logging practices, however, since any improvement in the present situation would have immediate benefits. The potential for the consumer nations to influence logging practices through certification is discussed below.

Regulating the exploitation of wildlife is another important step. Until recently, most attention was focused on reducing the exploitation of rare and endangered species, such as tigers, rhinoceroses, primates, and parrots. In countries with effective border controls, the control of international trade in these species and the products made from them has been the most effective approach (see below). However, for many species, the internal trade is now more important – witness, for example, the huge numbers of parrots and primates kept as pets in rain forest countries – while many rain forest countries, such as Indonesia, have borders that are far too long and porous for an effective enforcement of a trade ban.

The trade in bushmeat is now recognized as one of the most serious threat to rain forest animals (Fig. 9.16). Bushmeat is sold by weight, rather than rarity, so the bushmeat trade is a much broader threat to biodiversity than the trade in specific endangered species. Where bushmeat is the major source of protein for the local population, as it is in much of the African rain forest area, the provision of alternative sources of protein will also be essential. However, the

Fig. 9.16 Hunting threatens wildlife even in forest areas that appear intact in satellite pictures. This photograph shows three red-eared guenons (*Cercopithecus erythrotis*) and a crowned guenon (*Cercopithecus pogonias*) for sale in Malabo, Equatorial Guinea's bushmeat market, in 2005. (Courtesy of Jessica Weinberg.)

current importance of bushmeat protein in rural Africa is not an argument for uncontrolled exploitation, since current harvest levels are not sustainable (Fa & Brown 2009). Extraction rates are much lower in most of the Amazon region, making sustainability a more realistic target, although not necessarily any easier to achieve.

The role of consumers in rain forest conservation

The costs of rain forest conservation are met largely by local people in the tropical countries – the rural poor rather than the urban elite – while the benefits are shared globally (Whitten & Balmford 2006). A logical conclusion from this is that the developed world, as the wealthiest of the beneficiaries, should pay far more for tropical conservation. How can this best be done? As discussed above, the influence of the developed world in rain forest countries has been reduced by the growth of alternative markets and sources of capital. Moreover, poor governance in many tropical countries limits the effectiveness of aid and debt relief as a way of influencing conservation practices. The leaders and citizens of these countries are also often resentful of lectures from the developed world on the importance of saving the rain forest, seeing this unwanted advice as, at best, hypocritical, and at worst, neocolonial interference.

Stopping trade is the answer only for species that are directly threatened by harvesting for export. The multibillion dollar trade in endangered species

Fig. 9.17 Pangolins (*Manis* sp.) are being traded illegally in huge numbers from all over Southeast Asia to China, where they are sold as a health food and medicine. This photograph shows part of a large shipment of frozen pangolins from Indonesia and Malaysia seized in Hong Kong in 2005–6. (Photograph © AFCD, Hong Kong.)

and their products is controlled by the CITES convention – the Convention on International Trade in Endangered Species of Wild Fauna and Flora (Fig. 9.17). More than 175 countries are signatories to this convention. Signatories are committed to preventing or restricting trade in around 28,000 plant species and 5000 animal species, including elephants, rhinoceroses, tigers, and birds of paradise, as well as most rain forest primates. Periodic news reports of large seizures at ports and airports in the developed world show that there is still a big market there for wildlife products. Governments in the developed world can help reduce the impact of this trade by greater efforts in enforcement at home, by pressuring other importer nations to abide by the Convention, and by providing financial support and technical assistance to the source countries. In Southeast Asia, the Association of Southeast Asian Nations Wildlife Enforcement Network (ASEAN-WEN) coordinates the regional response to illegal trade in protected species and provides a single partner for collaboration with consumer nations, including China.

Boycotts of the major rain forest-related products, including rain forest timber and palm oil or soybeans grown on deforested land, are unlikely to be very effective, because of the many alternative markets for these products, and may even be counterproductive, by further reducing the influence of the developed world. Instead, the aim should be to work with the producers to reduce environmental damage. One way of doing this is by paying a premium for products

Fig. 9.18 The logo of the Forest Stewardship Council, an independent, non-governmental, not-for-profit organization, identifies wood products that meet a range of stringent but practical criteria relating to environmental and social sustainability.

that have been produced without damaging the rain forest. An example of an initiative that deserves support from consumers in the developed world is forest certification. Certification of a forest, and the wood products harvested from it, is intended to show the consumer that the forest is being managed sustainably, so that his or her purchase is not contributing to environmental degradation. Certification gives the producers the right to use a trademarked label. There are several international certification schemes, with the one run by the Forest Stewardship Council (FSC) (Fig. 9.18) most widely known, as well as a range of national schemes in consuming and producing countries. Requirements for certification vary between schemes, but all include reduced impact logging and compliance with local laws. An increasing number of both individual consumers and major industrial buyers of timber and wood products now insist on certification, and total FSC-labeled sales were estimated at over US$20 billion in 2010. Around 5% of the world's production forests are now certified, but less than one-tenth of these are in the tropics. More than a third of the FSC-certified tropical timber from natural forests comes from Bolivia, where a combination of compatible laws, strong enforcement, tax benefits, and large-scale, vertically integrated forestry operations has produced conditions particularly favorable to certification success (Ebeling & Yasué 2009). The major problem for most tropical producers is that the costs of meeting certification standards are rarely justified by the premium paid for certified products. An additional problem is that domestic markets for timber in many rain forest countries are far larger than export markets, so that international pressures have little effect. However, while forest certification is clearly not the complete answer to the problem of unsustainable logging of tropical rain forests, it can help by providing models for best forestry practice, which other producers can then be encouraged to copy.

Forest certification works because timber, furniture, and paper are easily recognizable rain forest products, so the consumer can be given a clear choice. However, many of the products most directly linked to rain forest loss are invisible in the end products purchased by the consumer, so mobilizing consumer support is a lot more difficult. Developed countries import huge amounts of palm oil and soybean, for instance, but few people knowingly buy either product at the supermarket. Palm oil is hidden in a huge variety of processed foods, as well as in animal feed, soaps, detergents, cosmetics, and candles. Rain forest soybeans reach consumers as chicken, pork, and beef, or in a similar variety of food and industrial products. In this case, the pressure needs to be on the private companies that import the raw products to insure that they have been produced in a way that does not contribute to deforestation. The Round Table on Sustainable Oil Palm (RSPO) is an association of producers, processors, traders, consumer goods manufacturers, retailers, banks, and NGOs that aims to develop and implement global standards for sustainable palm oil (www.rspo.org). Although a great deal of skepticism remains among conservationists (Laurance et al. 2010a), the RSPO does appear to be making a genuine effort to improve standards in the industry. Similar initiatives exist for soy and a number of other crops.

Money for carbon

Despite the setbacks at Copenhagen in 2009, the United Nations Framework Convention on Climate Change (UNFCC) may soon include a mechanism by which countries reducing their carbon emissions from deforestation and forest degradation (known by the acronym REDD) will receive financial compensation from industrial nations wishing to offset part of their own emissions. The precise mechanism for this compensation has not yet been worked out, but voluntary carbon markets are already funding REDD projects. REDD has the potential to not only provide far more funds for tropical rain forest conservation than are currently available, but also – and this is likely to prove more important than the amount of money – to link this funding to a rigorous assessment of each country's actual performance.

Five major types of management intervention are likely to qualify for REDD funding: slowing deforestation, reducing logging impacts (e.g. through RIL), preventing fires, restoring native forests, and expanding tree plantations (Stickler et al. 2009). The first four of these could provide direct conservation benefits, while the establishment of plantations could have a positive or negative impact, depending on how and where it is done. REDD funding for forest conservation in areas that are not currently threatened could be an additional option and one that would reward "high forest low deforestation countries," but unless these forests are still accumulating carbon, it will be hard to include this in a simple, carbon-based framework. REDD currently has numerous problems, many of them associated with verifying and measuring the reduction in emissions, but the tools being developed to overcome these problems, such as satellite monitoring of biomass, have themselves the potential for additional conservation benefits (Asner et al. 2010). Conservation would also benefit from the improvements in forest governance that will be needed. There have been concerns about the narrow focus on carbon, which could potentially favor fast-growing monoculture plantations over species-rich rain forests, but safeguards to prevent damage to forests of high conservation value are likely. Alternatively, there could be additional, non-REDD funding, for biodiversity, which would favor schemes that deliver conservation benefits (Harvey et al. 2010).

Restoring the rain forest

We have focused so far on ways to limit the continued clearance and over-exploitation of the world's remaining tropical rain forests. In an increasing proportion of the land previously covered in rain forest, however, deforestation and fragmentation have resulted in a landscape that can no longer sustain an intact rain forest community. In these areas, the conservation focus must be switched to restoration of tropical forest communities. It is not surprising that the region that has accumulated most experience in rain forest restoration is the highly fragmented Brazilian Atlantic Forest, where numerous projects are attempting to buffer and link fragments (Rodrigues et al. 2009). Although many problems remain, there has been enough success to encourage an increasing attention to reducing costs and increasing the rate of restoration, which currently lags the rate of deforestation by several orders of magnitude. Speed is important because there is a "time tax" on restoration, with forest fragments continuing to lose plant and animal species as long as they are embedded in a low diversity matrix of cattle pastures, agriculture, and secondary forests (Martínez-Garza and Howe 2003).

Planting nursery-raised seedlings is the commonest restoration technique (Fig. 9.19a,b), despite high costs, but a range of other options exist, including encouraging natural forest succession (Shono et al. 2007), sowing seeds directly at the site (Bonilla-Moheno & Holl 2009), planting live tree branches (c. 2 m tall) (Zahawi & Holl 2009), collecting wild seedlings for planting (Ådjers et al. 1998), and incorporating agricultural crops as a transitional phase (Vieira et al. 2009). In practice, since different techniques work better with different species, a combination of techniques is likely to be most efficient and successful. Restoration needs to be planned at the landscape level, with effort concentrated where it will do most good. It has sometimes been suggested that exotic plantation monocultures can act as "catalysts" for invasion by native forest species, but there is little evidence that this is an effective approach to restoration, unless the cash income from harvesting the exotics is a necessary incentive for reforestation.

The ecological restoration literature is dominated by botanical concerns, but the active reintroduction of locally extirpated animal species may be necessary after the forest structure has been restored. It may also be needed at sites where the vegetation is still intact but the fauna has been depleted by hunting. Reintroduction efforts in the tropics are often poorly planned, implemented, and documented, and in many cases confiscated or abandoned animals are simply dumped in the nearest forest. However, there have now been enough successful reintroduction projects worldwide to show that it is possible to reestablish populations of primates, ungulates, carnivores, and a variety of other organisms in areas from which they have been previously extirpated as long as the amount of habitat remaining is adequate, the cause of extirpation – usually hunting – is under control, and a wild or captive source population is available. The failure rate is still high, but persistence often pays off eventually. Successful reintroductions not only restore ecological processes, but can make a major contribution to conservation education and the development of local pride in wild species.

A good example of the role of reintroduction is the successful conservation program for the golden lion tamarin (*Leontopithecus rosalia*) (Fig. 9.20), which was critically endangered by the fragmentation of its rain forest habitat in the Brazilian Atlantic region. An estimated one-third of the 1000 or so wild individuals today are descendents of an international captive breeding program, while others were translocated from doomed populations in small and isolated forest fragments into larger populations (Kierulff et al. 2010). The reintroduction program has promoted forest protection and encouraged the creation of corridors and buffer zones by reforestation. This ultra-charismatic primate has also formed the basis of a successful public education program and encouraged community involvement in conservation.

(a)

(b)

Fig. 9.19 Reforestation of degraded areas has become an important tool in rain forest conservation. (a) Native tree nursery at Bawangling National Nature Reserve, Hainan, China, being used to raise seedlings for the reforestation of a corridor linking forest fragments inhabited by the critically endangered Hainan gibbon. (Courtesy of Billy Hau.) (b) Visiting Japanese students planting tree seedlings as part of a rain forest restoration project in northern Queensland, Australia. (Courtesy of Hiromi Kobori.)

Fig. 9.20 The golden lion tamarin (*Leontopithecus rosalia*) was the subject of a successful captive breeding and reintroduction program in the Brazilian Atlantic region. This ultra-charismatic primate has also formed the basis of a successful public education program and encouraged community involvement in conservation. (Courtesy of Tim Laman.)

Conclusions and future research directions

Each rain forest region has a unique flora and fauna that is a product of the biogeographical history of that region, and which represents an independent community-level response to the local climate, soils, and topography. Each rain forest community also faces different types and levels of current threats from increasing human activity. The enormous forest regions of the Amazon, the Congo basin, and New Guinea were protected until recently by their size, lack of development, and remoteness. The Amazon and New Guinea also had relatively low human populations. Large tracts of these forests will survive until at least the middle of this century, but their long-term future is in doubt due to planned road networks and the pressure to open new areas for logging, cash crops, and mineral exploration. Primary forests in most of everwet Southeast Asia have already disappeared because of increasing logging and agricultural development, and even some national parks are threatened. Madagascar may have stepped back from the brink, but the island's rain forests remain under extraordinary threat.

The situation is not yet hopeless. Even in the worst-hit regions, the majority of the rain forest biota still survives, in small protected areas, in unprotected fragments of primary forest on sites that are too steep, too wet, or too infertile to be worth clearing, in logged forests, in secondary forest on abandoned land, and in woody regrowth along streams and fences. Many of the more tolerant rain forest species can also make use of agricultural systems that include trees, such as shade coffee plantations, although few can survive in industrial mono-culture plantations. More species will survive if bigger areas are fully protected and if unprotected areas are managed sustainably.

International support is needed to insure that financing is available for the establishment and subsequent management of protected area systems and to encourage the use of best practices, such as RIL, in exploiting unprotected areas. It makes no sense to expect the world's poorest countries to pay for the protection of the world's richest ecosystems, when the benefits are global. Support for these conservation activities can come from countries of the developed world, conservation organizations, and multinational development banks, such as the World Bank. Carbon markets may eventually provide much-needed financing coupled with a rigorous assessment of conservation performance. Individual citizens can contribute to this effort by buying timber and other products obtained from sustainably managed forests, and by joining and donating funds to conservation organizations. And individual citizens and the conservation organizations to which they belong should insist that their governments contribute to international conservation efforts, including the control of greenhouse gas emissions.

The first priority of conservation efforts should be to protect representative examples of the most intact rain forest communities that remain, so that when the current, crazy round of forest clearing finishes, there will be core areas of forest out of which species can migrate and establish new forest communities. The ability of the forest to eventually heal following damage by logging, farming, and cattle is illustrated by the Maya forest region of Mexico and Central America (Primack et al. 1998). This area was highly fragmented and cleared over large areas by the Maya civilization during the first millennium to support large rural populations and growing cities. When the Maya civilization collapsed around 1000 AD, possibly due to some combination of exhaustion of its soil and forest resources and warfare, the human population declined drastically, and the forest was able to expand and recover from its remaining fragments. If one flies over the region today, one is struck by the extensive area of forest (Fig. 9.21). There is little evidence that any species went extinct in the region as a result of this transformation of the landscape by Maya agriculture. Although this example suggests that rain forests may be more resilient than is often thought, it is important to realize that recovery is a very slow process, taking centuries rather than decades. Moreover, a complete recovery will not be possible if native species have become extinct and exotic species have become established. The Maya had no guns, chain saws, or bulldozers and did not move species between continents.

In this book we have argued that comparative studies of rain forests in different parts of the world represent an important area for further research (see also Corlett & Primack 2006). We believe that in addition to their scientific importance such studies will help with conservation and management, by revealing key similarities and differences and thus the specific needs and vulnerabilities of each rain forest. Unfortunately, the pioneering pantropical comparisons of the 1950s, 1960s, and 1970s have largely given way to a more parochial approach in recent decades and few currently active rain forest researchers have experience in more than one major region. One notable exception has been the Center for Tropical Forest Science, which has established

Fig. 9.21 One thousand years ago, Mayan farms and cities occupied a wide area of the Central American lowlands, with no apparent loss of species. Today, the ruined cities are overgrown by tropical forests, as shown here by the ancient Mayan city of Tikal in Guatemala. (Courtesy of Harald Schuetz.)

a worldwide network of large (typically 25–50 ha, 50–125 acre) forest plots to look at patterns of tree species distribution and forest dynamics (www. ctfs.si.edu). Twenty-three of these plots are in tropical forests and they represent all the major rain forest regions, except Madagascar.

What is needed now are community-oriented studies that include multiple taxonomic levels: for example, to what extent do parrots, small primates, squirrels, and marsupials in each rain forest overlap in their ecology and compete with or substitute for each other? Similar questions could be asked of the large, vertebrate-consuming carnivores: in what ways are cats, snakes, and eagles competitors and ecological equivalents? Another key topic is determining how rain forests respond to increasing levels of human impact, such as logging, fragmentation, hunting, and global warming. How do logging impacts differ between forests with different dominant families of timber-producing trees? Are forest fragments less isolated in Africa and Asia, where many mammals are willing to cross open areas, than in the Neotropics, where most are not? What is the impact on the plant community when hunters remove all of the large animals from the forest, and does this impact differ between rain forest regions? Is there a resulting failure of seed dispersal and regeneration, and are Madagascan forests more vulnerable because they have fewer dispersal agents? What are the consequences of the contrasting bat-dispersal of most Neotropical pioneer trees and bird-dispersal of most of their Old World counterparts? Will the supra-annual reproductive cycles of Asian dipterocarp species be particularly vulnerable to disruption by global climate change?

   In many cases, experimental manipulations may be the best way to investigate the significance of observed differences between rain forests. Manipulating rain forest communities on the scale needed may seem a daunting task, but large-scale experiments have been successfully used in the study of rain forest fragmentation (Laurance et al. 2004). The same approach, with forest retained on part of a site that is otherwise being cleared for pasture or crops, could be used to compare the impacts of fragmentation in different rain forest regions. Experimental extirpations of native vertebrates from rain forests will rarely be acceptable, but reintroduction experiments in areas from which the species was previously extirpated can both answer ecological questions on the role of particular animal species and contribute to the development of restoration techniques. The availability in many areas of both defaunated forest fragments and surplus mammals and birds confiscated from hunters and traders suggests that such experiments would be practical. Invasive species provide another possible window into pantropical differences, and one that can also be manipulated, although in this case by removal experiments. Logging and other forms of exploitation are also open to manipulation, in collaboration with logging companies and legal (often indigenous) hunters.

   The world has many different rain forests, and the hope for each depends on a combination of the amount of forest remaining, the current threats to those forests, and the vulnerability of the species they still support. Protection of large areas of each rain forest type in well-managed national parks is the highest priority, followed by the sustainable management of forests outside these parks. If most species can survive until the present cycle of forest destruction ends, then rain forests will eventually expand from their remaining fragments and restore the damaged landscape. Biologists can play a role in these efforts by initiating comparative studies among rain forest regions that highlight the unique features of each area. Citizens can play a role by joining conservation organizations and buying rain forest products that have been produced sustainably. If biologists and informed leaders can convey the message that each forest has special qualities, the public will support efforts to protect representative examples of the "many tropical rain forests."

References

Aanen, D.K., Eggleton, P., Rouland-Lefèvre, C., Guldberg-Frøslev, T., Rosendahl, S. & Boomsma, J.J. (2002) The evolution of fungus-growing termites and their mutualistic fungal symbionts. *Proceedings of the National Academy of Sciences of the USA*, **99**, 14887–14892.

Abram, N.J., Gagan, M.K., Liu, Z., Hantoro, W.S., McCulloch, M.T. & Suwargadi, B.W. (2007) Seasonal characteristics of the Indian Ocean Dipole during the Holocene epoch. *Nature*, **445**, 299–302.

Adeney, J.M., Christensen, N.L., Jr. & Pimm, S.L. (2009) Reserves protect against deforestation fires in the Amazon. *PLoS One*, **4**, e5014.

Ådjers, G., Hadengganan, S., Kuusipalo, J., Otsamo, A. & Vesa, L. (1998) Production of planting stock from wildings of four *Shorea* species. *New Forests*, **16**, 185–197.

Adler, P.H. & Foottit, R.G. (2009) Introduction. In: *Insect Biodiversity: Science and Society* (eds R.G. Foottit & P.H. Adler), pp. 1–6. Blackwell, Hoboken, NJ.

Agnarsson, I., Kuntner, M. & May-Collado, L.J. (2010) Dogs, cats, and kin: a molecular species-level phylogeny of Carnivora. *Molecular Phylogenetics and Evolution*, **54**, 726–745.

Ali, J.R. & Huber, M. (2010) Mammalian biodiversity on Madagascar controlled by ocean currents. *Nature (London)*, **463**, 653.

Allmon, W.D. (1991) A plot study of forest floor litter frogs, Central Amazon, Brazil. *Journal of Tropical Ecology*, **7**, 503–522.

Altringham, J.D. & Fenton, M.B. (2003) Sensory ecology and communication in the Chiroptera. In: *Ecology of Bats* (eds T.H. Kunz & M.B. Fenton), pp. 90–27. University of Chicago Press, Chicago, IL.

An, S.Q., Zhu, X.L., Wang, Z.F., Campbell, D.G., Li, G.Q. & Chen, X.L. (1999) The plant species diversity in a tropical montane rain forest on Wuzhi Mountain, Hainan. *Acta Ecologica Sinica*, **19**, 803–809.

Andreone, F., Carpenter, A.I., Cox, N. et al. (2008) The challenge of conserving amphibian megadiversity in Madagascar. *PLoS Biology*, **6**, 943–946.

Arroyo-Rodríguez, V., Pineda, E., Escobar, F. & Benítez-Malvido, J. (2009) Value of small patches in the conservation of plant-species diversity in highly fragmented rainforest. *Conservation Biology*, **23**, 729–739.

Ashton, P.S. (2003) Floristic zonation of tree communities on wet tropical mountains revisited. *Perspectives in Plant Ecology Evolution and Systematics*, **6**, 87–104.

Ashton, P.S. (2010) Conservation of Borneo biodiversity: do small lowland parks have a role, or are big inland sanctuaries sufficient? Brunei as an example. *Biodiversity and Conservation*, **19**, 343–356.

Ashton, P.S. & CTFS-Working-Group (2004) Floristics and vegetation of the forest dynamics plots. In: *Tropical Forest Diversity and Dynamism: Findings from a Large-Scale Plot Network* (eds E.C. Losos & E.G. Leigh Jr). University of Chicago Press, Chicago, IL.

Asner, G.P., Powell, G.V.N., Mascaro, J. et al. (2010) High-resolution forest carbon stocks and emissions in the Amazon. *Proceedings of the National Academy of Sciences of the USA*, **107**, 16738–16742.

Asner, G.P., Rudel, T.K., Aide, T.M., Defries, R. & Emerson, R. (2009) A contemporary assessment of change in humid tropical forests. *Conservation Biology*, **23**, 1386–1395.

Athens, J.S. (2009) *Rattus exulans* and the catastrophic disappearance of Hawai'i's native lowland forest. *Biological Invasions*, **11**, 1489–1501.

Baker, R.R. (1978) *The Evolutionary Ecology of Animal Migration.* Hodder & Stoughton, London.

Balke, M., Gómez-Zurita, J., Ribera, I., Viloria, A., Zillikens, A., Steiner, J., García, M., Hendrich, L. & Vogler, A.P. (2008) Ancient associations of aquatic beetles and tank bromeliads in the Neotropical forest canopy. *Proceedings of the National Academy of Sciences of the USA*, **105**, 6356–6361.

Balslev, H., Renato, V., Paz y Miño, G., Christensen, H. & Nielsen, I. (1998) Species count of vascular plants in one hectare of humid lowland forest in Amazonian Ecuador. In: *Forest Biodiversity in North, Central and South America, and the Caribbean, Vol. 21* (eds F. Dallmeier & J.A. Comiskey), pp. 585–594. UNESCO, Paris.

Barlow, J. & Silveira, J.M. (2009) The consequences of fire for the fauna of humid tropical forests. In: *Tropical Fire Ecology: Climate Change, Land Use, and Ecosystem Dynamics*, pp. 543–556. Springer, Heidelberg, Germany.

Barona, E., Ramankutty, N., Hyman, G. & Coomes, O.T. (2010) The role of pasture and soybean in deforestation of the Brazilian Amazon. *Environmental Research Letters*, **5**.

Barrett, M.A. & Ratsimbazafy, J. (2009) Luxury bushmeat trade threatens lemur conservation. *Nature*, **461**, 470.

Bass, M.S., Finer, M.M., Jenkins, C.N. et al. (2010) Global conservation significance of Ecuador's Yasuní National Park. *PloS One*, **5**, article no. e8767.

Baussart, S., Korsoun, L. & Bels, V. (2007) Feeding mechanism in fruit-eating birds: toucans and hornbills. *Journal of Morphology*, **268**, 1048.

Beattie, A.J. & Hughes, L. (2002) Ant-plant interactions. In: *Plant-Animal Interactions: An Evolutionary Approach* (eds C.M. Herrera & O. Pellmyr), pp. 211–235. Blackwell Science, Oxford, UK.

Beehler, B.M. & Dumbacher, J.P. (1996) More examples of fruiting trees visited predominantly by birds of paradise. *Emu*, **96**, 81–88.

Beisiegel, B.M. (2001) Notes on the coati, *Nasua nasua* (Carnivora: Procyonidae) in an Atlantic forest area. *Brazilian Journal of Biology*, **61**, 689–692.

Bennett, B. (2000) Ethnobotany of the Bromeliaceae. In: *Bromeliaceae: Profile of an Adaptive Radiation* (ed D.H. Benzing), pp. 587–608. Cambridge University Press, Cambridge, UK.

Berghoff, S.M., Maschwitz, U. & Linsenmair, K.E. (2003) Influence of the hypogaeic army ant *Dorylus (Dichthadia) laevigatus* on tropical arthropod communities. *Oecologia*, **135**, 149–157.

Berry, N.J., Phillips, O.L., Lewis, S.L. et al. (2010) The high value of logged tropical forests: lessons from northern Borneo. *Biodiversity and Conservation*, **19**, 1–13.

Bezerra, B.M., Barnett, A.A., Souto, A. & Jones, G. (2009) Predation by the tayra on the common marmoset and the pale-throated three-toed sloth. *Journal of Ethology*, **27**, 91–96.

Biebouw, K. (2009) Home range size and use in *Allocebus trichotis* in Analamazaotra Special Reserve, Central Eastern Madagascar. *International Journal of Primatology*, **30**, 367–386.

Blake, S. & Fay, J.M. (1997) Seed production by *Gilbertiodendron dewevrei* in the Nouabalé-Ndoki National Park, Congo, and its implications for large mammals. *Journal of Tropical Ecology*, **13**, 885–891.

Blake, S., Deem, S.L., Strindberg, S. et al. (2008) Roadless wilderness area determines forest elephant movements in the Congo Basin. *PLoS One*, **3**, e3546.

Blake, S., Deem, S.L., Mossimbo, E., Maisels, F. & Walsh, P. (2009) Forest elephants: tree planters of the Congo. *Biotropica*, **41**, 459–468.

Blüthgen, N. & Fiedler, K. (2002) Interactions between weaver ants *Oecophylla smaragdina*, homopterans, trees and lianas in an Australian rain forest canopy. *Journal of Animal Ecology*, **71**, 793–801.

Blüthgen, N. & Reifenrath, K. (2003) Extrafloral nectaries in an Australian rainforest: Structure and distribution. *Australian Journal of Botany*, **51**, 515–527.

Boakes, E.H., Mace, G.M., McGowan, P.J.K. & Fuller, R.A. (2010) Extreme contagion in global habitat clearance. *Proceedings of the Royal Society Biological Sciences Series B*, **277**, 1081–1085.

Bobbink, R., Hicks, K., Galloway, J. et al. (2010) Global assessment of nitrogen deposition effects on terrestrial plant diversity: A synthesis. *Ecological Applications*, **20**, 30–59.

Bohannon, J. (2010) Madagascar's forests get a reprieve – but for how long? *Science*, **328**, 23–25.

Bonilla-Moheno, M. & Holl, K.D. (2009) Direct seeding to restore tropical mature-forest species in areas of slash-and-burn agriculture. *Restoration Ecology*, doi: 10.1111/j.1526-100X.2009.00580.x.

Bostwick, K.S., Elias, D.O., Mason, A. & Montealegre-Z.F. (2010) Resonating feathers produce courtship song. *Proceedings of the Royal Society Biological Sciences Series B*, **277**, 835–841.

Bourguignon, T., Sobotnik, J., Lepoint, G., Martin, J.-M. & Roisin, Y. (2009) Niche differentiation among neotropical soldierless soil-feeding termites revealed by stable isotope ratios. *Soil Biology & Biochemistry*, **41**, 2038–2043.

Boyer, A.G. (2008) Extinction patterns in the avifauna of the Hawaiian islands. *Diversity and Distributions*, **14**, 509–517.

Braby, M.F. (2000) *Butterflies of Australia: Their Identification, Biology and Distribution, Vol. 1.* CSIRO Publishing, Collingwood, Victoria.

Braby, M.F. & Pierce, N.E. (2007) Systematics, biogeography and diversification of the Indo-Australian genus *Delias* Hubner (Lepidoptera: Pieridae): phylogenetic evidence supports an "out-of-Australia" origin. *Systematic Entomology*, **32**, 2–25.

Bradford, M.G., Dennis, A.J. & Westcott, D.A. (2008) Diet and dietary preferences of the southern cassowary (*Casuarius casuarius*) in North Queensland, Australia. *Biotropica*, **40**, 338–343.

Brady, S.G. (2003) Evolution of the army ant syndrome: The origin and long-term evolutionary stasis of a complex of behavioral and reproductive adaptations. *Proceedings of the National Academy of Sciences of the USA*, **100**, 6575–6579.

Bravo, S.P. (2009) Implications of behavior and gut passage for seed dispersal quality: the case of black and gold howler monkeys. *Biotropica*, **41**, 751–758.

Brochu, C.A. (2007) Morphology, relationships, and biogeographical significance of an extinct horned crocodile (Crocodylia, Crocodylidae) from the Quaternary of Madagascar. *Zoological Journal of the Linnean Society*, **150**, 835–863.

Brodmann, J., Twele, R., Francke, W., Luo, Y.B., Song, X.Q. & Ayasse, M. (2009) Orchid mimics honey bee alarm pheromone in order to attract hornets for pollination. *Current Biology*, **19**, 1368–1372.

Brown, E.D. & Hopkins, M.J.G. (1996) How New Guinea rainforest flower resources vary in time and space: Implications for nectarivorous birds. *Australian Journal of Ecology*, **21**, 363–378.

Brown, J.H. & Lomolino, M.V. (1998) *Biogeography.* Sinauer Associates, Sunderland, MA.

Brown, L. & Amadon, D. (1968) *Eagles, Hawks and Falcons of the World.* Country Life Books, Feltham, UK.

Brown, N., Press, M. & Bebber, D. (1999) Growth and survivorship of dipterocarp seedlings: Differences in shade persistence create a special case of dispersal limitation. *Philosophical Transactions of the Royal Society of London B Biological Sciences*, **354**, 1847–1855.

Brühl, C.A. & Eltz, T. (2010) Fuelling the biodiversity crisis: species loss of ground-dwelling forest ants in oil palm plantations in Sabah, Malaysia (Borneo). *Biodiversity and Conservation*, **19**, 519–529.

Brühl, C.A., Gunsalam, G. & Linsenmair, K.E. (1998) Stratification of ants (Hymenoptera, Formicidae) in a primary rain forest in Sabah, Borneo. *Journal of Tropical Ecology*, **14**, 285–297.

Brühl, C.A., Eltz, T. & Linsenmair, K.E. (2003) Size does matter: Effects of tropical rainforest fragmentation on the leaf litter ant community in Sabah, Malaysia. *Biodiversity and Conservation*, **12**, 1371–1389.

Brumfield, R.T., Tello, J.G., Cheviron, Z.A., Carling, M.D., Crochet, N. & Rosenberg, K.V. (2007) Phylogenetic conservatism and antiquity of a tropical specialization: Army-ant-following in the typical antbirds (Thamnophilidae). *Molecular Phylogenetics and Evolution*, **45**, 1–13.

Bühler, P. (1997) The visual peculiarities of the toucan's bill and their principal biological role (Ramphastidae, Aves). In: *Tropical Biodiversity and Systematics* (ed H. Ulrich), pp. 305–310. Zoologisches Forschungsinstitut und Museum Alexander Koenig, Bonn, Germany.

Burgess, N.D., Butynski, T.M., Cordeiro, N.J. et al. (2007) The biological importance of the Eastern Arc Mountains of Tanzania and Kenya. *Biological Conservation*, **134**, 209–231.

Burney, D.A., Robinson, G.S. & Burney, L.P. (2003) *Sporormiella* and the late Holocene extinctions in Madagascar. *Proceedings of the National Academy of Sciences of the USA*, **100**, 10800–10805.

Bush, A. (2007) Extratropical influences on the El Niño–Southern Oscillation through the late Quaternary. *Journal of Climate*, **20**, 788–800.

Butchart, S.H.M. & Bird, J.P. (2010) Data deficient birds on the IUCN Red List: What don't we know and why does it matter? *Biological Conservation*, **143**, 239–247.

Byrne, M.M. (1994) Ecology of twig-dwelling ants in a wet lowland tropical forest. *Biotropica*, **26**, 61–72.

Byrnes, G., Lim, N.T.L. & Spence, A.J. (2008) Take-off and landing kinetics of a free-ranging gliding mammal, the Malayan colugo (*Galeopterus variegatus*). *Proceedings of the Royal Society Biological Sciences Series B*, **275**, 1007–1013.

Caine, N.G., Osorio, D. & Mundy, N.I. (2010) A foraging advantage for dichromatic marmosets (*Callithrix geoffroyi*) at low light intensity. *Biology Letters*, **6**, 36–38.

Camargo, J.M.F. & Pedro, S.R.M. (2002) Mutualistic association between a tiny Amazonian stingless bee and a wax-producing scale insect. *Biotropica*, **34**, 446–451.

Cameron, S.A. (2004) Phylogeny and biology of the neotropical orchid bees (Euglossini). *Annual Review of Entomology*, **49**, 377–404.

Campos-Arceiz, A., Lin, T.Z., Htun, W., Takatsuki, S. & Leimgruber, P. (2008) Working with mahouts to explore the diet of work elephants in Myanmar (Burma). *Ecological Research*, **23**, 1057–1064.

Cannon, C.H., Morley, R.J. & Bush, A.B.G. (2009) The current refugial rainforests of Sundaland are unrepresentative of their biogeographic past and highly vulnerable to disturbance. *Proceedings of the National Academy of Sciences of the USA*, **106**, 11188–11193.

Caujapé-Castells, J., Tye, A., Crawford, D.J. et al. (2010) Conservation of oceanic island floras: Present and future global challenges. *Perspectives in Plant Ecology, Evolution and Systematics*, **12**, 107–129.

Changizi, M.A., Zhang, Q. & Shimojo, S. (2006) Bare skin, blood and the evolution of primate colour vision. *Biology Letters*, **2**, 217–221.

Chapman, C.A. & Russo, S.E. (2007) Primate seed dispersal: linking behavioral ecology with forest community structure. In: *Primates in Perspective* (eds C.J. Campbell, A.F. Fuentes, K.C. MacKinnon, M. Panger & S.K. Bearder), pp. 510–525. Oxford University Press, Oxford, UK.

Cheke, A.S. & Hume, J. (2008) *Lost Land of the Dodo: An Ecological History of Mauritius, Réunion & Rodrigues*. T. & A.D. Poyser, London, UK.

Chen, I.C., Shiu, H.-J., Benedick, S. et al. (2009) Elevation increases in moth assemblages over 42 years on a tropical mountain. *Proceedings of the National Academy of Sciences of the USA*, **106**, 1479–1483.

Chippaux, J.P. (1998) Snake-bites: Appraisal of the global situation. *Bulletin of the World Health Organization*, **76**, 515–524.

Clark, C.J., Poulsen, J.R., Malonga, R. & Elkan, P.W., Jr. (2009) Logging concessions can extend the conservation estate for Central African tropical forests. *Conservation Biology*, **23**, 1281–1293.

Clark, D.B., Clark, D.A. & Oberbauer, S.F. (2010) Annual wood production in a tropical rain forest in NE Costa Rica linked to climatic variation but not to increasing CO_2. *Global Change Biology*, **16**, 747–759.

Cochrane, M.A. (2009) *Tropical Fire Ecology: Climate Change, Land Use, and Ecosystem Dynamics*. Praxis Publishing, Chichester, UK.

Cody, S., Richardson, J.E., Rull, V., Ellis, C. & Pennington, R.T. (2010) The Great American Biotic Interchange revisited. *Ecography*, **33**, 326–332.

Colwell, R.K., Brehm, G., Cardelus, C.L., Gilman, A.C. & Longino, J.T. (2008) Global warming, elevational range shifts, and lowland biotic attrition in the wet tropics. *Science*, **322**, 258–261.

Compton, S.G., Grehan, L. & Van Noort, S. (2009) A fig crop pollinated by three or more species of agaonid fig wasps. *African Entomology*, **17**, 215–222.

Corlett, R.T. (1998) Frugivory and seed dispersal by vertebrates in the oriental (Indomalayan) region. *Biological Reviews*, **73**, 413–448.

Corlett, R.T. (2004) Flower visitors and pollination in the Oriental (Indomalayan) Region. *Biological Reviews*, **79**, 497–532.

Corlett, R.T. (2007a) What's so special about Asian tropical forests? *Current Science*, **93**, 1551–1557.

Corlett, R.T. (2007b) The impact of hunting on the mammalian fauna of tropical Asian forests. *Biotropica*, **39**, 292–303.

Corlett, R.T. (2009a) *The Ecology of Tropical East Asia.* Oxford University Press, Oxford, UK.

Corlett, R.T. (2009b) Seed dispersal distances and plant migration potential in tropical East Asia. *Biotropica*, **41**, 592–598.

Corlett, R.T. (2010) Megafaunal extinctions and their consequences in the tropical Indo-Pacific. *Terra Australis*, **32**, 117–131.

Corlett, R.T. & Primack, R.B. (2006) Tropical rain-forests and the need for cross-continental comparisons. *Trends in Ecology & Evolution*, **21**, 104–110.

Cowie, R.H. (1995) Variation in species diversity and shell shape in Hawaiian land snails: In situ speciation and ecological relationships. *Evolution*, **49**, 1191–1202.

Cristoffer, C. (1987) Body size differences between New World and Old World, arboreal, tropical vertebrates: cause and consequences. *Journal of Biogeography*, **14**, 165–172.

Cristoffer, C. & Peres, C.A. (2003) Elephants versus butterflies: The ecological role of large herbivores in the evolutionary history of two tropical worlds. *Journal of Biogeography*, **30**, 1357–1380.

Crowe, T.M., Bowie, R.C.K., Bloomer, P. et al. (2006) Phylogenetics, biogeography and classi-fication of, and character evolution in, gamebirds (Aves: Galliformes): effects of character exclusion, data partitioning and missing data. *Cladistics*, **22**, 495–532.

Curran, L.M. & Leighton, M. (2000) Vertebrate responses to spatiotemporal variation in seed production of mast-fruiting Dipterocarpaceae. *Ecological Monographs*, **70**, 101–128.

Curran, L.M., Trigg, S.N., McDonald, A.K. et al. (2004) Lowland forest loss in protected areas of Indonesian Borneo. *Science*, **303**, 1000–1003.

Dallimer, M., King, T. & Atkinson, R.J. (2009) Pervasive threats within a protected area: con-serving the endemic birds of São Tomé, West Africa. *Animal Conservation*, **12**, 209–219.

Darwin, C.R. (1862) *On the Various Contrivances by which British and Foreign Orchids are Fertilised by Insects, and on the Good Effects of Intercrossing*. John Murray, London, UK.

Davidson, D.W., Cook, S.C., Snelling, R.R. & Chua, T.H. (2003) Explaining the abundance of ants in lowland tropical rainforest canopies. *Science*, **300**, 969–972.

Davidson, D.W., Lessard, J.-P., Bernau, C.R. & Cook, S.C. (2007) The tropical ant mosaic in a primary Bornean rain forest. *Biotropica*, **39**, 468–475.

Davies, R.G., Eggleton, P., Jones, D.T., Gathorne-Hardy, F.J. & Hernández, L.M. (2003) Evolution of termite functional diversity: Analysis and synthesis of local ecological and regional influences on local species richness. *Journal of Biogeography*, **30**, 847–877.

Davies, S.J.J.F. (2002) *Ratites and Tinamous: Tinamidae, Rheidae, Dromaiidae, Casuariidae, Apterygidae, Struthionidae*. Oxford University Press, Oxford, UK.

Davis, A.L.V. & Philips, T.K. (2009) Regional fragmentation of rain forest in West Africa and its effect on local dung beetle assemblage structure. *Biotropica*, **41**, 215–220.

Davis, B.W., Li, G. & Murphy, W.J. (2010) Supermatrix and species tree methods resolve phylogenetic relationships within the big cats, *Panthera* (Carnivora: Felidae). *Molecular Phylogenetics and Evolution*, **56**, 64–76.

De Gouvenain, R.C. & Silander, J.A., Jr. (2003) Do tropical storm regimes influence the structure of tropical lowland rain forests? *Biotropica*, **35**, 166–180.

Dejean, A., Grangier, J., Leroy, C. & Orivel, J. (2009) Predation and aggressiveness in host plant protection: a generalization using ants from the genus *Azteca*. *Naturwissenschaften*, **96**, 57–63.

de la Rosa, C.L. & Nocke, C.C. (2000) *A Guide to the Carnivora of Central America: Natural History, Ecology, and Conservation*. University of Texas Press, Austin, TX.

de Oliveira, F.B., Molina, E.C. & Marroig, G. (2009) Paleogeography of the South Atlantic: a route for primates and rodents into the New World? In: *South American Primates: Comparative Perspectives in the Study of Behavior, Ecology, and Conservation* (eds P.A. Garber, A. Estrada, J.C. Bicca-Marques, E.W. Heymann & K.B. Strier), pp. 55–68. Springer, Heidelberg, Germany.

de Oliveira, M.L. & Cunha, J.A. (2005) Do Africanized honeybees explore resources in the Amazonian forest? *Acta Amazonica*, **35**, 389–394.

Dekker, R.W.R.J. (2007) Distribution and speciation of megapodes (Megapodiidae) and subsequent development of their breeding behaviour. In: *Biogeography, Time, and Place: Distributions, Barriers, and Islands* (ed W. Renema), pp. 93–102. Springer, Dordrecht, Netherlands.

Delissio, L.J., Primack, R.B., Hall, P. & Lee, H.S. (2002) A decade of canopy-tree seedling survival and growth in two Bornean rain forests: Persistence and recovery from suppression. *Journal of Tropical Ecology*, **18**, 645–658.

den Tex, R.J., Thorington, R., Maldonado, J.E. & Leonard, J.A. (2010) Speciation dynamics in the SE Asian tropics: Putting a time perspective on the phylogeny and biogeography of Sundaland tree squirrels, *Sundasciurus*. *Molecular Phylogenetics and Evolution*, **55**, 711–720.

Dennis, A.J. (2003) Scatter-hoarding by musky rat-kangaroos, *Hypsiprymnodon moschatus*, a tropical rain-forest marsupial from Australia: Implications for seed dispersal. *Journal of Tropical Ecology*, **19**, 619–627.

Deufel, A. & Cundall, D. (2006) Functional plasticity of the venom delivery system in snakes with a focus on the poststrike prey release behavior. *Zoologischer Anzeiger*, **245**, 249–267.

Deutsch, C.A., Tewksbury, J.J., Huey, R.B. et al. (2008) Impacts of climate warming on terrestrial ectotherms across latitude. *Proceedings of the National Academy of Sciences of the USA*, **105**, 6668–6672.

DeVries, P.J. (2001) Butterflies. In: *Encyclopedia of Biodiversity, Vol. 1* (ed S.A. Levin), pp. 559–573. Academic Press, San Diego, CA.

Dick, C.W., Etchelecu, G. & Austerlitz, F. (2003) Pollen dispersal of tropical trees (*Dinizia excelsa*: Fabaceae) by native insects and African honeybees in pristine and fragmented Amazonian rainforest. *Molecular Ecology*, **12**, 753–764.

Dick, C.W. & Heuertz, M. (2008) The complex biogeographic history of a widepread tree species. *Evolution*, **62**, 2760–2774.

Donovan, D., de Jong, W. & Abe, K.-I. (2007) Tropical forests and extreme conflict. In: *Extreme Conflict and Tropical Forests* (eds W. de Jong, D. Donovan & K.I. Abe), pp. 1–15. Springer, Berlin.

Dransfield, J. & Beentje, H. (1995) *The Palms of Madagascar*. Royal Botanic Gardens, Kew, UK.

Ducousso, M., Bena, G., Bourgeois, C. et al. (2004) The last common ancestor of Sarcolaenaceae and Asian dipterocarp trees was ectomycorrhizal before the India-Madagascar separation, about 88 million years ago. *Molecular Ecology*, **13**, 231–236.

Dudley, R., Byrnes, G., Yanoviak, S.P., Borrell, B., Brown, R.M. & McGuire, J.A. (2007) Gliding and the functional origins of flight: Biomechanical novelty or necessity? *Annual Review of Ecology, Evolution, and Systematics*, **38**, 179–201.

Duellman, W.E. & Pianka, E.R. (1990) Biogeography of nocturnal insectivores: Historical events and ecological filters. *Annual Review of Ecology and Systematics*, **21**, 57–68.

Dumbacher, J.P., Deiner, K., Thompson, L. & Fleischer, R.C. (2008) Phylogeny of the avian genus *Pitohui* and the evolution of toxicity in birds. *Molecular Phylogenetics and Evolution*, **49**, 774–781.

Dumont, E.R. (2003) Bats and fruit: an ecomorphological approach. In: *Ecology of Bats* (eds T.H. Kunz & M.B. Fenton), pp. 398–429 University of Chicago Press, Chicago, IL.

Dumont, E.R., Herrel, A., Medellín, R.A., Vargas-Contreras, J.A. & Santana, S.E. (2009) Built to bite: cranial design and function in the wrinkle-faced bat. *Journal of Zoology*, **279**, 329–337.

Dyer, F.C. (2002) The biology of the dance language. *Annual Review of Entomology*, **47**, 917–949.

Eaton, M.J., Martin, A., Thorbjarnarson, J. & Amato, G. (2009) Species-level diversification of African dwarf crocodiles (Genus *Osteolaemus*): A geographic and phylogenetic perspective. *Molecular Phylogenetics and Evolution*, **50**, 496–506.

Ebeling, J. & Yasué, M. (2009) The effectiveness of market-based conservation in the tropics: Forest certification in Ecuador and Bolivia. *Journal of Environmental Management*, **90**, 1145–1153.

Eberhard, S.H., Nemeschkal, H.L. & Krenn, H.W. (2009) Biometrical evidence for adaptations of the salivary glands to pollen feeding in *Heliconius* butterflies (Lepidoptera: Nymphalidae). *Biological Journal of the Linnean Society*, **97**, 604–612.

Edwards, K.A., Doescher, L.T., Kaneshiro, K.Y. & Yamamoto, D. (2007) A database of wing diversity in the Hawaiian Drosophila. *PLoS One*, **2**, e487.

Eggleton, P. & Davies, R.G. (2003) Isoptera, termites. In: *The Natural History of Madagascar* (eds S.M. Goodman & J.P. Benstead), pp. 654–660. University of Chicago Press, Chicago, IL.

Eggleton, P., Homathevi, R., Jones, D.T. et al. (1999) Termite assemblages, forest disturbance and greenhouse gas fluxes in Sabah, East Malaysia. *Philosophical Transactions of the Royal Society of London B Biological Sciences*, **354**, 1791–1802.

Ehrlich, P.R. & Goulder, L.H. (2007) Is current consumption excessive? A general framework and some indications for the United States. *Conservation Biology*, **21**, 1145–1154.

EIA & Global Witness (2010) *Investigation into the Global Trade in Malagasy Precious Woods*. Environmental Investigation Agency and Global Witness, Washington, DC, USA.

Ellwood, M.D.F. & Foster, W.A. (2004) Doubling the estimate of invertebrate biomass in a rainforest canopy. *Nature*, **429**, 549–551.

Emmons, L.H. (2000) *Tupai: A Field Study of Bornean Tree-Shrews*. University of Califonia Press, Berkeley, CA.

Engel, M.S., Grimaldi, D.A. & Krishna, K. (2009a) Termites (Isoptera): Their phylogeny, classification, and rise to ecological dominance. *American Museum Novitates*, **3650**, 1–27.

Engel, M.S., Hinojosa-Diaz, I.A. & Rasnitsyn, A.P. (2009b) A honey bee from the Miocene of Nevada and the biogeography of *Apis* (Hymenoptera: Apidae: Apini). *Proceedings of the California Academy of Sciences*, **60**, 23–38.

Epstein, J.H., Olival, K.J., Pulliam, J.R.C. et al. (2009) *Pteropus vampyrus*, a hunted migratory species with a multinational home-range and a need for regional management. *Journal of Applied Ecology*, **46**, 991–1002.

Ericson, P.G.P. & Johansson, U.S. (2003) Phylogeny of Passerida (Aves: Passeriformes) based on nuclear and mitochondrial sequence data. *Molecular Phylogenetics and Evolution*, **29**, 126–138.

Ewers, R.M. & Didham, R.K. (2006) Confounding factors in the detection of species responses to habitat fragmentation. *Biological Reviews*, **81**, 117–142.

Fa, J.E. & Brown, D. (2009) Impacts of hunting on mammals in African tropical moist forests: a review and synthesis. *Mammal Review*, **39**, 231–264.

Fairbairn, A.S., Hope, G.S. & Summerhayes, G.R. (2006) Pleistocene occupation of New Guinea's highland and subalpine environments. *World Archaeology*, **38**, 371–386.

Fayle, T.M., Dumbrell, A.J., Eggleton, P. & Foster, W.A. (2009) Rainforest canopy architecture differentially affects the distribution of two species of epiphytic fern (*Asplenium* spp.). *Biotropica*, **41**, 676–681.

Fearnside, P.M. (2008) The roles and movements of actors in the deforestation of Brazilian Amazonia. *Ecology and Society*, **13**, 23.

Federle, W., Maschwitz, U., Fiala, B., Riederer, M. & Hölldobler, B. (1997) Slippery ant-plants and skillful climbers: Selection and protection of specific ant partners by epicuticular wax blooms in *Macaranga* (Euphorbiaceae). *Oecologia*, **112**, 217–224.

Feeley, K.J. & Silman, M.R. (2010) Biotic attrition from tropical forests correcting for truncated temperature niches. *Global Change Biology*, **16**, 1830–1836.

Ferguson-Lees, J. & Christie, D.A. (2001) *Raptors of the World*. Christopher Helm, London.

Fernández-Marin, H., Zimmerman, J.K., Nash, D.R., Boomsma, J.J. & Wcislo, W.T. (2009) Reduced biological control and enhanced chemical pest management in the evolution of fungus farming in ants. *Proceedings of the Royal Society Biological Sciences Series B*, **276**, 2263–2269.

Fiala, B., Maschwitz, U., Pong, T.Y. & Helbig, A.J. (1989) Studies of a South East Asian ant-plant association: protection of *Macaranga* trees by *Crematogaster borneensis*. *Oecologia*, **79**, 463–470.

Fietz, J. & Ganzhorn, J.U. (1999) Feeding ecology of the hibernating primate *Cheirogaleus medius*: how does it get so fat? *Oecologia*, **121**, 157–164.

Finer, M. & Orta-Martínez, M. (2010) A second hydrocarbon boom threatens the Peruvian Amazon: trends, projections, and policy implications. *Environmental Research Letters*, **5**, 1–10.

Finer, M., Jenkins, C.N., Pimm, S.L., Keane, B. & Ross, C. (2008) Oil and gas projects in the Western Amazon: Threats to wilderness, biodiversity, and indigenous peoples. *PLoS One*, **3**, e2932.

Fisher, B.L. (2010) Biogeography. In: *Ant Ecology* (eds L. Lach, C.L. Parr & K.L. Abbott), pp. 18–31. Oxford University Press, Oxford, UK.

Fittkau, E.J. & Klinge, H. (1973) On biomass and trophic structure of the central Amazonian rain forest ecosystem. *Biotropica*, **5**, 2–14.

Flannery, T.F. (1995a) *Mammals of New Guinea*. Reed Books, Chatswood, New South Wales.

Flannery, T.F. (1995b) *Mammals of the South West Pacific and Moluccan Islands*. Reed Books, Chatswood, New South Wales.

Fleagle, J.G. (1999) *Primate Adaptation and Evolution, 2nd edn*. Academic Press, San Diego, CA.

Fleischer, R.C. & McIntosh, C.E. (2001) Molecular systematics and biogeography of the Hawaiian avifauna. *Studies in Avian Biology*, **22**, 51–60.

Fleischer, R.C., James, H.F. & Olson, S.L. (2008) Convergent evolution of Hawaiian and Australo-Pacific honeyeaters from distant songbird ancestors. *Current Biology*, **18**, 1927–1931.

Fleming, T.H. & Muchhala, N. (2008) Nectar-feeding bird and bat niches in two worlds: pantropical comparisons of vertebrate pollination systems. *Journal of Biogeography*, **35**, 764–780.

Floren, A. & Linsenmair, K.E. (2000) Do ant mosaics exist in pristine lowland rain forests? *Oecologia*, **123**, 129–137.

Forget, P.-M. & Vander Wall, S.B. (2001) Scatter-hoarding rodents and marsupials: Convergent evolution on diverging continents. *Trends in Ecology and Evolution*, **16**, 65–67.

Franks, N.R., Sendova-Franks, A.B. & Anderson, C. (2001) Division of labour within teams of New World and Old World army ants. *Animal Behaviour*, **62**, 635–642.

Fredriksson, G.M. (2005) Predation on sun bears by reticulated python in East Kalimantan, Indonesian Borneo. *Raffles Bulletin of Zoology*, **53**, 165–168.

Frith, C.B. & Beehler, B.M. (1998) *The Birds of Paradise*. Oxford University Press, Oxford, UK.

Frith, C.B., Frith, D.W. & Barnes, E. (2004) *The Bowerbirds*. Oxford University Press, Oxford, UK.

Fry, B.G., Vidal, N., Norman, J.A. et al. (2006) Early evolution of the venom system in lizards and snakes. *Nature*, **439**, 584–588.

Ganzhorn, J.U., Arrigo-Nelson, S., Boinski, S. et al. (2009) Possible fruit protein effects on primate communities in Madagascar and the Neotropics. *PLoS One*, **4**, e8253.

Garbutt, N. (1999) *Mammals of Madagascar*. Pica Press, East Sussex, UK.

Gartrell, B.D. & Jones, S.M. (2001) Eucalyptus pollen grain emptying by two Australian nectarivorous psittacines. *Journal of Avian Biology*, **32**, 224–230.

Gathorne-Hardy, F.J., Jones, D.T. & Syaukani (2002) A regional perspective on the effects of human disturbance on the termites of Sundaland. *Biodiversity and Conservation*, **11**, 1991–2006.

Gaubert, P. & Veron, G. (2003) Exhaustive sample set among Viverridae reveals the sister-group of felids: The linsangs as a case of extreme morphological convergence within Feliformia. *Proceedings of the Royal Society Biological Sciences Series B*, **270**, 2523–2530.

Gaveau, D.L.A., Linkie, M., Suyadi, Levang, P. & Leader-Williams, N. (2009) Three decades of deforestation in southwest Sumatra: Effects of coffee prices, law enforcement and rural poverty. *Biological Conservation*, **142**, 597–605.

Gayot, M., Henry, O., Dubost, G. & Sabatier, D. (2004) Comparative diet of the two forest cervids of the genus *Mazama* in French Guiana. *Journal of Tropical Ecology*, **20**, 31–43.

Gelang, M., Cibois, A., Pasquet, E., Olsson, U., Alström, P. & Ericson, P.G.P. (2009) Phylogeny of babblers (Aves, Passeriformes): major lineages, family limits and classification. *Zoologica Scripta*, **38**, 225–236.

Gillespie, R.G. (1999) Comparison of rates of speciation in web-building and non-web-building groups within a Hawaiian spider radiation. *Journal of Arachnology*, **27**, 79–85.

Givnish, T.J. & Renner, S.S. (2004) Tropical intercontinental disjunctions: Gondwana breakup, immigration from the boreotropics, and trans-oceanic dispersal. *International Journal of Plant Sciences*, **165**, S1–S6.

Givnish, T.J., Millam, K.C., Berry, P.E. & Sytsma, K.J. (2007) Phylogeny, adaptive radiation, and historical biogeography of Bromeliaceae inferred from ndhF sequence data. *Aliso*, **23**, 3–26.

Givnish, T.J., Millam, K.C., Mast, A.R. et al. (2009) Origin, adaptive radiation and diversification of the Hawaiian lobeliads (Asterales: Campanulaceae). *Proceedings of the Royal Society Biological Sciences Series B*, **276**, 407–416.

Glen, M.E. & Cords, M. (2002) *The Guenons: Diversity and Adaptation in African Monkeys*. Springer, New York, NY.

Golden, C.D. (2009) Bushmeat hunting and use in the Makira Forest, north-eastern Madagascar: a conservation and livelihoods issue. *Oryx*, **43**, 386–392.

Goodman, S.M. & Ganzhorn, J.U. (1997) Rarity of figs (*Ficus*) on Madagascar and its relationship to a depauperate frugivore community. *Revue d'Ecologie*, **52**, 321–329.

Goodman, S.M., Kerridge, F.J. & Ralisoamalala, R.C. (2003) A note on the diet of *Fossa fossana* (Carnivora) in the central eastern humid forests of Madagascar. *Mammalia*, **67**, 595–598.

Goodman, S.M., Rasoloarison, R.M. & Ganzhorn, J.U. (2004) On the specific identification of subfossil Cryptoprocta (Mammalia, Carnivora) from Madagascar. *Zoosystema*, **26**, 129–143.

Gotwald, W.H. (1995) *Army Ants: the Biology of Social Predation*. Cornell University Press, Ithaca, NY.

Goulding, M. (1989) *Amazon: the Flooded Forest*. BBC Books, London.

Grace, M.S., Church, D.R., Kelly, C.T., Lynn, W.F. & Cooper, T.M. (1999) The python pit organ: Imaging and immunocytochemical analysis of an extremely sensitive natural infrared detector. *Biosensors and Bioelectronics*, **14**, 53–59.

Graham, A. (1999) The Tertiary history of the northern temperate element in the northern Latin American biota. *American Journal of Botany*, **86**, 32–38.

Graham, C.H., Parra, J.L., Rahbek, C. & McGuire, J.A. (2009) Phylogenetic structure in tropical hummingbird communities. *Proceedings of the National Academy of Sciences*, **106**, 19673–19678.

Graham, E.A., Mulkey, S.S., Kitajima, K., Phillips, N.G. & Wright, S.J. (2003) Cloud cover limits net CO_2 uptake and growth of a rainforest tree during tropical rainy seasons. *Proceedings of the National Academy of Sciences of the USA*, **100**, 572–576.

Grogan, J., Blundell, A.G., Landis, R.M. et al. (2010) Over-harvesting driven by consumer demand leads to population decline: big-leaf mahogany in South America. *Conservation Letters*, **3**, 12–20.

Gross, C.L. (2005) A comparison of the sexual systems in the trees from the Australian tropics with other tropical biomes – More monoecy but why? *American Journal of Botany*, **92**, 907–919.

Groves, C.P. & Schaller, G.B. (2000) The phylogeny and biogeography of the newly discovered Annamite artiodactyls. In: *Antelopes, Deer, and Relatives* (eds E.S. Vrba & G.B. Schaller). Yale University Press, New Haven, CT.

Grubb, P.J. (2003) Interpreting some outstanding features of the flora and vegetation of Madagascar. *Perspectives in Plant Ecology Evolution and Systematics*, **6**, 125–146.

Guedes, P.G. & Salles, L.O. (2005) New insights on the phylogenetic relationships of the two giant extinct New World monkeys (Primates, Platyrrhini). *Arquivos do Museu Nacional, Rio de Janeiro*, **63**, 147–159.

Hackett, S.J., Kimball, R.T., Reddy, S. et al. (2008) A phylogenomic study of birds reveals their evolutionary history. *Science*, **320**, 1763–1768.

Hansen, M.C., Roy, D.P., Lindquist, E., Adusei, B., Justice, C.O. & Altstatt, A. (2008a) A method for integrating MODIS and Landsat data for systematic monitoring of forest cover and change in the Congo Basin. *Remote Sensing of Environment*, **112**, 2495–2513.

Hansen, M.C., Stehman, S.V., Potapov, P.V. et al. (2008b) Humid tropical forest clearing from 2000 to 2005 quantified by using multitemporal and multiresolution remotely sensed data. *Proceedings of the National Academy of Sciences of the USA*, **105**, 9439–9444.

Hanson, T., Brooks, T.M., Da Fonseca, G.A.B. et al. (2009) Warfare in biodiversity hotspots. *Conservation Biology*, **23**, 578–587.

Happold, D.C. (1996) Mammals of the Guinea-Congo rain forest, West Africa. In: *Essays on the Ecology of the Guinea-Congo Rain Forest* (eds I.J. Alexander, M.D. Swaine & R. Watling), pp. 243–284. Royal Society of Edinburgh, Edinburgh, UK.

Harbaugh, D.T., Wagner, W.L., Percy, D.M., James, H.F. & Fleischer, R.C. (2009) Genetic structure of the polymorphic *Metrosideros* (Myrtaceae) complex in the Hawaiian Islands using nuclear microsatellite data. *PLoS One*, **4**, e4698.

Harper, G.J., Steininger, M.K., Tucker, C.J., Juhn, D. & Hawkins, F. (2007) Fifty years of deforestation and forest fragmentation in Madagascar. *Environmental Conservation*, **34**, 325–333.

Harrison, J.F., Fewell, J.H., Anderson, K.E. & Loper, G.M. (2006) Environmental physiology of the invasion of the Americas by Africanized honeybees. *Integrative and Comparative Biology*, **46**, 1110–1122.

Harrison, M.E., Page, S.E. & Limin, S.H. (2009) The global impact of Indonesian forest fires. *Biologist*, **56**, 156–163.

Harrison, R.D. (2001) Drought and the consequences of El Niño in Borneo: A case study of figs. *Population Ecology*, **43**, 63–75.

Harrison, R.D., Hamid, A.A., Kenta, T. et al. (2003) The diversity of hemi-epiphytic figs (*Ficus*; Moraceae) in a Bornean lowland rain forest. *Biological Journal of the Linnean Society*, **78**, 439–455.

Harshman, J., Braun, E.L., Braun, M.J. et al. (2008) Phylogenomic evidence for multiple losses of flight in ratite birds. *Proceedings of the National Academy of Sciences of the USA*, **105**, 13462–13467.

Hart, T.B. (1995) Seed, seedling and subcanopy survival in momodominant and mixed forests of the Ituri Forest, Africa. *Journal of Tropical Ecology*, **11**, 443–459.

Harvey, C.A., Dickson, B. & Kormos, C. (2010) Opportunities for achieving biodiversity conservation through REDD. *Conservation Letters*, **3**, 53–61.

Haugaasen, T. & Peres, C.A. (2005) Mammal assemblage structure in Amazonian flooded and unflooded forests. *Journal of Tropical Ecology*, **21**, 133–145.

Head, J.J., Bloch, J.I., Hastings, A.K. et al. (2009) Giant boid snake from the Palaeocene Neotropics reveals hotter past equatorial temperatures. *Nature*, **457**, 715.

Heaney, L.R. (2004) Conservation biogeography in oceanic archipelagoes. In: *Frontiers of Biogeography: New Directions in the Geography of Nature* (eds M.V. Lomolino & L.R. Heaney), pp. 345–360. Sinauer Associates, Sunderland, MA.

Hedin, L.O., Brookshire, E.N.J., Menge, D.N.L. & Barron, A.R. (2009) The nitrogen paradox in tropical forest ecosystems. *Annual Review of Ecology Evolution and Systematics*, **40**, 613–635.

Heil, M., Fiala, B., Linsemair, K.E., Zotz, G., Menke, P. & Maschwitz, U. (1997) Food body production in *Macaranga triloba* (Euphorbiaceae): A plant investment in anti-herbivore defence via symbiotic ant partners. *Journal of Ecology*, **85**, 847–861.

Henkel, T.W. (2003) Monodominance in the ectomycorrhizal *Dicymbe corymbosa* (Caesalpiniaceae) from Guyana. *Journal of Tropical Ecology*, **19**, 417–437.

Heywood, V.H., Brummit, R.K., Culham, A. & Seberg, O. (2007) *Flowering Plant Families of the World*. Royal Botanic Gardens, Kew, UK.

Hnatiuk, R.J., Smith, J.M.B. & McVean, D.N. (1976) *The Climate of Mount Wilhelm*. Australian National University, Canberra.

Hocknull, S., Zhao, J., Feng, Y. & Webb, G. (2007) Responses of Quaternary rainforest vertebrates to climate change in Australia. *Earth and Planetary Science Letters*, **264**, 317–331.

Hodgkison, R. & Kunz, T.H. (2006) *Balionycteris maculata*. *Mammalian Species*, **793**, 1–3.

Holland, R.A., Wikelski, M., Kuemmeth, F. & Bosque, C. (2009) The secret life of oilbirds: new insights into the movement ecology of a unique avian frugivore. *PLoS One*, **4**, e8264.

Hooijer, A., Page, S., Canadell, J.G., Silvius, M., Kwadijk, J., Wösten, H. & Jauhiainen, J. (2010) Current and future CO_2 emissions from drained peatlands in Southeast Asia. *Biogeosciences*, **7**, 1505–1514.

Horn, H.G. (2004) *Varanus salvadorii*. In: *Varanoid Lizards of the World* (eds E.R. Pianka, D. King & R.A. King), pp. 234–244. Indiana University Press, Bloomington, IN.

Horner, M.A., Fleming, T.H. & Sahley, C.T. (1998) Foraging behaviour and energetics of nectar-feeding bat, *Leptonycteris curasoae* (Chiroptera: Phyllostomidae). *Journal of Zoology*, **244**, 575–586.

Horvath, J.E., Weisrock, D.W., Embry, S.L. et al. (2008) Development and application of a phylogenomic toolkit: Resolving the evolutionary history of Madagascar's lemurs. *Genome Research*, **18**, 489–499.

Houlton, B.Z., Wang, Y.P., Vitousek, P.M. & Field, C.B. (2008) A unifying framework for dinitrogen fixation in the terrestrial biosphere. *Nature*, **454**, 327–330.

Houston, D.C. (1994) Family Cathartidae (New World vultures). In: *Handbook of the Birds of the World, Vol. 2, New World Vultures to Guineafowl* (eds J. del Hoyo, A. Elliott & J. Sargatal), pp. 24–41. Lynx Edicions, Barcelona, Spain.

Huey, R.B., Deutsch, C.A., Tewksbury, J.J. et al. (2009) Why tropical forest lizards are vulnerable to climate warming. *Proceedings of the Royal Society Biological Sciences Series B*, **276**, 1939–1948.

Hume, I.D. (1999) *Marsupial Nutrition*. Cambridge University Press, Cambridge, UK.

Hunt, R.M. (1996) Biogeography of the order Carnivora. In: *Carnivore behavior, ecology, and evolution* (ed J.L. Gittleman), pp. 485–541. Cornell University Press, Ithaca, NY.

Inger, R.F. (1980) Densities of floor-dwelling frogs and lizards in lowland forests of southeast Asia and central America. *American Naturalist*, **115**, 761–770.

Inui, Y., Tanaka, H.O., Hyodo, F. & Itioka, T. (2009) Within-nest abundance of a tropical cockroach *Pseudoanaplectinia yumotoi* associated with *Crematogaster* ants inhabiting epiphytic fern domatia in a Bornean dipterocarp forest. *Journal of Natural History*, **43**, 1139–1145.

IPPC (2007) *Climate Change 2007: the Physical Science Basis.* Cambridge University Press, New York.

Itioka, T., Inoue, T., Kaliang, H. et al. (2001) Six-year population fluctuation of the giant honey bee *Apis dorsata* (Hymenoptera: Apidae) in a tropical lowland dipterocarp forest in Sarawak. *Annals of the Entomological Society of America*, **94**, 545–549.

Iwaniuk, A.N., Olson, S.L. & James, H.F. (2009) Extraordinary cranial specialization in a new genus of extinct duck (Aves: Anseriformes) from Kauai, Hawaiian Islands. *Zootaxa*, **2296**, 47–67.

Jacobs, G.H. & Nathans, J. (2009) The evolution of primate color vision. *Scientific American*, **300**, 56–63.

Jaeger, J.-J., Beard, K.C., Chaimanee,Y. et al. (2010) Late middle Eocene epoch of Libya yields earliest known radiation of African anthropoids. *Nature*, **467**, 1095–1098.

Jaeggi, A.V., Dunkel, L.P., Noordwijk, M.A.V., Wich, S.A., Sura, A.A.L. & Schaik, C.P.V. (2010) Social learning of diet and foraging skills by wild immature Bornean orangutans: implications for culture. *American Journal of Primatology*, **72**, 62–71.

James, H.F. & Burney, D.A. (1997) The diet and ecology of Hawaii's extinct flightless waterfowl: Evidence from coprolites. *Biological Journal of the Linnean Society*, **62**, 279–297.

Janis, C. (2008) An evolutionary history of browsing and grazing ungulates. In: *The Ecology of Browsing and Grazing* (eds I.J. Gordon & H.H.T. Prins), pp. 21–45. Springer, New York.

Janzen, D.H. (1974) Tropical blackwater rivers, animals and mast fruiting by the Dipterocarpaceae. *Biotropica*, **6**, 69–103.

Janzen, D.H. & Martin, P.S. (1982) Neotropical anachronisms: The fruits the gomphotheres ate. *Science*, **215**, 19–27.

Jaramillo, M.A., Callejas, R., Davidson, C., Smith, J.F., Stevens, A.C. & Tepe, E.J. (2008) A phylogeny of the tropical genus *Piper* using ITS and the chloroplast intron psbJ-petA. *Systematic Botany*, **33**, 647–660.

Johansson, U.S., Bowie, R.C.K., Hackett, S.J. & Schulenberg, T.S. (2008a) The phylogenetic affinities of Crossley's babbler (*Mystacornis crossleyi*): adding a new niche to the vanga radiation of Madagascar. *Biology Letters*, **4**, 677–680.

Johansson, U.S., Fjeldså, J. & Bowie, R.C.K. (2008b) Phylogenetic relationships within Passerida (Aves: Passeriformes): A review and a new molecular phylogeny based on three nuclear intron markers. *Molecular Phylogenetics and Evolution*, **48**, 858–876.

Johnsgard, P.A. (1994) *Arena Birds: Sexual Selection and Behavior.* Smithsonian Institution Press, Washington, DC.

Jones, J.P.G., Andriamarovololona, M.M. & Hockley, N. (2008) The importance of taboos and social norms to conservation in Madagascar. *Conservation Biology*, **22**, 976–986.

Jones, M.E. & Stoddart, D.M. (1998) Reconstruction of the predatory behaviour of the extinct marsupial thylacine (*Thylacinus cynocephalus*). *Journal of Zoology*, **246**, 239–246.

Jones, P.J. (1994) Biodiversity in the Gulf of Guinea: an overview. *Biodiversity and Conservation*, **3**, 772–784.

Joppa, L.N., Loarie, S.R. & Pimm, S.L. (2008) On the protection of "protected areas". *Proceedings of the National Academy of Sciences of the USA*, **105**, 6673–6678.

Joron, M., Papa, R., Beltran, M. et al. (2006) A conserved supergene locus controls colour pattern diversity in Heliconius butterflies. *PLoS Biology*, **4**, 1831–1840.

Jullien, M. & Thiollay, J.-M. (1998) Multi-species territoriality and dynamic of neotropical forest understorey bird flocks. *Journal of Animal Ecology*, **67**, 227–252.

Junqueira, A.B., Shepard Jr, G.H. & Clement, C.R. (2010) Secondary forests on anthropogenic soils in Brazilian Amazonia conserve agrobiodiversity. *Biodiversity and Conservation*, **19**, 1–29.

Kappeler, P.M. (2000) Lemur origins: Rafting by groups of hibernators? *Folia Primatologica*, **71**, 422–425.

Karl, T.R. (2006) Written statement for an oversight hearing: Introduction to Climate Change before the Committee on Government Reform, US House of Representatives, Washington, DC.

Karr, J.R. (1990) Birds of tropical rainforest: comparative biogeography and ecology. In: *Biogeography and Ecology of Forest Bird Communities* (ed A. Keast), pp. 215–228. SPB Academic Publishing, The Hague, Netherlands.

Kaufmann, E. & Maschwitz, U. (2006) Ant-gardens of tropical Asian rainforests. *Naturwissenschaften*, **93**, 216–227.

Kay, R.F. & Madden, R.H. (1997) Paleogeography and paleoecology. In: *Vertebrate Paleontology in the Neotropics: The Miocene Fauna of La Venta, Colombia* (eds R.F. Kay, R.H. Madden, R.L. Cifelli & J.J. Flynn), pp. 520–550. Smithsonian Institution Press, Washington, DC.

Kemp, N.J. & Burnett, J.B. (2007) A non-native primate (*Macaca fascicularis*) in Papua: implications for biodiversity. In: *The Ecology of Papua. Part Two* (eds A.J. Marshall & B.M. Beehler), pp. 1348–1364. Periplus Editions, Singapore.

Kendrick, E.L., Shipley, L.A., Hagerman, A.E. & Kelley, L.M. (2009) Fruit and fibre: the nutritional value of figs for a small tropical ruminant, the blue duiker (*Cephalophus monticola*). *African Journal of Ecology*, **47**, 556–566.

Kenfack, D., Thomas, D.W., Chuyong, G. & Condit, R. (2007) Rarity and abundance in a diverse African forest. *Biodiversity and Conservation*, **16**, 2045–2074.

Keppel, G., Buckley, Y.M. & Possingham, H.P. (2010) Drivers of lowland rain forest community assembly, species diversity and forest structure on islands in the tropical South Pacific. *Journal of Ecology*, **98**, 87–95.

Kierulff, M.C.M., Rylands, A.B. & de Oliveira, M.M. (2010) *Leontopithecus rosalia*. *IUCN Red List of Threatened Species. Version 2009.2.* www.iucnredlist.org (accessed November 22, 2010).

Kingdon, J. (1997) *The Kingdon Field Guide to African Mammals.* Princeton University Press, Princeton, NJ.

Kinnaird, M.F. & O'Brian, T.G. (2007) *The Ecology and Conservation of Asian Hornbills.* University of Chicago Press, Chicago.

Kinoshita, S., Yoshioka, S. & Miyazaki, J. (2008) Physics of structural colors. *Reports on Progress in Physics*, **71**, 1–30.

Kirchman, J.J. & Steadman, D.W. (2007) New species of extinct rails (Aves: Rallidae) from archaeological sites in the Marquesas Islands, French Polynesia. *Pacific Science*, **61**, 145–163.

Kirkpatrick, R.C. (20007) The Asian colobines: diversity among leaf-eating monkeys. In: *Primates in Perspective* (eds C.J. Campbell, A.F. Fuentes, K.C. MacKinnon, M. Panger & S. Bearder), pp. 186–200. Oxford University Press, Oxford, UK.

Knogge, C., Heymann, E.W. & Tirado Herrera, E.R. (1998) Seed dispersal of *Asplundia peruviana* (Cyclanthaceae) by the primate *Saguinus fuscicollis*. *Journal of Tropical Ecology*, **14**, 99–102.

Koch, P.L. & Barnosky, A.D. (2006) Late quaternary extinctions: State of the debate. *Annual Review of Ecology Evolution and Systematics*, **37**, 215–250.

Koepfli, K.-P., Gompper, M.E., Eizirik, E. et al. (2007) Phylogeny of the Procyonidae (Mammalia: Carnivora): Molecules, morphology and the Great American Interchange. *Molecular Phylogenetics and Evolution*, **43**, 1076–1095.

Koltunov, A., Ustin, S.L., Asner, G.P. & Fung, I. (2009) Selective logging changes forest phenology in the Brazilian Amazon: Evidence from MODIS image time series analysis. *Remote Sensing of Environment*, **113**, 2431–2440.

Konoplyova, A., Petropoulou, Y., Yiotis, C., Psaras, G.K. & Manetas, Y. (2008) The fine structure and photosynthetic cost of structural leaf variegation. *Flora (Jena)*, **203**, 653–662.

Kress, W.J. & Beach, J.H. (1994) Flowering plant reproductive systems. In: *La Selva: Ecology and Natural History of a Neotropical Rain Forest* (eds L.A. McDade, K.S. Bawa, H.A. Hespenheide & G.S. Hartshorn), pp. 161–182.

Kress, W.J., Schatz, G.E., Andrianifahanana, M. & Morland, H.S. (1994) Pollination of *Ravenala madagascariensis* (Strelitziaceae) by lemurs in Madagascar – evidence for an archaic coevolutionary system. *American Journal of Botany*, **81**, 542–551.

Kricher, J. (1997) *A Neotropical Companion, 2nd edn.* Princeton University Press, Princeton, NJ.

Kronauer, D.J.C. (2009) Recent advances in army ant biology (Hymenoptera: Formicidae). *Myrmecological News*, **12**, 51–65.

LaFrankie, J.V., Ashton, P.S., Chuyong, G.B. (2006) Contrasting structure and composition of the understory in species-rich tropical rain forests. *Ecology*, **87**, 2298–2305.

Lähteenoja, O., Ruokolainen, K., Schulman, L. & Oinonen, M. (2009) Amazonian peatlands: An ignored C sink and potential source. *Global Change Biology*, **15**, 2311–2320.

Laman, T. (2000) Gliders: the creatures of Borneo's rain forest go airborne. *National Geographic*, **198**, 68–85.

Lan, G.-Y., Hu, Y.-H., Cao, M. et al. (2008) Establishment of Xishuangbanna tropical forest dynamics plot: Species compositions and spatial distribution patterns. *Zhiwu Shengtai Xuebao*, **32**, 287–298.

Lapola, D.M., Schaldach, R., Alcamo, J. et al. (2010) Indirect land-use changes can overcome carbon savings from biofuels in Brazil. *Proceedings of the National Academy of Sciences*, **107**, 3388–3393.

Laurance, W.F. (2005) When bigger is better: the need for Amazonian megareserves. *Trends in Ecology & Evolution*, **20**, 645–648.

Laurance, W.F. (2008) The need to cut China's illegal timber imports. *Science*, **319**, 1184.

Laurance, W.F. (2010) Better governance to save rainforests. *Nature*, **467**, 789.

Laurance, W.F., Camargo, J.L.C., Luizão, R.C.C. et al. (2010b) The fate of Amazonian forest fragments: a 32-year investigation. *Biological Conservation*, doi: 10.1016/j.biocon.2010.09.021.

Laurance, W.F., Lovejoy, T.E., Vasconcelos, H.L. et al. (2002) Ecosystem decay of Amazonian forest fragments: A 22-year investigation. *Conservation Biology*, **16**, 605–618.

Laurance, W.F., Goosem, M. & Laurance, S.G.W. (2009) Impacts of roads and linear clearings on tropical forests. *Trends in Ecology & Evolution*, **24**, 659–669.

Laurance, W.F., Koh, L.P., Butler, R. et al. (2010a) Improving the performance of the Roundtable on Sustainable Palm Oil for nature conservation. *Conservation Biology*, **24**, 377–381.

Lee, D.W. (2001) Leaf colour in tropical plants: Some progress and much mystery. *Malayan Nature Journal*, **55**, 117–131.

Lee, H.S., Davies, S.J., Lafrankie, J.V., Tan, S. et al. (2002) Floristic and structural diversity of mixed dipterocarp forest in Lambir Hills National Park, Sarawak, Malaysia. *Journal of Tropical Forest Science*, **14**, 379–400.

Leigh Jr, E.G., Hladik, A., Hladik, C.M. & Jolly, A. (2007) The biogeography of large islands, or how does the size of the ecological theater affect the evolutionary play? *Revue d'Ecologie (La Terre et la Vie)*, **62**, 105–168.

Levey, D.J., Moermond, T.C. & Denslow, J.S. (1994) Frugivory: an overview. In: *La Selva: Ecology and Natural History of a Neotropical Rain Forest* (eds L.A. McDade, K.S. Bawa, H.A. Hespenheide & G.S. Hartshorn), pp. 287–294. University of Chicago Press, Chicago, IL.

Lewis, S.L., Lopez-Gonzalez, G., Sonke, B. et al. (2009) Increasing carbon storage in intact African tropical forests. *Nature*, **457**, 1003.

Liu, W.J., Wang, P.Y., Liu, W.Y., Li, J.T. & Li, P.J. (2008) The importance of radiation fog in the tropical seasonal rain forest of Xishuangbanna, south-west China. *Hydrology Research*, **39**, 79–87.

Lobova, T.A., Geiselman, C.K. & Mori, S.A. (2009) *Seed Dispersal by Bats in the Neotropics.* New York Botanic Garden, New York.

Lomolino, M.V., Riddle, B.R. & Brown, J.H. (2006) *Biogeography.* Sinauer Associates, Sunderland, MA.

Long, J.L. (2003) *Introduced Mammals of the World.* CSIRO Publishing, Collingwood, Australia.

Long, J., Archer, M., Flannery, T. & Hand, S. (2002) *Prehistoric Mammals of Australia and New Guinea: One Hundred Million Years of Evolution.* John Hopkins University Press, Baltimore, MD.

Lopez, J.E. & Vaughan, C. (2007) Food niche overlap among neotropical frugivorous bats in Costa Rica. *Revista de Biologia Tropical*, **55**, 301–313.

Lopez, O.R. & Kursar, T.A. (2007) Interannual variation in rainfall, drought stress and seedling mortality may mediate monodominance in tropical flooded forests. *Oecologia*, **154**, 35–43.

Lopez-Vaamonde, C., Wikström, N., Kjer, K.M. et al. (2009) Molecular dating and biogeography of fig-pollinating wasps. *Molecular Phylogenetics and Evolution*, **52**, 715–726.

Losos, J.B. (2009) *Lizards in an Evolutionary Tree: Ecology and Adaptive Radiation of Anoles.* University of California Press, Berkeley, CA.

Lugo, A.E. (2008) Visible and invisible effects of hurricanes on forest ecosystems: an international review. *Australian Ecology*, **33**, 368–398.

Lycett, S.J., Collard, M. & McGrew, W.C. (2007) Phylogenetic analyses of behavior support existence of culture among wild chimpanzees. *Proceedings of the National Academy of Sciences of the USA*, **104**, 17588–17592.

MacArthur, R.H. & Wilson, E.O. (1967) *The Theory of Island Biogeography.* Princeton University Press, Princeton, NJ.

MacDonald, D.W. & Sillero-Zubiri, C. (2004) *The Biology and Conservation of Wild Canids.* Oxford University Press, Oxford, UK.

Mack, A.L. (1993) The sizes of vertebrate-dispersed fruits: A neotropical-paleotropical comparison. *American Naturalist*, **142**, 840–856.

Mack, A.L. & Jones, J. (2003) Low-frequency vocalizations by cassowaries (*Casuarius* spp.). *Auk*, **120**, 1062–1068.

Mack, A.L. & Wright, D.D. (1998) The vulturine parrot, *Psittrichas fulgidus*, a threatened New Guinea endemic: notes on its biology and conservation. *Bird Conservation International*, **8**, 185–194.

Mackessy, S.P., Sixberry, N.A., Heyborne, W.H. & Fritts, T. (2006) Venom of the brown treesnake, *Boiga irregularis*: ontogenetic shifts and taxa-specific toxicity. *Toxicon*, **47**, 537–548.

MacKinnon, J. (2000) New mammals in the 21st century? *Annals of the Missouri Botanical Garden*, **87**, 63–66.

MacPhee, R.D.E. & Horovitz, I. (2002) Extinct Quaternary playrrhines of the Great Antilles and Brazil. In: *The Primate Fossil Record* (ed W.C. Hartwig), pp. 189–200. Cambridge University Press, Cambridge, UK.

MacPhee, R.D.E., Iturralde-Vinent, M.A. & Jimenez Vázquez, O. (2007) Prehistoric sloth extinctions in Cuba: Implications of a new "Last" appearance date. *Caribbean Journal of Science*, **43**, 94–98.

Magnusson, W.E. & Lima, A.P. (1991) The ecology of a cryptic predator, *Paleosuchus trigonatus*, in a tropical rainforest. *Journal of Herpetology*, **25**, 41–48.

Magnusson, W.E., Lima, A.P. & Sampaio, R.M. (1985) Sources of heat for nests of *Paleosuchus trigonatus* and a review of crocodilian nest temperatures. *Journal of Herpetology*, **19**, 199–207.

Maisels, F. (2004) Defoliation of a monodominant rain-forest tree by a noctuid moth in Gabon. *Journal of Tropical Ecology*, **20**, 239–241.

Malekian, M., Cooper, S.J.B., Norman, J.A., Christidis, L. & Carthew, S.M. (2010) Molecular systematics and evolutionary origins of the genus *Petaurus* (Marsupialia: Petauridae) in Australia and New Guinea. *Molecular Phylogenetics and Evolution*, **54**, 122–135.

Maley, J. (2001) The impact of arid phases on the African rain forest through geological history. In: *African Rain Forest Ecology and Conservation: An Interdisciplinary Perspective* (eds W. Weber, L.J.T. White, A. Vedder & L. Naughton-Treves), pp. 69–87. Yale University Press, New Haven, CT.

Mandujano, S. & Naranjo, E.J. (2010) Ungulate biomass across a rainfall gradient: a comparison of data from neotropical and palaeotropical forests and local analyses in Mexico. *Journal of Tropical Ecology*, **26**, 13–23.

Marcot, J.D. (2007) Molecular phylogeny of terrestrial artiodactyls. In: *The Evolution of Artiodactyls* (eds D.R. Prothero & S.E. Foss), pp. 4–18. John Hopkins University Press, Baltimore, MD.

Marshall, A.G. (1983) Bats, flowers and fruit: Evolutionary relationships in the Old World. *Biological Journal of the Linnean Society*, **20**, 115–135.

Martin, C. (1991) *The Rainforests of West Africa: Ecology, Threats, Conservation.* Birkhauser Verlag, Basel, Switzerland.

Martin, P.S. & Steadman, D.W. (1999) Prehistoric extinctions on islands and continents. In: *Extinctions in Near Time: Causes, Contexts, and Consequences* (ed R.D.E. MacPhee), pp. 17–55. Kluwer, New York.

Martin, T.E. & Blackburn, G.A. (2009) The effectiveness of a Mesoamerican "paper park" in conserving cloud forest avifauna. *Biodiversity and Conservation*, **18**, 3841–3859.

Martínez-Garza, C. & Howe, H.F. (2003) Restoring tropical diversity: Beating the time tax on species loss. *Journal of Applied Ecology*, **40**, 423–429.

Martius, C. (1994) Diversity and ecology of termites in Amazonian forests. *Pedobiologia*, **38**, 407–428.

Matsubayashi, H., Lagan, P., Majalap, N., Tangah, J., Sukor, J.R.A. & Kitayama, K. (2007) Importance of natural licks for the mammals in Bornean inland tropical rain forests. *Ecological Research*, **22**, 742–748.

McCain, C.M. (2009) Vertebrate range sizes indicate that mountains may be "higher" in the tropics. *Ecology Letters*, **12**, 550–560.

McClure, H.E. (1974) *Migration and Survival of the Birds of Asia.* Applied Scientific Research Corporation of Thailand, Bangkok.

McConkey, K.R. & Chivers, D.J. (2007) Influence of gibbon ranging patterns on seed dispersal distance and deposition site in a Bornean forest. *Journal of Tropical Ecology*, **23**, 269–275.

McConkey, K.R. & Drake, D.R. (2006) Flying foxes cease to function as seed dispersers long before they become rare. *Ecology*, **87**, 271–276.

McGinley, M. (2008) Biological diversity in the Caribbean Islands. In: *Encyclopedia of Earth* (ed C.J. Cleveland). Environmental Information Coalition, Washington, DC.

McGraw, K.J. (2006) Mechanics of uncommon colors: pterins, porphyrins, and psittacofulvins. In: *Bird Coloration: Vol. 1 Mechanisms and Measurements* (eds G.E. Hill & K.J. McGraw). Harvard University Press, Cambridge, MA.

McGregor, G.R. & Nieuwolt, S. (1998) *Tropical Climatology: An Introduction to the Climates of Low Latitudes.* Wiley, Chichester, UK.

McGuire, J.A. & Dudley, R. (2005) The cost of living large: comparative gliding performance in flying lizards (Agamidae: *Draco*). *American Naturalist*, **166**, 93–106.

McGuire, K.L. (2008) Ectomycorrhizal associations function to maintain tropical monodominance. In: *Mycorrhizae: Sustainable Agriculture and Forestry* (eds Z.A. Siddiqui, M.S. Akhtar & K. Futai), pp. 287–302. Springer, Dordrecht, Netherlands.

Meir, P. & Woodward, F.I. (2010) Amazonian rain forests and drought: response and vulnerability. *New Phytologist*, **187**, 553–557.

Meiri, S., Meijaard, E., Wich, S.A., Groves, C.P. & Helgen, K.M. (2008) Mammals of Borneo – small size on a large island. *Journal of Biogeography*, **35**, 1087–1094.

Melo, F.P.L., Rodriguez-Herrera, B., Chazdon, R.L., Medellin, R.A. & Ceballos, G.G. (2009) Small tent-roosting bats promote dispersal of large-seeded plants in a Neotropical forest. *Biotropica*, **41**, 737–743.

Mercader, J. (2003) *Under the Canopy: The Archaeology of Tropical Rainforests.* Rutgers University Press, New Brunswick, NJ.

Mercader, J., Panger, M. & Boesch, C. (2002) Excavation of a chimpanzee stone tool site in the African rainforest. *Science*, **296**, 1452–1455.

Meredith, R.W., Westerman, M. & Springer, M.S. (2009) A phylogeny of Diprotodontia (Marsupialia) based on sequences for five nuclear genes. *Molecular Phylogenetics and Evolution*, **51**, 554–571.

Meyer, C.F.J. & Kalko, E.K.V. (2008) Assemblage-level responses of phyllostomid bats to tropical forest fragmentation: land-bridge islands as a model system. *Journal of Biogeography*, **35**, 1711–1726.

Meyer, S.T., Leal, I.R. & Wirth, R. (2009) Persisting hyper-abundance of leaf-cutting ants (*Atta* spp.) at the edge of an old Atlantic forest fragment. *Biotropica*, **41**, 711–716.

Micheneau, C., Fournel, J., Warren, B.H. et al. (2010) Orthoptera, a new order of pollinator. *Annals of Botany*, **105**, 355–364.

Michener, C.D. (2007) *The Bees of the World*. John Hopkins University Press, Baltimore, MA.

Mickleburgh, S., Waylen, K. & Racey, P. (2009) Bats as bushmeat: a global review. *Oryx*, **43**, 217–234.

Mikheyev, A.S., Mueller, U.G. & Abbot, P. (2010) Comparative dating of attine ant and lepiotaceous cultivar phylogenies reveals coevolutionary synchrony and discord. *American Naturalist*, **175**, E126–E133.

Milne, R.I. (2006) Northern hemisphere plant disjunctions: A window on tertiary land bridges and climate change? *Annals of Botany (London)*, **98**, 465–472.

Mittermeier, R.A., Gill, P.R., Hoffman, M. et al. (2005) *Hotspots Revisited: Earth's Biologically Richest and Most Endangered Terrestrial Ecoregions*. University of Chicago Press, Chicago, IL.

Miura, T., Roisin, Y. & Matsumoto, T. (1998) Developmental pathways and polyethism of neuter castes in the processional nasute termite *Hospitalitermes medioflavus* (Isoptera: Termitidae). *Zoological Science*, **15**, 843–848.

Moffett, M.W. (1987) Division of labor and diet in the extremely polymorphic ant *Pheidologeton diversus*. *National Geographic Research*, **3**, 282–304.

Molleman, F., Grunsven, R.H.A., Liefting, M., Zwaan, B.J. & Brakefield, P.M. (2005) Is male puddling behaviour of tropical butterflies targeted at sodium for nuptial gifts or activity? *Biological Journal of the Linnean Society*, **86**, 345–361.

Momose, K., Yumoto, T., Nagamitsu, T. et al. (1998) Pollination biology in a lowland dipterocarp forest in Sarawak, Malaysia. I. Characteristics of the plant-pollinator community in a lowland dipterocarp forest. *American Journal of Botany*, **85**, 1477–1501.

Morgan, G.S. & Woods, C.A. (1986) Extinction and the zoogeography of West Indian land mammals. *Biological Journal of the Linnean Society*, **28**, 167–203.

Morley, R.J. (2000) *Origin and Evolution of Tropical Rain Forests*. Wiley, Chichester, UK.

Morley, R.J. (2003) Interplate dispersal paths for megathermal angiosperms. *Perspectives in Plant Ecology Evolution and Systematics*, **6**, 5–20.

Morley, R.J. (2007) Cretaceous and Tertiary climate change and the past distribution of megathermal rainforests. In: *Tropical Rainforest Responses to Climatic Change* (eds M.B. Bush & J.R. Flenley), pp. 1–378. Praxis Publishing, Chichester, UK.

Moyle, R.G. & Marks, B.D. (2006) Phylogenetic relationships of the bulbuls (Aves: Pycnonotidae) based on mitochondrial and nuclear DNA sequence data. *Molecular Phylogenetics and Evolution*, **40**, 687–695.

Moyle, R.G., Filardi, C.E., Smith, C.E. & Diamond, J. (2009) Explosive Pleistocene diversification and hemispheric expansion of a "great speciator". *Proceedings of the National Academy of Sciences of the USA*, **106**, 1863–1868.

Moynihan, M. (1976) *The New World Primates*. Princeton University Press, Princeton, NJ.

Muchhala, N. & Thomson, J.D. (2010) Fur versus feathers: pollen delivery by bats and hummingbirds and consequences for pollen production. *American Naturalist*, **175**, 717–726.

Muellner, A.N., Savolainen, V., Samuel, R. & Chase, M.W. (2006) The mahogany family "out-of-Africa": Divergence time estimation, global biogeographic patterns inferred from plastid rbcL DNA sequences, extant, and fossil distribution of diversity. *Molecular Phylogenetics and Evolution*, **40**, 236–250.

Munn, C.A. (1985) Permanent canopy and understorey flocks in Amazonia: species composition and population density. *Ornithological Monographs*, **36**, 683–712.

Muscarella, R. & Fleming, T.H. (2007) The role of frugivorous bats in tropical forest succession. *Biological Reviews*, **82**, 573–590.

Neall, V.E. & Trewick, S.A. (2008) The age and origin of the Pacific islands: a geological overview. *Philosophical Transactions of the Royal Society of London B Biological Sciences*, **363**, 3293–3308.

Nellemann, C., Redmond, I. & Refisch, J. (2010) *The Last Stand of the Gorilla – Environmental Crime and Conflict in the Congo Basin*. United Nations Environmental Programme, GRID-Arendal, Norway.

Newbery, D.M., Chuyong, G.B. & Zimmermann, L. (2006) Mast fruiting of large ectomycorrhizal African rain forest trees: importance of dry season intensity, and the resource-limitation hypothesis. *New Phytologist*, **170**, 561–579.

Newton, I. & Brockie, K. (2008) *The Migration Ecology of Birds*. Academic Press, London, UK.

Nieder, J., Prosperi, J. & Michaloud, G. (2001) Epiphytes and their contribution to canopy diversity. *Plant Ecology*, **153**, 51–63.

Nieh, J.C. (2004) Recruitment communication in stingless bees (Hymenoptera, Apidae, Meliponini). *Apidologie*, **35**, 159–182.

Nijman, V. (2010) An overview of international wildlife trade from Southeast Asia. *Biodiversity and Conservation*, **19**, 1101–1114.

Nilsson, L.A. (1998) Deep flowers for long tongues. *Trends in Ecology and Evolution*, **13**, 259–260.

Nobre, C. & Aanen, D.K. (2010) Dispersion and colonisation by fungus-growing termites. *Communicative & Integrative Biology*, **3**, 1–3.

Nobre, T., Eggleton, P. & Aanen, D.K. (2010) Vertical transmission as the key to the colonization of Madagascar by fungus-growing termites? *Proceedings of the Royal Society Biological Sciences Series B*, **277**, 359–365.

Nogueira, M.R., Monteiro, L.R., Peracchi, A.L. & de Araujo, A.F.B. (2005) Ecomorphological analysis of the masticatory apparatus in the seed-eating bats, genus *Chiroderma* (Chiroptera: Phyllostomidae). *Journal of Zoology*, **266**, 355–364.

Nogueira-Filho, S.L.G., Nogueira, S.S.C. & Fragoso, J.M.V. (2009) Ecological impacts of feral pigs in the Hawaiian Islands. *Biodiversity and Conservation*, **18**, 3677–3683.

Normand, E., Ban, S.D. & Boesch, C. (2009) Forest chimpanzees (*Pan troglodytes verus*) remember the location of numerous fruit trees. *Animal Cognition*, **12**, 797–807.

Nortcliff, S. (2010) Soils of the tropics. In: *Soil Biology and Agriculture in the Tropics*. (ed P. Dion), pp. 1–15. Springer-Verlag, Berlin, Germany.

Novotny, V., Tonner, M. & Spitzer, K. (1991) Distribution and flight behaviour of the junglequeen butterfly, *Stichophthalma louisa* (Lepidoptera: Nymphalidae), in an Indochinese montane rainforest. *Journal of Research on the Lepidoptera*, **30**, 279–288.

Novotny, V. (2010) Rain forest conservation in a tribal world: why forest dwellers prefer loggers to conservationists. *Biotropica*, **42**, 546–549.

Nowak, R.M. (1999) *Walker's Mammals of the World, 6th edn*. John Hopkins University Press, Baltimore, MD.

Ohkuma, M. (2008) Symbioses of flagellates and prokaryotes in the gut of lower termites. *Trends in Microbiology*, **16**, 345–352.

Oldroyd, B.P. & Nanork, P. (2009) Conservation of Asian honey bees. *Apidologie*, **40**, 296–312.

Oliver, P.M. & Sanders, K.L. (2009) Molecular evidence for Gondwanan origins of multiple lineages within a diverse Australasian gecko radiation. *Journal of Biogeography*, **36**, 2044–2055.

Orivel, J. & Leroy, C. (2010) The diversity and ecology of ant gardens (Hymenoptera: Formicidae; Spermatophyta, Angiospermae). *Myrmecological News*, **14**, 73–85.

Orlando, L., Calvignac, S., Schnebelen, C., Douady, C.J., Godfrey, L.R. & Hanni, C. (2008) DNA from extinct giant lemurs links archaeolemurids to extant indriids. *BMC Evolutionary Biology*, **8**, 121.

Osada, N., Takeda, H., Furukawa, A. & Awang, M. (2001) Fruit dispersal of two dipterocarp species in a Malaysian rain forest. *Journal of Tropical Ecology*, **17**, 911–917.

O'Shea, M. (2007) *Boas and Pythons of the World*. Princeton University Press, Princeton, NJ.

Owen-Smith, R.N. (1992) *Megaherbivores: The Influence of Very Large Body Size on Ecology*. Cambridge University Press, Cambridge, UK.

Paijmans, K. (1976) *New Guinea Vegetation*. Australian National University Press, Canberra, Australia.

Parry, L., Peres, C.A., Day, B. & Amaral, S. (2010) Rural-urban migration brings conservation threats and opportunities to Amazonian watersheds. *Conservation Letters*, **3**, 251–259.

Patel, E.R. (2007) Logging of rare rosewood and palisandre (*Dalbergia* spp.) within Morojejy National Park, Madagascar. *Madagascar Conservation and Development*, **2**, 11–16.

Paul, G.S. & Yavitt, J.B. (2010) Tropical vine growth and the effects on forest succession: a review of the ecology and management of tropical climbing plants. *Botanical Review*, doi: 10.1007/s12229-010-9059-3.cv.

Pearson, D.L. (1977) A pantropical comparison of bird community structure in six lowland forest sites. *Condor*, **79**, 232–244.

Peay, K.G., Kennedy, P.G., Davies, S.J., Tan, S. & Bruns, T.D. (2010) Potential link between plant and fungal distributions in a dipterocarp rainforest: community and phylogenetic structure of tropical ectomycorrhizal fungi across a plant and soil ecotone. *New Phytologist*, **185**, 529–542.

Pedersen, L.B. & Kress, W.J. (1999) Honeyeater (Meliphagidae) pollination and the floral biology of Polynesian *Heliconia* (Heliconiaceae). *Plant Systematics and Evolution*, **216**, 1–21.

Pemberton, R.W. (2010) Biotic resource needs of specialist orchid pollinators. *Botanical Review*, **76**, 275–292.

Pennington, R.T. & Dick, C.W. (2004) The role of immigrants in the assembly of the South American rainforest tree flora. *Philosophical Transactions of the Royal Society of London B Biological Sciences*, **359**, 1611–1622.

Pereira, S.L., Johnson, K.P., Clayton, D.H. & Baker, A.J. (2007) Mitochondrial and nuclear DNA sequences support a Cretaceous origin of Columbiformes and a dispersal-driven radiation in the Paleogene. *Systematic Biology*, **56**, 656–672.

Peres, C.A. (2005) Why we need megareserves in Amazonia. *Conservation Biology*, **19**, 728–733.

Perry, G.H. & Dominy, N.J. (2009) Evolution of the human pygmy phenotype. *Trends in Ecology & Evolution*, **24**, 218–225.

Peters, M.K. (2010) Ant-following and the prevalence of blood parasites in birds of African rainforests. *Journal of Avian Biology*, **41**, 105–110.

Peters, M.K., Likare, S. & Kraemer, M. (2008) Effects of habitat fragmentation and degradation on flocks of African ant-following birds. *Ecological Applications*, **18**, 847–858.

Pfeiffer, M. & Linsenmair, K.E. (2001) Territoriality in the Malaysian giant ant *Camponotus gigas* (Hymenoptera / Formicidae). *Journal of Ethology*, **19**, 75–85.

Phillips, O.L., Lewis, S.L., Baker, T.R., Chao, K.-J. & Higuchi, N. (2008) The changing Amazon forest. *Philosophical Transactions of the Royal Society of London B Biological Sciences*, **363**, 1819–1827.

Pinto-Tomás, A.A., Anderson, M.A., Suen, G. et al. (2009) Symbiotic nitrogen fixation in the fungus gardens of leaf-cutter ants. *Science*, **326**, 1120–1123.

Plana, V. (2004) Mechanisms and tempo of evolution in the African Guineo-Congolian rainforest. *Philosophical Transactions of the Royal Society B: Biological Sciences*, **359**, 1585–1594.

Polasky, S. (2008) Why conservation planning needs socioeconomic data. *Proceedings of the National Academy of Sciences of the USA*, **105**, 6505–6506.

Potapov, P., Yaroshenko, A., Turubanova, S. et al. (2008) Mapping the world's intact forest landscapes by remote sensing. *Ecology and Society*, **13**, 51.

Potts, M.D. (2003) Drought in a Bornean everwet rain forest. *Journal of Ecology*, **91**, 467–474.

Powell, G.V.N. (1985) Sociobiology and adaptive significance of interspecific foraging flocks in the Neotropics. *Ornithological Monographs*, **36**, 713–732.

Powell, G.V.N. & Bjork, R.D. (2004) Habitat linkages and the conservation of tropical biodiversity as indicated by seasonal migrations of three-wattled bellbirds. *Conservation Biology*, **18**, 500–509.

Powell, S. & Clark, E. (2004) Combat between large derived societies: A subterranean army ant established as a predator of mature leaf-cutting ant colonies. *Insectes Sociaux*, **51**, 342–351.

Power, M.L. & Myers, E.W. (2009) Digestion in the common marmoset (*Callithrix jacchus*), a gummivore-frugivore. *American Journal of Primatology*, **71**, 957–963.

Price, S., Donovan, D. & De Jong, W. (2007) Confronting conflict timber. In: *Extreme Conflict and Tropical Forests* (eds W. de Jong, D. Donovan & K.I. Abe), pp. 117–132. Springer, Berlin.

Pridgeon, A.M. (1994) The realm of wonder. *Proceedings of the 14th World Orchid Conference*, pp. 5–12. HMSO, Edinburgh, UK.

Primack, R.B. (1987) Relationships among flowers, fruits, and seeds. *Annual Review of Ecology and Systematics*, **18**, 409–430.

Primack, R.B. (2010) *Essentials of Conservation Biology, 5th edn*. Sinauer Associates, Sunderland, MA.

Primack, R.B., Bray, D., Galleti, H.A. & Ponciano, I. (1998) *Timber, Tourists, and Temples: Conservation and Development in the Maya Forest of Belize, Guatemala, and Mexico*. Island Press, Washington, DC.

Primack, R.B., Chai, E.O.K. & Lee, H.S. (1989) Relative performance of dipterocarp trees in natural forest, managed forest, logged forest and plantations throughout Sarawak, East Malaysia. In: *Growth and Yield in Tropical Mixed/Moist Forests* (eds M. Wan Razali, H.T. Chan & S. Appanah), pp. 161–175. Forest Research Institute, Kuala Lumpur, Malaysia.

Proctor, J., Haridasan, K. & Smith, G.W. (1998) How far north does lowland evergreen tropical rain forest go? *Global Ecology and Biogeography Letters*, **7**, 141–146.

Proctor, J., Brearley, F.Q., Dunlop, H., Proctor, K., Supramono & Taylor, D. (2001) Local wind damage in Barito Ulu, Central Kalimantan: A rare but essential event in a lowland dipterocarp forest? *Journal of Tropical Ecology*, **17**, 473–475.

Prum, R.O. (1994) Phylogenetic analysis of the evolution of alternative social behavior in the manakins (Aves: Pipridae). *Evolution*, **48**, 1657–1675.

Putz, F.E., Zuidema, P.A., Pinard, M.A. et al. (2008) Improved tropical forest management for carbon retention. *PLoS Biology*, **6**, e166.

Rabeling, C., Brown, J.M. & Verhaagh, M. (2008) Newly discovered sister lineage sheds light on early ant evolution. *Proceedings of the National Academy of Sciences of the USA*, **105**, 14913–14917.

Rabinowitz, A., Myint, T., Khaing, S.T. & Rabinowitz, S. (1999) Description of the leaf deer (*Muntiacus putaoensis*), a new species of muntjac from northern Myanmar. *Journal of Zoology*, **249**, 427–435.

Raghuram, H., Thangadurai, C., Gopukumar, N., Nathar, K. & Sripathi, K. (2009) The role of olfaction and vision in the foraging behaviour of an echolocating megachiropteran fruit bat, *Rousettus leschenaulti* (Pteropodidae). *Mammalian Biology*, **74**, 9–14.

Ramirez, S.R., Gravendeel, B., Singer, R.B., Marshall, C.R. & Pierce, N.E. (2007) Dating the origin of the Orchidaceae from a fossil orchid with its pollinator. *Nature*, **448**, 1042–1045.

Rasingam, L. & Parathasarathy, N. (2009) Tree species diversity and population structure across major forest formations and disturbance categories in Little Andaman Island, India. *Tropical Ecology*, **50**, 89–102.

Rasmussen, C. & Cameron, S.A. (2010) Global stingless bee phylogeny supports ancient divergence, vicariance, and long distance dispersal. *Biological Journal of the Linnean Society*, **99**, 206–232.

Rasmussen, D. & Sussman, R. (2007) Parallelisms among primates and possums. In: *Primate Origins: Adaptations and Evolution.* (eds M.J. Ravosa & M. Dagosto), pp. 775–803. Plenum Press, New York.

Rawlings, L.H., Rabosky, D.L., Donnellan, S.C. & Hutchinson, M.N. (2008) Python phylogenetics: inference from morphology and mitochondrial DNA. *Biological Journal of the Linnean Society*, **93**, 603–619.

Rawlins, D.R. & Handasyde, K.A. (2002) The feeding ecology of the striped possum *Dactylopsila trivirgata* (Marsupialia: Petauridae) in far north Queensland, Australia. *Journal of Zoology*, **257**, 195–206.

Reid, N. (1991) Coevolution of mistletoes and frugivorous birds? *Australian Journal of Ecology*, **16**, 457–469.

Renner, S.S. & Schaefer, H. (2010) The evolution and loss of oil-offering flowers: new insights from dated phylogenies for plants and bees. *Philosophical Transactions of the Royal Society of London B Biological Sciences*, **365**, 423–435.

Rex, K., Kelm, D.H., Wiesner, K., Kunz, T.H. & Voigt, C.C. (2008) Species richness and structure of three Neotropical bat assemblages. *Biological Journal of the Linnean Society*, **94**, 617–629.

Ribeiro, M.C., Metzger, J.P., Martensen, A.C., Ponzoni, F.J. & Hirota, M.M. (2009) The Brazilian Atlantic Forest: How much is left, and how is the remaining forest distributed? Implications for conservation. *Biological Conservation*, **142**, 1141–1153.

Richter, H.V. & Cumming, G.S. (2008) First application of satellite telemetry to track African straw-coloured fruit bat migration. *Journal of Zoology*, **275**, 172–176.

Ricketts, T.H., Soares-Filho, B., da Fonseca, G.A.B. et al. (2010) Indigenous lands, protected areas, and slowing climate change. *PLoS Biology*, **8**, e100331.

Ricklefs, R. & Bermingham, E. (2008) The West Indies as a laboratory of biogeography and evolution. *Philosophical Transactions of the Royal Society of London B Biological Sciences*, **363**, 2393–2413.

Rinderknecht, A. & Blanco, R.E. (2008) The largest fossil rodent. *Proceedings of the Royal Society Biological Sciences Series B*, **275**, 923–928.

Rodrigues, R.R., Lima, R.A.F., Gandolfi, S. & Nave, A.G. (2009) On the restoration of high diversity forests: 30 years of experience in the Brazilian Atlantic Forest. *Biological Conservation*, **142**, 1242–1251.

Roubik, D.W. (1989) *Ecology and Natural History of Tropical Bees.* Cambridge University Press, Cambridge, UK.

Roubik, D.W. (2009) Ecological impact on native bees by the invasive africanized honey bee. *Acta Biologica Colombica*, **14**, 115–124.

Roubik, D.W. & Villanueva-Gutierrez, R. (2009) Invasive Africanized honey bee impact on native solitary bees: a pollen resource and trap nest analysis. *Biological Journal of the Linnean Society*, **98**, 152–160.

Rubinoff, D. (2008) Phylogeography and ecology of an endemic radiation of Hawaiian aquatic case-bearing moths (*Hyposmocoma*: Cosmopterigidae). *Philosophical Transactions of the Royal Society of London B Biological Sciences*, **363**, 3459–3465.

Russell, J.A., Moreau, C.S., Goldman-Huertas, B., Fujiwara, M., Lohman, D.J. & Pierce, N.E. (2009) Bacterial gut symbionts are tightly linked with the evolution of herbivory in ants. *Proceedings of the National Academy of Sciences*, **106**, 21236–21241.

Rust, J., Singh, H., Rana, R.S. et al. (2010) Biogeographic and evolutionary implications of a diverse paleobiota in amber from the early Eocene of India. *Proceedings of the National Academy of Sciences of the USA*, doi: 10.1073/pnas.1007407107.

Safer, A.B. & Grace, M.S. (2004) Infrared imaging in vipers: differential responses of crotaline and viperine snakes to paired thermal targets. *Behavioural Brain Research*, **154**, 55–61.

Sakai, S., Harrison, R.D., Momose, K. et al. (2006) Irregular droughts trigger mass flowering in aseasonal tropical forests in Asia. *American Journal of Botany*, **93**, 1134–1139.

Sallam, H.M., Seiffert, E.R., Steiper, M.E. & Simons, E.L. (2009) Fossil and molecular evidence constrain scenarios for the early evolutionary and biogeographic history of hystricognathous rodents. *Proceedings of the National Academy of Sciences of the USA*, **106**, 16722–16727.

Sanz, C.M., Schöning, C. & Morgan, D.B. (2010) Chimpanzees prey on army ants with specialized tool set. *American Journal of Primatology*, **72**, 17–24.

Saverschek, M., Herz, H., Wagner, M. & Roces, F. (2010) Avoiding plants unsuitable for the symbiotic fungus: learning and long-term memory in leaf-cutting ants. *Animal Behaviour*, **79**, 689–698.

Savolainen, P., Leitner, T., Wilton, A.N., Matisoo-Smith, E. & Lundeberg, J. (2004) A detailed picture of the origin of the Australian dingo, obtained from the study of mitochondrial DNA. *Proceedings of the National Academy of Sciences of the USA*, **101**, 12387–12390.

Sax, D.F. & Gaines, S.D. (2008) Species invasions and extinction: The future of native biodiversity on islands. *Proceedings of the National Academy of Sciences of the USA*, **105**, 11490–11497.

Schatz, G.E. (2001) *Generic Tree Flora of Madagascar.* Royal Botanic Gardens, Kew, UK.

Schöning, C., Humle, T., Möbius, Y. & McGrew, W.C. (2008) The nature of culture: Technological variation in chimpanzee predation on army ants revisited. *Journal of Human Evolution*, **55**, 48–59.

Schuiteman, A. & de Vogel, E.F. (2007) Orchidaceae of Papua. In: *The Ecology of Papua, Part One* (eds A.J. Marshall & B.M. Beehler), pp. 435–456. Periplus Editions, Singapore.

Schultz, T.R. & Brady, S.G. (2008) Major evolutionary transitions in ant agriculture. *Proceedings of the National Academy of Sciences of the USA*, **105**, 5435–5440.

Scopece, G., Cozzolino, S., Johnson, S.D. & Schiestl, F.P. (2010) Pollination efficiency and the evolution of specialized deceptive pollination systems. *American Naturalist*, **175**, 98–105.

Scott, J.J., Budsberg, K.J., Suen, G., Wixon, D.L., Balser, T.C. & Currie, C.R. (2010) Microbial community structure of leaf-cutter ant fungus gardens and refuse dumps. *PLoS One*, **5**, e9922.

Secor, S.M. (2008) Digestive physiology of the Burmese python: broad regulation of integrated performance. *Journal of Experimental Biology*, **211**, 3767–3774.

Setchell, J.M., Charpentier, M.J.E., Abbott, K.M., Wickings, E.J. & Knapp, L.A. (2009) Is brightest best? Testing the Hamilton-Zuk hypothesis in mandrills. *International Journal of Primatology*, **30**, 825–844.

Shanahan, M., So, S., Compton, S.G. & Corlett, R. (2001) Fig-eating by vertebrate frugivores: A global review. *Biological Reviews*, **76**, 529–572.

Shapiro, B., Sibthorpe, D., Rambaut, A. et al. (2002) Flight of the dodo. *Science*, **295**, 1683.

Shearman, P.L., Ash, J., Mackey, B., Bryan, J.E. & Lokes, B. (2009) Forest conversion and degradation in Papua New Guinea 1972–2002. *Biotropica*, **41**, 379–390.

Sheldon, F.H., Styring, A. & Hosner, P.A. (2010) Bird species richness in a Bornean exotic tree plantation: A long-term perspective. *Biological Conservation*, **143**, 399–407.

Shine, R., Harlow, P.S., Keogh, J.S. & Boeadi (1996) Commercial harvesting of giant lizards: The biology of water monitors *Varanus salvator* in southern Sumatra. *Biological Conservation*, **77**, 125–134.

Shono, K., Cadaweng, E.A. & Durst, P.B. (2007) Application of assisted natural regeneration to restore degraded tropical forestlands. *Restoration Ecology*, **15**, 620–626.

Slik, J.W.F., Aiba, S.-I., Brearley, F.Q. et al. (2010) Environmental correlates of tree biomass, basal area, wood specific gravity and stem density gradients in Borneo's tropical forests. *Global Ecology and Biogeography*, **19**, 50–60.

Smedmark, J.E.E. & Anderberg, A.A. (2007) Boreotropical migration explains hybridization between geographically distant lineages in the pantropical clade Sideroxyleae (Sapotaceae). *American Journal of Botany*, **94**, 1491–1505.

Smith, A.P. & Ganzhorn, J.U. (1996) Convergence in community structure and dietary adaptation in Australian possums and gliders and Malagasy lemurs. *Australian Journal of Ecology*, **21**, 31–46.

Smith, S.Y., Collinson, M.E. & Rudall, P.J. (2008) Fossil *Cyclanthus* (Cyclanthaceae, Pandanales) from the Eocene of Germany and England. *American Journal of Botany*, **95**, 688–699.

Socha, J.J., O'Dempsey, T. & LaBarbera, M. (2005) A 3-D kinematic analysis of gliding in a flying snake, *Chrysopelea paradisi*. *Journal of Experimental Biology*, **208**, 1817–1833.

Sodhi, N., Lee, T., Sekercioglu, C. et al. (2010) Local people value environmental services provided by forested parks. *Biodiversity and Conservation*, **19**, 1175–1188.

Spergel, B. & Taieb, P. (2008) *Rapid Review of Conservation Trust Funds, 2nd edn.* Conservation Finance Alliance.

Srygley, R.B. & Penz, C.M. (1999) Lekking in Neotropical owl butterflies, *Caligo illioneus* and *C. oileus* (Lepidoptera: Brassolinae). *Journal of Insect Behavior*, **12**, 81–103.

Steadman, D.W. (2006) *Extinction & Biogeography of Tropical Pacific Birds.* University of Chicago Press, Chicago.

Stein, B. (1992) Sicklebill hummingbirds, ants, and flowers. *Bioscience*, **42**, 27–33.

Stein, B.A., Kutner, L.S. & Adams, J.S. (2000) *Precious Heritage: The Status of Biodiversity in the United States.* Oxford University Press, New York.

Stickler, C.M., Nepstad, D.C., Coe, M.T. et al. (2009) The potential ecological costs and cobenefits of REDD: a critical review and case study from the Amazon region. *Global Change Biology*, **15**, 2803–2824.

Struhsaker, T.T. & Leakey, M. (1990) Prey selectivity by crowned hawk-eagles on monkeys in the Kibale Forest, Uganda. *Behavioral Ecology and Sociobiology*, **26**, 435–443.

Styring, A.R. & Ickes, K. (2003) Woodpeckers (Picidae) at Pasoh: Foraging ecology, flocking and the impacts of logging on abundance and diversity. In: *Pasoh: Ecology of a Lowland Rain Forest in Southeast Asia* (eds T. Okuda, N. Manokaran, Y. Matsumoto, K. Niiyama, S.C. Thomas & P.S. Ashton), pp. 547–557. Springer, New York.

Sunquist, M. & Sunquist, F. (2002) *Wild Cats of the World*. University of Chicago Press, Chicago, IL.

Swapna, N., Radhakrishna, S., Gupta, A.K. & Kumar, A. (2010) Exudativory in the Bengal slow loris (*Nycticebus bengalensis*) in Trishna Wildlife Sanctuary, Tripura, northeast India. *American Journal of Primatology*, **72**, 113–121.

Swartz, M.B. (2001) Bivouac checking, a novel behavior distinguishing obligate from opportunistic species of army-ant-following birds. *Condor*, **103**, 629–633.

Sweet, S.S. & Pianka, E.R. (2003) The lizard kings. *Natural History*, **112**, 40–45.

Tanaka, H.O., Inui, Y. & Itioka, T. (2009) Anti-herbivore effects of an ant species, *Crematogaster difformis*, inhabiting myrmecophytic epiphytes in the canopy of a tropical lowland rainforest in Borneo. *Ecological Research*, **24**, 1393–1397.

Tanaka, H.O., Yamane, S. & Itioka, T. (2010) Within-tree distribution of nest sites and foraging areas of ants on canopy trees in a tropical rainforest in Borneo. *Population Ecology*, **52**, 147–157.

Tattersall, G.J., Andrade, D.V. & Abe, A.S. (2009) Heat exchange from the toucan bill reveals a controllable vascular thermal radiator. *Science*, **325**, 468–470.

Teeling, E.C., Springer, M.S., Madsen, O., Bates, P., O'Brien, S.J. & Murphy, W.J. (2005) A molecular phylogeny for bats illuminates biogeography and the fossil record. *Science*, **307**, 580–584.

Tello, J.G. (2003) Frugivores at a fruiting *Ficus* in south-eastern Peru. *Journal of Tropical Ecology*, **19**, 717–721.

Temeles, E.J., Koulouris, C.R., Sander, S.E. & Kress, W.J. (2009) Effect of flower shape and size on foraging performance and trade-offs in a tropical hummingbird. *Ecology*, **90**, 1147–1161.

Terborgh, J., Lopez, L., Nuñez, P.V. et al. (2001) Ecological meltdown in predator-free forest fragments. *Science*, **294**, 1923–1926.

Ter Steege, H., Pitman, N., Sabatier, D. et al. (2003) A spatial model of tree alpha-diversity and tree density for the Amazon. *Biodiversity and Conservation*, **12**, 2255–2277.

Thiollay, J.M. (1991) Foraging, home range use and social behaviour of a group-living rainforest raptor, the red-throated caracara *Daptrius americanus*. *Ibis*, **133**, 382–393.

Thiollay, J.-M. (2003) Comparative foraging behavior between solitary and flocking insectivores in a neotropical forest: Does vulnerability matter? *Ornitologia Neotropical*, **14**, 47–65.

Thompson, J., Brokaw, N., Zimmerman, J.K. et al. (2002) Land use history, environment, and tree composition in a tropical forest. *Ecological Applications*, **12**, 1344–1363.

Tschapka, M. & Dressler, S. (2002) Chiropterophily: On bat-flowers and flower bats. *Curtis's Botanical Magazine*, **19**, 114–125.

Tsuji, Y., Yangozene, K. & Sakamaki, T. (2010) Estimation of seed dispersal distance by the bonobo, *Pan paniscus*, in a tropical forest in Democratic Republic of Congo. *Journal of Tropical Ecology*, **26**, 115–118.

Turner, I.M. (2001a) Rainforest ecosystems, plant diversity. In: *Encyclopedia of Biodiversity* (ed S.A. Levin), pp. 13–23. Academic Press, San Diego, CA.

Turner, I.M. (2001b) *The Ecology of Trees in the Tropical Rain Forest*. Cambridge University Press, Cambridge, UK.

Ueda, S., Quek, S.-P., Itioka, T., Murase, K. & Itino, T. (2010) Phylogeography of the *Coccus* scale insects inhabiting myrmecophytic *Macaranga* plants in Southeast Asia. *Population Ecology*, **52**, 137–146.

Valentine, P.S. & Hill, R. (2008) The establishment of a World Heritage Area. In: *Living in a Dynamic Tropical Forest Landscape* (eds N.E. Stork & S.M. Turton). Blackwell, Malden, MA.

Van Bael, S.A., Brawn, J.D. & Robinson, S.K. (2003) Birds defend trees from herbivores in a Neotropical forest canopy. *Proceedings of the National Academy of Sciences of the USA*, **100**, 8304–8307.

van der Werf, G.R., Morton, D.C., DeFries, R.S. et al. (2009) CO_2 emissions from forest loss. *Nature Geoscience*, **2**, 737–738.

Vanhooydonck, B., Meulepas, G., Herrel, A. et al. (2009) Ecomorphological analysis of aerial performance in a non-specialized lacertid lizard, *Holaspis guentheri*. *Journal of Experimental Biology*, **212**, 2475–2482.

van Schaik, C.P. (2009) Geographical variation in the behavior of wild great apes: is it really cultural? In: *The Question of Animal Culture* (eds K.N. Laland & B.G. Galef), pp. 70–98. Harvard University Press, Cambridge, MA.

Vidal, N., Marin, J., Morini, M. et al. (2010) Blindsnake evolutionary tree reveals long history on Gondwana. *Biology Letters*, **6**, 558–561.

Vieira, D.L.M., Holl, K.D. & Peneireiro, F.M. (2009) Agro-successional restoration as a strategy to facilitate tropical forest recovery. *Restoration Ecology*, **17**, 451–459.

Vinyard, C.J., Wall, C.E., Williams, S.H. & Hylander, W.L. (2003) Comparative functional analysis of skull morphology of tree-gouging primates. *American Journal of Physical Anthropology*, **120**, 153–170.

Vitousek, P.M., Porder, S., Houlton, B.Z. & Chadwick, O.A. (2010) Terrestrial phosphorus limitation: Mechanisms, implications, and nitrogen-phosphorus interactions. *Ecological Applications*, **20**, 5–15.

Voirin, J.B., Kays, R., Lowman, M.D. & Wikelski, M. (2009) Evidence for three-toed sloth (*Bradypus variegatus*) predation by spectacled owl (*Pulsatrix perspicillata*). *Edentata*, **8–10**, 15–20.

Walker, R., Moore, N.J., Arima, E. et al. (2009) Protecting the Amazon with protected areas. *Proceedings of the National Academy of Sciences of the USA*, **106**, 10582–10586.

Wallace, A.R. (1859) *On the Zoological Geography of the Malay Archipelago*. Linnean Society, London.

Wallace, R.B., Painter, R.L.E. & Saldania, A. (2002) An observation of bush dog (*Speothos venaticus*) hunting behaviour. *Mammalia*, **66**, 309–311.

Walsh, R.P.D. (1996) Climate. In: *The Tropical Rain Forest: An Ecological Study* (eds P.W. Richards, R.P.D. Walsh, I.C. Baillie & P. Greig-Smith), pp. 159–205. Cambridge University Press, Cambridge, UK.

Walsh, R.P.D. & Newbery, D.M. (1999) The ecoclimatology of Danum, Sabah, in the context of the world's rainforest regions, with particular reference to dry periods and their impact. *Philosophical Transactions of the Royal Society of London B Biological Sciences*, **354**, 1869–1883.

Wang, H., Moore, M.J., Soltis, P.S. et al. (2009) Rosid radiation and the rapid rise of angiosperm-dominated forests. *Proceedings of the National Academy of Sciences of the USA*, **106**, 3853–3858.

Wanntorp, L. & Kunze, H. (2009) Identifying synapomorphies in the flowers of *Hoya* and *Dischidia* – towards phylogenetic understanding. *International Journal of Plant Sciences*, **170**, 331–342.

Ward, P.S. (2010) Taxonomy, phylogenetics, and evolution. In: *Ant Ecology* (eds L. Lach, C.L. Parr & K.L. Abbott), pp. 3–17. Oxford University Press, Oxford, UK.

Watkins Jr, J.E., Cardelus, C.L. & Mack, M.C. (2008) Ants mediate nitrogen relations of an epiphytic fern. *New Phytologist*, **180**, 5–8.

Webb, S.D. (1997) The great American faunal interchange. In: *Central America: A Natural and Cultural History* (ed A.G. Coates), pp. 97–122. Yale University Press, New Haven, CT.

Weir, J.T., Bermingham, E. & Schluter, D. (2009) The Great American Biotic Interchange in birds. *Proceedings of the National Academy of Sciences of the USA*, **106**, 21737–21742.

Weisrock, D.W., Rasoloarison, R.M., Fiorentino, I. et al. (2010) Delimiting species without nuclear monophyly in Madagascar's mouse lemurs. *PLoS One*, **5**, e9883.

Wet Tropics Management Authority (2008) *Climate Change in the Wet Tropics: Impacts and Responses*. Cairns, Australia.

White, J.L. & MacPhee, R.D.E. (2001) The sloths of the West Indies: a systematic and phylogenetic review. In: *Biogeography of the West Indies: Patterns and Perspectives* (eds C.A. Woods & F.E. Sergile), pp. 201–235. CRC Press, Boca Raton, FL.

White, L.J.T., Tutin, C.E.G. & Fernandez, M. (1993) Group composition and diet of forest elephants, *Loxodonta africana cyclotis* Matschie 1900, in the Lopé Reserve, Gabon. *African Journal of Ecology*, **31**, 181–199.

Whitmore, T.C. (1998) *An Introduction to Tropical Rain Forests, 2nd edn*. Oxford University Press, Oxford, UK.

Whitmore, T.C. & Burslem, D.F.R.P. (1998) Major disturbances in tropical rainforests. In: *Dynamics of Tropical Communities* (eds D.M. Newbery, H.H.T. Prins & N.D. Brown), pp. 549–565. Blackwell Science, Oxford, UK.

Whitten, T. & Balmford, A. (2006) Who should pay for tropical forest conservation, and how should the costs be met? In: *Emerging Threats to Tropical Forests* (eds W.F. Laurance & C.A. Peres), pp. 317–336. Chicago University Press, Chicago.

Wiederholt, R. & Post, E. (2010) Tropical warming and the dynamics of endangered primates. *Biology Letters*, **6**, 257–260.

Wiens, F., Zitzmann, A. & Hussein, N.A. (2006) Fast food for slow lorises: Is low metabolism related to secondary compounds in high-energy plant diet? *Journal of Mammalogy*, **87**, 790–798.

Williams, D.D. (2006) *The Biology of Temporary Waters*. Oxford University Press, Oxford, UK.

Williams, G.A. & Adam, P. (1997) The composition of the bee (Apoidea: Hymenoptera) fauna visiting flowering trees in New South Wales lowland subtropical rainforest remnants. *Proceedings of the Linnean Society of New South Wales*, **118**, 69–95.

Williamson, G.B., Laurance, W.F., Oliveira, A.A. et al. (2000) Amazonian tree mortality during the 1997 El Niño drought. *Conservation Biology*, **14**, 1538–1542.

Wilson, E.E., Mullen, L.M. & Holway, D.A. (2009) Life history plasticity magnifies the ecological effects of a social wasp invasion. *Proceedings of the National Academy of Sciences of the USA*, **106**, 12809–12813.

Wink, M., Heidrich, P., Sauer-Gurth, H., Elsayed, A.-A. & Gonzalez, J. (2009) Molecular phylogeny and systematics of owls (Strigiformes). In: *Owls of the World* (eds C. König, F. Weick & J.-H.

Becking), pp. 39–57. Yale University Press, New Haven, CT.

Winston, M.L. (1992) The biology and management of Africanized honey bees. *Annual Review of Entomology*, **37**, 173–194.

Wirth, R., Herz, H., Ryel, R.J., Beyschlag, W. & Hölldobler, B. (2003) *Herbivory of Leaf-cutting Ants: A Case Study on Atta colombica in the Tropical Rainforest of Panama.* Springer, Heidelberg, Germany.

Worthy, T.H., Anderson, A.J. & Molnar, R.E. (1999) Megafaunal expression in a land without mammals: The first fossil faunas from terrestrial deposits in Fiji (Vertebrata: Amphibia, Reptilia, Aves). *Senckenbergiana Biologica*, **79**, 237–242.

Wrege, P.H., Wikelski, M., Mandel, J.T., Rassweiler, T. & Couzin, I.D. (2005) Antbirds parasitize foraging army ants. *Ecology*, **86**, 555–559.

Wright, D.D., Jessen, J.H., Burke, P. & Gómez De Silva Garza, H. (1997) Tree and liana enumeration and diversity on a one-hectare plot in Papua New Guinea. *Biotropica*, **29**, 250–260.

Wright, P.C. (1998) Impact of predation risk on the behaviour of *Propithecus diadema edwardsi* in the rain forest of Madagascar. *Behaviour*, **135**, 483–512.

Wright, S.J. (2002) Plant diversity in tropical forests: A review of mechanisms of species coexistence. *Oecologia*, **130**, 1–14.

Wright, S.J. & Calderón, O. (2006) Seasonal, El Niño and longer term changes in flower and seed production in a moist tropical forest. *Ecology Letters*, **9**, 35–44.

Wright, S.J. & Muller-Landau, H.C. (2006) The future of tropical forest species. *Biotropica*, **38**, 287–301.

Wright, S.J., Carrasco, C., Calderón, O. & Paton, S. (1999) The El Niño Southern Oscillation, variable fruit production, and famine in a tropical forest. *Ecology*, **80**, 1632–1647.

Wright, S.J., Stoner, K.E., Beckman, N. et al. (2007) The plight of large animals in tropical forests and the consequences for plant regeneration. *Biotropica*, **39**, 289–291.

Wright, T.F., Schirtzinger, E.E., Matsumoto, T. et al. (2008) A multilocus molecular phylogeny of the parrots (Psittaciformes): Support for a Gondwanan origin during the Cretaceous. *Molecular Biology and Evolution*, **25**, 2141–2156.

Yack, J.E., Otero, L.D., Dawson, J.W., Surlykke, A. & Fullard, J.H. (2000) Sound production and hearing in the blue cracker butterfly *Hamadryas feronia* (Lepidoptera, Nymphalidae) from Venezuela. *Journal of Experimental Biology*, **203**, 3689–3702.

Yamagishi, S., Honda, M., Eguchi, K. & Thorstrom, R. (2001) Extreme endemic radiation of the Malagasy vangas (Aves: Passeriformes). *Journal of Molecular Evolution*, **53**, 39–46.

Yanoviak, S.P. (2010) The directed aerial descent of arboreal ants. In: *Ant Ecology* (eds L. Lach, C.L. Parr & K.L. Abbott), pp. 223–224. Oxford University Press, Oxford.

Yatabe, Y., Shinohara, W., Matsumoto, S. & Murakami, N. (2009) Patterns of hybrid formation among cryptic species of bird-nest fern, *Asplenium nidus* complex (Aspleniaceae), in West Malesia. *Botanical Journal of the Linnean Society*, **160**, 42–63.

Yoder, A.D., Burns, M.M., Zehr, S. et al. (2003) Single origin of Malagasy Carnivora from an African ancestor. *Nature*, **421**, 734–737.

Youngsteadt, E., Nojima, S., Haeberlein, C., Schulz, S. & Schal, C. (2008) Seed odor mediates an obligate ant-plant mutualism in Amazonian rainforests. *Proceedings of the National Academy of Sciences of the USA*, **105**, 4571–4575.

Youngsteadt, E., Alvarez Baca, J., Osborne, J. & Schal, C. (2009) Species-specific seed dispersal in an obligate ant-plant mutualism. *PLoS One*, **4**, e4335.

Yu, D.W., Levi, T. & Shepard, G.H. (2010) Conservation in low-governance environments. *Biotropica*, **42**, 569–571.

Yumoto, T. (1999) Seed dispersal by Salvin's curassow, *Mitu salvini* (Cracidae), in a tropical forest of Colombia: Direct measurements of dispersal distance. *Biotropica*, **31**, 654–660.

Yumoto, T., Maruhashi, T., Yamagiwa, J. & Mwanza, N. (1995) Seed-dispersal by elephants in a tropical rain forest in Kahuzi-Biéga National Park, Zaire. *Biotropica*, **27**, 526–530.

Zahawi, R.A. & Holl, K.D. (2009) Comparing the performance of tree stakes and seedlings to restore abandoned tropical pastures. *Restoration Ecology*, **17**, 854–864.

Zhang, P. & Wake, M.H. (2009) A mitogenomic perspective on the phylogeny and biogeography of living caecilians (Amphibia: Gymnophiona). *Molecular Phylogenetics and Evolution*, **53**, 479–491.

Zhu, H. (1997) Ecological and biogeographical studies on the tropical rain forest of south Yunnan, SW China with a special reference to its relation with rain forests of tropical Asia. *Journal of Biogeography*, **24**, 647–662.

Zimmermann, Y., Roubik, D.W. & Eltz, T. (2006) Species-specific attraction to pheromonal analogues in orchid bees. *Behavioral Ecology and Sociobiology*, **60**, 833–843.

Zimmermann, Y., Roubik, D.W., Quezada-Euan, J.J.G., Paxton, R.J. & Eltz, T. (2009) Single mating in orchid bees (Euglossa, Apinae): implications for mate choice and social evolution. *Insectes Sociaux*, **56**, 241–249.

Index

Page numbers in *italics* refer to Figures; those in **bold** to Tables.

Tropical Rain Forests: An Ecological and Biogeographical Comparison, Second edition.
© Richard T. Corlett and Richard B. Primack. Published 2011 by Blackwell Publishing Ltd.